Jianhua Zheng

Value Distribution of Meromorphic Functions

Jianhua Zheng

Value Distribution of Meromorphic Functions

TSINGHUA UNIVERSITY PRESS

Springer

Author
Prof. Jianhua Zheng
Department of Mathematical Sciences
Tsinghua University
Beijing 100084, P. R. China
Email: jzheng@math.tsinghua.edu.cn

ISBN 978-7-302-22329-0
Tsinghua University Press, Beijing

ISBN 978-3-642-12908-7 e-ISBN 978-3-642-12909-4
Springer Heidelberg Dordrecht London New York

Library of Congress Control Number: 2010927409

Cover design: Frido Steinen-Broo, EStudio Calamar, Spain

Printed on acid-free paper

Springer is part of Springer Science+Business Media (www.springer.com)

Preface

This book is devoted to the study of value distribution of functions which are meromorphic on the complex plane or in an angular domain with vertex at the origin. We characterize such meromorphic functions in terms of distribution of some of their value points. The study, together with certain related topics, is known as theory of value distribution of meromorphic functions. The theory is too vast to be justified within a single work. Therefore we selected and organized the content based on their significant importance to our understanding and interests in this book. I gladly acknowledge my indebtedness in particular to the books of M. Tsuji, A. A. Goldberg and I. V. Ostrovskii, Yang L. and the papers of A. Eremenko.

An outline of the book is provided below. The introduction of the Nevanlinna characteristic to the study of meromorphic functions is a new starting symbol of the theory of value distribution. The Nevanlinna characteristic is powerful, and its thought has been used to produce various characteristics such as the Nevanlinna characteristic and Tsuji characteristic for an angular domain. And from geometric point of view, namely the Ahlfors theory of covering surfaces, the Ahlfors-Shimizu characteristic have also been introduced. These characteristics are real-valued functions defined on the positive real axis. Therefore, in the first chapter, we collect the basic results about positive real functions that are often used in the study of meromorphic function theory. Some of these results are distributed in other books, some in published papers, and some have been newly established in order to serve our specific objectives in the book.

In the present book, we discuss value distribution not only in the complex plane, but also in an angular domain. Therefore, we introduce, in the second chapter, various characteristics of a meromorphic function: The Nevanlinna characteristic for a disk, the Nevanlinna characteristic for an angle, the Tsuji characteristic and the Ahlfors-Shimizu characteristic for an angle. Although they were distributed in another books, we collected all of them, and more importantly, we carefully compared them with one another to reveal their relations that enabled us to produce new results and applications. We establish the first and second fundamental theorems for the various characteristics and the corresponding integrated counting functions, and provide an estimate of the error term related to the Nevanlinna characteristic for an angle in terms of the Nevanlinna characteristic in a larger angle. We discuss in an angle the growth order of a meromorphic function and exponent of convergence of its a-points by means of the Ahlfors-Shimizu characteristic. We establish unique theorems in an angular domain with the help of the Tsuji characteristic, which is a new topic, because this has never been touched before while only the case of the whole complex plane was discussed.

After providing a brief overview of the characteristics in Chapter 2, we carefully investigate, in the third chapter, a new singular direction of a meromorphic function called T direction, which is different from the Julia, Borel and Nevanlinna directions. A singular direction is characterized essentially with the help of a property that in any angle containing it, the function assumes abundantly any value possibly except at most two values. The word "abundantly" is expressed by "infinitely often" for the Julia directions and by the growth order of the function for the Borel directions. The definition of T directions is to compare the integrated counting function in an angle to the characteristic and so it does not depend on the growth order, which is different from the Borel directions. So we can naturally consider T directions of meromorphic functions with zero order or infinite order. The second fundamental theorem of Nevanlinna is considered as the background of T directions. The following inequality

$$\limsup_{r\to\infty}\frac{N(r,\mathbb{C},f=a)}{T(r,f)}>0$$

always holds for all but at most two values of a. For a T direction, we consider the above inequality in any angle containing it instead of the whole complex plane. First we discuss the existence of T directions including the case of small functions in our consideration, next do relationship with the Borel directions, then common T directions of the function and its derivatives including the Hayman T directions. The singular directions of meromorphic solutions of linear differential equations possess some special properties, which are carefully studied and finally, we survey the results on the uniqueness and singular directions of an algebroid function.

The book includes discussion of argument distribution as well as modulo distribution and their relations. In the fourth chapter, we reveal relations between the numbers of deficient values and T directions. The results established there are new and unpublished elsewhere. The essential idea for discussion of this topic comes from observation that if the function assumes two values a and b at few points and is in close proximity to a complex number $c\neq a,b$ at enough points in a bounded domain, then it is close to c in the whole domain possibly outside a small set and that if the function is analytic, in view of the two constant theorem for the harmonic measure, we can use the modulo of the function on some part of the boundary of the domain to control the function modulo inside the domain. In the final section, we make a survey on this topic.

In the fifth chapter, we discuss the growth of the meromorphic functions that have two radially distributed values and a Nevanlinna deficient value. We first consider the growth of the meromorphic functions without any restriction imposed on their order and then those with the finite lower order. We attain our purpose in terms of the Nevanlinna characteristic for an angle, as Goldberg and Ostrovskii did, but our starting point is to establish an estimate of the Nevanlinna characteristic for a disk centered at the origin in terms of $B_{\alpha,\beta}(r,f)$ under an observation of the Nevanlinna deficient value, and then $B_{\alpha,\beta}(r,f)$ is estimated by two $C_{\alpha,\beta}(r,*)$ which may deal with the derivatives with help of fundamental inequalities for the Nevanlinna characteristic for an angle, and finally, $C_{\alpha,\beta}(r,*)$ are replaced by the integrated counting functions $N(r,\Omega,*)$ in terms of the relations between them. Thus the Nevanlinna characteristic for a disk can be estimated by two $N(r,\Omega,*)$ and we reduce the study of this subject to estimation of $B_{\alpha,\beta}$ in terms of $C_{\alpha,\beta}$. However, this comes from the study of fundamental inequality for the Nevanlinna characteristic for an angle. As we know, most of the fundamental inequalities for a disk can be validly and easily transferred to the case of an angle and therefore, we give out a simple and elementary approach to the discussions of this subject. When the function is of the finite lower order, we use the Baernstein spread relation to discuss the estimation of the Nevanlinna characteristic for a disk in terms of $B_{\alpha,\beta}(r,f)$ and hence we can attain deeper results for this subject.

In the sixth chapter, we collect and develop results about singularities of the inverse of a meromorphic function. A transcendental meromorphic function is equipped with a parabolic simply connected Riemann surface. The boundary points of the Riemann surface correspond to transcendental singularities of the inverse of the function, that is, asymptotic values of the function, and vice versa. We discuss relationships between the number of direct singularities and the growth (lower) order. The isolated transcendental singularity is logarithmic, and hence we observe that an asymptotic value over which the singularity is not logarithmic is a limit of other singular values. For a meromorphic function of finite order, such an asymptotic value is a limit point of critical values , which is the Bergweiler-Eremenko's result. We show Eremenko's construction of a transcendental meromorphic function with the finite given order which has every value on the extended complex plane as its asymptotic value, and next discuss the fixed points of bounded-type meromorphic functions, that is, meromorphic functions whose singular value set are bounded, from which we obverse that meromorphic functions possess special characters if their singular values are suitably restricted.

The final chapter is mainly devoted to the Eremenko's proof of the famous F. Nevanlinna conjecture on meromorphic functions with maximum total sum of Nevanlinna deficiencies. The conjecture was proved first by David Drasin, but his proof is very complicated. A. Eremenko used the potential theory to give a simple proof to the conjecture, from which we see the power of the potential theory in the study of value distribution of meromorphic functions. The theory to study subharmonic functions is the potential theory. The defence of two subharmonic functions is called δ-subharmonic. The logarithm of modulo of a meromorphic function is a δ-subharmonic function. Therefore, some problems about value distribution of meromorphic functions can be transferred to those about subharmonic functions. And the limit functions of a sequence of subharmonic functions produced by the subharmonic function in question are easier to be characterized than the meromorphic functions. The property or behavior of the limit functions can be used to describe the meromorphic functions. This is one of the approaches in which the potential theory are used to discuss problems about meromorphic functions.

For the benefit of readers, and for our intent to introduce and develop the potential theory in value distributions, we introduce and gather the basic knowledge about the potential theory including the normality of subharmonic function family in the sense of $\mathscr{L}_{\mathrm{loc}}$ and the Nevanlinna theory of subharmonic functions which consist of works of Anderson, Baernstein, Eremenko, Sodin, and others. The works of these mathematicians are very special and very important, and in our opinion, represent one aspect of value distribution theory which is worth further investigating and developing.

The first draft of this book was finished at the end of 2006, and main content of the book, except the seventh chapter was lectured in the summer course for postgraduated students held at Jiang Xi Normal University in the summer of 2007. I am indebted to Professor Yi Caifeng for her organizing the summer school, to Professor He Yuzhan for his comments and offering me some important materials, and to Professor Ye Zhuang for his support of this book. I would like to send many thanks to others including my students who pointed out some mistakes or some tough statements in the original draft when they read. This book has been partially supported by the National Natural Science Foundation of China.

Jianhua Zheng
Beijing,
December, 2009

Contents

Chapter 1
Preliminaries of Real Functions

Jianhua Zheng
Department of Mathematical Sciences, Tsinghua University, Beijing 100084, P. R. China
jzheng@math.tsinghua.edu.cn

Abstract: The various characteristics of meromorphic functions are main tool in the study of value distribution of meromorphic functions this book will introduce. They are real-valued functions defined on the positive real axis. In this chapter, we discuss certain properties of such real functions for application in later chapters. We begin with the order and the lower order of such functions which include the proximate order and the type function. We discuss the existence of the Pólya peak sequence. Also, we identify a sequence of positive numbers with some of the Pólya peak properties. We mainly introduce a result of Edrei and Fuchs for the regularity, thereby, improving the lemma of Borel and quasi-invariance of inequalities of two real functions under differentiation and integration. Finally, we exhibit the Green formula and collect several integral inequalities.

Key words: Real functions, Proximate order, Pólya peak, Regularity, Quasi-invariance

1.1 Functions of a Real Variable

In investigation of theory of meromorphic functions, we often meet the study of some properties of functions of a real variable, because various characteristics of meromorphic functions are such functions. Therefore, in this section, we collect the main properties of such functions which will be frequently used in the sequel.

1.1.1 The Order and Lower Order of a Real Function

Let $T(r)$ be a non-negative continuous function on $[r_0,\infty)$ for some $r_0 \geqslant 0$ and define $\log^+ x = \log\max\{1,x\}$. For $T(r)$, we define its lower order μ and order λ in turn as follows:

$$\mu = \mu(T) = \liminf_{r\to\infty} \frac{\log^+ T(r)}{\log r}$$

and

$$\lambda = \lambda(T) = \limsup_{r\to\infty} \frac{\log^+ T(r)}{\log r}.$$

We concentrate mainly on the function $T(r)$ which tends to infinity as r does. The order of a positive increasing continuous function can be characterized in term of an integral value.

Lemma 1.1.1. *Let $T(r)$ be a continuous, non-decreasing and positive function on $[r_0,\infty)$. Then for each $\rho < \lambda(T)$, we have*

$$\int^{\infty} \frac{T(t)}{t^{\rho+1}} \mathrm{d}t = \infty;$$

Conversely, if the above equation holds for certain ρ, then $\lambda(T) \geqslant \rho$.

Proof. Suppose that the integral is finite, and then for all $r \geqslant r_0$,

$$K > \int_r^{2r} \frac{T(t)}{t^{\rho+1}} \mathrm{d}t \geqslant \frac{T(r)}{(2r)^{\rho+1}} r = 2^{-\rho-1} T(r) r^{-\rho},$$

where $K = \int_{r_0}^{\infty} \frac{T(t)}{t^{\rho+1}} \mathrm{d}t$. This immediately deduces $\lambda(T) \leqslant \rho$ and the former half part of the lemma follows.

If $\lambda(T) < \rho$, then for each s with $\lambda(T) < s < \rho$, we have $T(r) < r^s$ for all sufficiently large r. Thus $T(r)r^{-\rho-1} < r^{-(\rho-s)-1}$, which yields the integral $\int^{\infty} \frac{T(t)}{t^{\rho+1}} \mathrm{d}t$ is convergent.

This completes the proof of Lemma 1.1.1. □

A continuous function may be too complicated to grasp, and thus sometime it is necessary to modify it by preserving, roughly speaking, only the values of r at which $T(r)$ can be approximately written into r^{λ}. The precise statement is as under

Theorem 1.1.1. *Let $T(r)$ be a continuous and positive function for $r \geqslant r_0 > 0$ and tend to infinity as $r \to \infty$ with $\lambda = \lambda(T) < \infty$. Then, there exists a function $\lambda(r)$ with the following properties:*

(1) *$\lambda(r)$ is a monotone and piecewise continuous differentiable function for $r \geqslant r_0$ with $\lim_{r\to\infty} \lambda(r) = \lambda$;*

(2) $\lim_{r\to\infty} \lambda'(r) r \log r = 0$;

(3) $\limsup_{r\to\infty} \frac{T(r)}{r^{\lambda(r)}} = 1$;

(4) *for each positive number d,*

$$\lim_{r\to\infty} \frac{U(dr)}{U(r)} = d^{\lambda}, \qquad U(r) = r^{\lambda(r)}. \tag{1.1.1}$$

We shall call the function $\lambda(r)$ the proximate order of $T(r)$ and the function $U(r)$ the type function of $T(r)$. It is obvious that the proximate order and the type function of a real function are not unique. As $\lambda > 0$, $U(r) = e^{\lambda(r)\log r}$ is increasing for all larger r. A simple calculation implies that a monotone increasing function $T(r)$ satisfying (1.1.1) must have $\mu(T) = \lambda(T) = \lambda$. The formula (1.1.1) is the key point of Theorem 1.1.1 and it makes sense essentially for the limit being finite. This explains the necessity for the condition that a function $T(r)$ in question is of finite order. However, in the case of infinite order, we have the following

Theorem 1.1.2. *Let $T(r)$ be a continuous and positive function for $r \geqslant r_0 > 0$ and tend to infinity as $r \to \infty$ with $\lambda = \lambda(T) = \infty$. Assume that $\omega(r)$ is a positive, continuous and non-increasing function with $\int_1^\infty \frac{\omega(t)}{t}\mathrm{d}t < +\infty$.*

Then, there exists a function $\lambda(r)$ with the following properties

(1) $\lambda(r)$ is non-decreasing and continuous and tends to infinity as $r \to \infty$;

(2) $\limsup\limits_{r\to\infty} \frac{T(r)}{r^{\lambda(r)}} = 1$;

(3) Set $U(r) = r^{\lambda(r)}$ and

$$\lim_{r\to\infty} \frac{U(r+\omega(U(r)))}{U(r)} = 1. \tag{1.1.2}$$

The proofs of Theorem 1.1.1 and Theorem 1.1.2 can be found in Chuang [2]. The following result will be used often in the next chapters.

Lemma 1.1.2. *Let $T(r)$ be a non-negative and non-decreasing function in $0 < r < \infty$. If*

$$\liminf_{r\to\infty} \frac{T(dr)}{T(r)} \geqslant c > 1$$

for some $d > 1$, then

$$\int_1^r \frac{T(t)}{t}\mathrm{d}t \leqslant \frac{2c\log d}{c-1}T(r) + O(1);$$

If

$$\liminf_{r\to\infty} \frac{T(dr)}{T(r)} > d^\omega$$

for some $d > 1$ and $\omega > 0$, then

$$\int_1^r \frac{T(t)}{t^{\omega+1}}\mathrm{d}t \leqslant K\frac{T(r)}{r^\omega} + O(1),$$

where K is a positive constant.

Proof. Write $s = \frac{c+1}{2}$ and we can find a natural number N such that for $r \geqslant r_0 = d^N$, we have $T(d^{-1}r) < s^{-1}T(r)$. Then for each $r \geqslant r_0 = d^N$, we have $n \geqslant N$ such that $d^n \leqslant r < d^{n+1}$, and let us estimate the following integral

$$\begin{aligned}\int_{r_0}^{r}\frac{T(t)}{t}\mathrm{d}t &= \sum_{k=N}^{n-1}\int_{d^k}^{d^{k+1}}\frac{T(t)}{t}\mathrm{d}t+\int_{d^n}^{r}\frac{T(t)}{t}\mathrm{d}t\\ &\leqslant \sum_{k=N}^{n-1}T(d^{k+1})\log d+T(r)\log d\\ &= T(d^n)\log d\sum_{k=N}^{n-1}\frac{T(d^{k+1})}{T(d^n)}+T(r)\log d\\ &< T(d^n)\log d\sum_{k=0}^{\infty}s^{-k}+T(r)\log d\\ &\leqslant \frac{2c\log d}{c-1}T(r).\end{aligned}$$

This yields the first desired inequality.

Now, we come to the proof of the second part of Lemma 1.1.2. Under the given assumption, for $r\geqslant r_0=d^N$ and some $\varepsilon>0$ we have $T(d^{-1}r)<(d+\varepsilon)^{-\omega}T(r)$. Thus, it follows that

$$\begin{aligned}\int_{r_0}^{r}\frac{T(t)}{t^{\omega+1}}\mathrm{d}t &= \sum_{k=N}^{n-1}\int_{d^k}^{d^{k+1}}\frac{T(t)}{t^{\omega+1}}\mathrm{d}t+\int_{d^n}^{r}\frac{T(t)}{t^{\omega+1}}\mathrm{d}t\\ &\leqslant \sum_{k=N}^{n-1}T(d^{k+1})\frac{1}{\omega}\left(\frac{1}{d^{k\omega}}-\frac{1}{d^{(k+1)\omega}}\right)+T(r)\frac{1}{\omega}\left(\frac{1}{d^{n\omega}}-\frac{1}{r^{\omega}}\right)\\ &< \frac{1}{\omega}T(d^n)\sum_{k=N}^{n-1}(d+\varepsilon)^{-\omega(n-k-1)}\left(\frac{1}{d^{k\omega}}-\frac{1}{d^{(k+1)\omega}}\right)+\frac{1}{\omega}\frac{T(r)}{d^{n\omega}}\\ &< \frac{d^{\omega}-1}{\omega}\frac{T(d^n)}{(d+\varepsilon)^{n\omega}}\frac{\left(\frac{d+\varepsilon}{d}\right)^{(n+1)\omega}-1}{\left(\frac{d+\varepsilon}{d}\right)^{\omega}-1}+\frac{1}{\omega}\frac{T(r)}{d^{n\omega}}\\ &\leqslant K_0\frac{T(r)}{d^{n\omega}}<K_0d^{\omega}\frac{T(r)}{r^{\omega}},\end{aligned}$$

where $K_0=\frac{d^{\omega}-1}{\omega}\frac{(d+\varepsilon)^{\omega}}{(d+\varepsilon)^{\omega}-d^{\omega}}+\frac{1}{\omega}$.

This completes the proof of Lemma 1.1.2. □

1.1.2 The Pólya Peak Sequence of a Real Function

In this subsection, we consider the Pólya peak for a $T(r)$, which was first introduced by Edrei [6].

Definition 1.1.1. *A sequence of positive numbers $\{r_n\}$ is called a sequence of Pólya peaks of order β for $T(r)$ (outside a set E) provided that there exist four sequences $\{r'_n\}$, $\{r''_n\}$, $\{\varepsilon_n\}$ and $\{\varepsilon'_n\}$ such that*

(1) $r_n\notin E,\ r'_n\to\infty,\ \frac{r_n}{r'_n}\to\infty,\ \frac{r''_n}{r_n}\to\infty,\ \varepsilon_n\to 0,\ \varepsilon'_n\to 0\ (n\to\infty)$;

(2) $\liminf_{n\to\infty} \frac{\log T(r_n)}{\log r_n} \geqslant \beta$;

(3) $T(t) < (1+\varepsilon_n)\left(\frac{t}{r_n}\right)^{\beta} T(r_n)$, $t \in [r'_n, r''_n]$;

(4) $T(t)/t^{\beta-\varepsilon'_n} \leqslant KT(r_n)/r_n^{\beta-\varepsilon'_n}$, $1 \leqslant t \leqslant r''_n$ *and for a positive constant* K.

Actually, it is easy to see that (2) follows from (4). It is obvious that any subsequence of a Pólya peak sequence is still a sequence of the Pólya peak. Please note that the above definition of the Pólya peaks has some differences from that in other literatures where a sequence of Pólya peak is only required to satisfy (1) and (3) listed in Definition 1.1.1. The sequence $\{r_n\}$ is called a sequence of relaxed Pólya peaks of order β for a constant $C > 1$, provided that (1), (2) and (4) in Definition 1.1.1 hold and (3) does for C in place of "$(1+\varepsilon_n)$". It is easily seen that for a sequence $\{r_n\}$ of Pólya peak and $d \geqslant 1$, $\{dr_n\}$ must be a sequence of the relaxed Pólya peak.

The following is a modifying version of well-known result which can be found in Section 8.1 of Yang [12].

Theorem 1.1.3. *Let $T(r)$ be a non-negative and non-decreasing continuous function in $0 < r < \infty$ with $0 \leqslant \mu(T) < \infty$ and $0 < \lambda(T) \leqslant \infty$. Then for arbitrary finite and positive number β satisfying $\mu \leqslant \beta \leqslant \lambda$ and a set F with finite logarithmic measure, i.e., $\int_F t^{-1}\mathrm{d}t < \infty$, there exists a sequence of the Pólya peaks of order β for $T(r)$ outside F.*

Proof. We choose a sequence of positive numbers $\{\varepsilon_n\}$ with $\varepsilon_n \to 0$ as $n \to \infty$. By induction, we seek the desired Pólya peak sequence $\{r_n\}$. Suppose we have r_{n-1} and want to find r_n.

First of all consider the case when $\beta = \lambda(T) < \infty$. It is easy to see that for n,

$$\limsup_{t\to\infty} \frac{T(t)}{t^{\beta-\varepsilon_n}} = \infty \quad \text{and} \quad \lim_{t\to\infty} \frac{T(t)}{t^{\beta+\varepsilon_n}} = 0.$$

Therefore, we can find a real number $u > \max\{n\varepsilon_n^{-1}, r_{n-1}\}$ such that

$$T(u)u^{-\beta+\varepsilon_n} = \max_{1\leqslant t\leqslant u} \{T(t)t^{-\beta+\varepsilon_n}\}$$

and a $v \geqslant u$ such that

$$T(v)v^{-\beta-\varepsilon_n} = \max_{t\geqslant u}\{T(t)t^{-\beta-\varepsilon_n}\}.$$

We choose r_n with $u \leqslant r_n \leqslant v$ such that

$$T(r_n)r_n^{-\beta+\varepsilon_n} = \max_{u\leqslant t\leqslant v} \{T(t)t^{-\beta+\varepsilon_n}\} \geqslant T(u)u^{-\beta+\varepsilon_n}.$$

Thus for $t \leqslant v$, we have

$$T(r_n)r_n^{-\beta+\varepsilon_n} \geqslant T(t)t^{-\beta+\varepsilon_n} \tag{1.1.3}$$

and for $t \geqslant r_n$

$$T(t)t^{-\beta-\varepsilon_n} \leqslant T(v)v^{-\beta-\varepsilon_n} \leqslant T(r_n)r_n^{-\beta+\varepsilon_n}v^{-2\varepsilon_n} \leqslant T(r_n)r_n^{-\beta-\varepsilon_n}$$

and, therefore, for $r_n \leqslant t \leqslant r_n/\varepsilon_n$,

$$\begin{aligned} T(t)t^{-\beta+\varepsilon_n} &= T(t)t^{-\beta-\varepsilon_n}t^{2\varepsilon_n} \leqslant T(r_n)r_n^{-\beta-\varepsilon_n}t^{2\varepsilon_n} \\ &= T(r_n)r_n^{-\beta+\varepsilon_n}\left(\frac{t}{r_n}\right)^{2\varepsilon_n} \\ &\leqslant \left(\frac{1}{\varepsilon_n}\right)^{2\varepsilon_n} T(r_n)r_n^{-\beta+\varepsilon_n}. \end{aligned} \tag{1.1.4}$$

Combining (1.1.3) and (1.1.4) deduces that r_n satisfies (4) for $r_n'' = r_n/\varepsilon_n$. This also immediately yields

$$T(t) \leqslant \mathrm{e}^{-2\varepsilon_n \log \varepsilon_n}\left(\frac{t}{r_n}\right)^{\beta} T(r_n) \text{ for } \varepsilon_n r_n \leqslant t \leqslant \varepsilon_n^{-1} r_n. \tag{1.1.5}$$

Now let us consider the case when $\mu \leqslant \beta < \lambda$. Assume without any loss of generalities that $\varepsilon_n < \lambda - \beta$. Then

$$\limsup_{t\to\infty} \frac{T(t)}{t^{\beta+\varepsilon_n}} = \infty \text{ and } \liminf_{t\to\infty} \frac{T(t)}{t^{\beta+\varepsilon_n/2}} = 0.$$

Application of a theorem of Edrei [6] deduces the existence of r_n with $r_n > \max\{r_{n-1}, \varepsilon_n^{-\frac{2\beta+\varepsilon_n}{\varepsilon_n}}\}$ such that

$$T(t) \leqslant \left(\frac{t}{r_n}\right)^{\beta+\varepsilon_n} T(r_n)$$

for $1 \leqslant t \leqslant r_n^{\frac{\beta+\varepsilon_n}{\beta+\varepsilon_n/2}}$. This immediately implies (1.1.5) and r_n satisfies (4), because for $1 \leqslant t \leqslant \varepsilon_n^{-1} r_n (< r_n^{\frac{\beta+\varepsilon_n}{\beta+\varepsilon_n/2}})$,

$$\left(\frac{t}{r_n}\right)^{2\varepsilon_n} \leqslant \mathrm{e}^{2\varepsilon_n|\log \varepsilon_n|}$$

and the quantity on the right side is bounded and tends to 1.

Thus, we have gotten a sequence $\{r_n\}$ satisfying (1.1.5) and (4) in Definition 1.1.1.

Put $d_n = 1 + 1/n$ and $V = \cup_{n=1}^{\infty}[r_n, d_n r_n]$. V has the infinite logarithmic measure and, therefore, there exist a subsequence of $\{[r_n, d_n r_n]\}$, each member of which contains at least a point outside F. Without any loss of generalities we can assume for each n a $\hat{r}_n \in [r_n, d_n r_n] \setminus F$. Then for $\hat{\varepsilon}_n \hat{r}_n \leqslant t \leqslant \hat{r}_n/\hat{\varepsilon}_n$ with $\hat{\varepsilon}_n = d_n \varepsilon_n$, we have

$$\begin{aligned} T(t) &\leqslant \left(\frac{t}{r_n}\right)^{\beta+\varepsilon_n} T(r_n) \leqslant (d_n)^{\beta+\varepsilon_n} \left(\frac{t}{\hat{r}_n}\right)^{\beta+\varepsilon_n} T(\hat{r}_n) \\ &\leqslant (d_n)^{\beta+\varepsilon_n} \left(\frac{1}{\hat{\varepsilon}_n}\right)^{2\varepsilon_n} \left(\frac{t}{\hat{r}_n}\right)^{\beta} T(\hat{r}_n), \end{aligned}$$

this implies that $\{\hat{r}_n\}$ satisfies (3) in Definition 1.1.1. It is easy to show $\{\hat{r}_n\}$ satisfies other conditions of the Pólya peak.

This completes the proof of Theorem 1.1.3. □

Chuang considered in [4] the type function and in [3] the Pólya peak sequence of a continuous real function and revealed some relations between the type function and the Pólya peak sequence by demonstrating their existence simultaneously starting from a basic theorem, that is, Theorem 1 of [3] or Lemma 4.4 of [4]. In fact, we easily obtain a sequence of the Pólya peak of order $\lambda(T)$ from the type function, for an example, a careful calculation implies that a sequence of positive real numbers $\{r_n\}$ with $U(r_n) = (1+o(1))T(r_n)$ must be a Pólya peak sequence of $T(r)$ of order $\lambda(T)$. Drasin and Shea [5] obtained a necessary and sufficient condition for existence of a sequence of Pólya peaks of order β which satisfies only (1) and (3) listed in Definition 1.1.1. Set

$$\lambda^*(T) = \sup\left\{\tau : \limsup_{x,A\to\infty} \frac{T(Ax)}{A^{\tau}T(x)} = \infty\right\}$$

and

$$\mu_*(T) = \inf\left\{\tau : \liminf_{x,A\to\infty} \frac{T(Ax)}{A^{\tau}T(x)} = 0\right\}.$$

It is proved in [5] that $\mu_*(T) \leqslant \mu(T) \leqslant \lambda(T) \leqslant \lambda^*(T)$ and if $\mu_* < \infty$, then a sequence of Pólya peaks of order β satisfying only (1) and (3) listed in Definition 1.1.1 exists if and only if $\mu_* \leqslant \beta \leqslant \lambda^*$ and $\beta < \infty$. However, we do not know if this condition is sufficient to the existence of our Pólya peak sequence. Usually, we call λ^* and μ_* respectively the Pólya order and Pólya lower order of $T(r)$.

Generally, there exists no Pólya peak sequence of $T(r)$ whose lower order is of infinite order. However, we have the following, which will be often used in the sequel.

Lemma 1.1.3. *Let $T(r)$ be an increasing and non-negative continuous function with the infinite order and F a set of positive real numbers having finite logarithmic measure. Then given a sequence $\{s_n\}$ of positive real numbers, there exists an unbounded sequence $\{r_n\}$ of positive real numbers outside F such that*

$$\frac{T(t)}{t^{s_n}} \leqslant \mathrm{e}\frac{T(r_n)}{r_n^{s_n}}, \qquad 1 \leqslant t \leqslant r_n.$$

Proof. Since $T(r)$ is of infinite order, for a fixed s_n we have

$$\limsup_{t\to\infty}\frac{T(t)}{t^{s_n}}=\infty$$

and it is easy to see that we can find a sequence $\{\hat{r}_m\}$ such that $\hat{r}_m > 2^{nm}$ and $\hat{r}_{m+1} > e^{1/s_n}\hat{r}_m$ and

$$\frac{T(t)}{t^{s_n}}\leqslant\frac{T(\hat{r}_m)}{\hat{r}_m^{s_n}},\qquad 1\leqslant t\leqslant\hat{r}_m.$$

Set

$$F_n=\bigcup_{m=1}^{\infty}[\hat{r}_m,e^{1/s_n}\hat{r}_m].$$

Then

$$\int_{F_n}\frac{dt}{t}=\sum_{m=1}^{\infty}\int_{\hat{r}_m}^{e^{1/s_n}\hat{r}_m}\frac{dt}{t}=\sum_{m=1}^{\infty}\frac{1}{s_n}=\infty$$

so that $F_n\setminus F$ has the infinite logarithmic measure. We can find a $r_n\in F_n\setminus F$ such that for some m, $\hat{r}_m\leqslant r_n\leqslant e^{1/s_n}\hat{r}_m$ and choose a r'_n in $[\hat{r}_m,r_n]$ such that

$$\frac{T(r'_n)}{r'^{s_n}_n}=\max\left\{\frac{T(t)}{t^{s_n}}:\hat{r}_m\leqslant t\leqslant r_n\right\}.$$

Thus for $1\leqslant t\leqslant r_n$, we have

$$\frac{T(t)}{t^{s_n}}\leqslant\frac{T(r'_n)}{r'^{s_n}_n}\leqslant\left(\frac{r_n}{r'_n}\right)^{s_n}\frac{T(r_n)}{r_n^{s_n}}\leqslant e\frac{T(r_n)}{r_n^{s_n}}.$$

The desired sequence $\{r_n\}$ has been attained. □

1.1.3 The Regularity of a Real Function

We first of all consider the density and the logarithmic density of a Lebesgue measurable set on the positive real axis. However, we begin with a general case, which will bring us some benefits.

An absolutely continuous function $\psi(r)$ on an interval $[a,b]$ has finite derivative almost everywhere in the sense of Lebesgue and $\psi'(r)\in L^1([a,b])$ and for each $r\in[a,b]$

$$\psi(r)=\psi(a)+\int_a^r\psi'(t)dt$$

and an indefinite integral of a function in $L^1([a,b])$ is absolutely continuous. A convex function is absolutely continuous and its right (left) derivative is non-decreasing. We say that an increasing function $\psi(r)$ is a convex function of another increasing $\varphi(r)$ if the right (left) derivative $d\psi(t)/d\varphi(t)$ exists and is non-decreasing.

We denote by m the Lebesgue measure on the positive real axis. Let E be a Lebesgue measurable subset of the positive real axis and $\psi(r)$ a positive and absolutely continuous function of r for $r \geqslant r_0$. Following Barry [1], we define the ψ-measure of $E(r) = E \cap [r_0, r]$ by

$$\psi\_m(E(r)) = \int_{E(r)} \psi'(t)\mathrm{d}t$$

and the upper and lower ψ-densities, respectively, of E by

$$\psi\_\overline{\underline{\mathrm{dens}}}E = \lim_{r\to\infty} \begin{matrix}\sup\\ \inf\end{matrix} \frac{\psi\_m(E(r))}{\psi(r)}.$$

When $\psi(r)$ is taken to be r, we obtain the definition of the upper and lower densities of E, denoted by $\overline{\mathrm{dens}}E$ and $\underline{\mathrm{dens}}E$ and when $\psi(r)$ is $\log r$, we have the upper and lower logarithmic densities of E, denoted by $\overline{\log\mathrm{dens}}E$ and $\underline{\log\mathrm{dens}}E$. When $\psi\_\overline{\mathrm{dens}}E = \psi\_\underline{\mathrm{dens}}E$, it is said that E has a ψ-density and we use notation $\psi\_\mathrm{dens}E$ to denote the common value and in this case, specially we have the definition of the density and logarithmic density of a set.

It is easy to see that for a set E on the positive real axis with the finite logarithmic measure, i.e., $\int_E t^{-1}\mathrm{d}t < \infty$, we have $\mathrm{dens}E = 0$. Actually, it follows from the following equation

$$m(E(r)) = m(E(\sqrt{r})) + m(E \cap [\sqrt{r}, r]) \leqslant \sqrt{r} + r\int_{E\cap[\sqrt{r},r]} t^{-1}\mathrm{d}t = o(r).$$

The following is Lemma 1 of Barry [1].

Lemma 1.1.4. *Let $\psi(r)$ and $\varphi(r)$ be positive, increasing, unbounded and absolutely continuous functions of r, and $\psi(r)$ a convex function of $\varphi(r)$ for $r \geqslant r_0$. Then*

$$\psi\_\underline{\mathrm{dens}}E \leqslant \varphi\_\underline{\mathrm{dens}}E \leqslant \varphi\_\overline{\mathrm{dens}}E \leqslant \psi\_\overline{\mathrm{dens}}E.$$

Proof. According to the definition of the upper ψ-density of a set, given arbitrarily $\varepsilon > 0$, for $t \geqslant r_1(\varepsilon) > r_0$, we have

$$\psi\_m(E(t)) < (\psi\_\overline{\mathrm{dens}}E + \varepsilon)\psi(t).$$

Noticing that $\mathrm{d}\psi(t)/\mathrm{d}\varphi(t)$ is non-decreasing in t, in view of the formula for integration by parts, we have for $r > r_1$

$$\begin{aligned}\varphi\_m(E(r)) &= \int_{E(r)} \mathrm{d}\varphi(t) = \int_{E(r)} \left(\frac{\mathrm{d}\psi(t)}{\mathrm{d}\varphi(t)}\right)^{-1} \mathrm{d}\psi(t)\\ &= \int_{r_0}^{r} \left(\frac{\mathrm{d}\psi(t)}{\mathrm{d}\varphi(t)}\right)^{-1} \mathrm{d}\psi\_m(E(t))\\ &= \psi\_m(E(t))\left(\frac{\mathrm{d}\psi(t)}{\mathrm{d}\varphi(t)}\right)^{-1}\Big|_{r_0}^{r} + \int_{r_0}^{r} \psi\_m(E(t))\mathrm{d}\left[-\left(\frac{\mathrm{d}\psi(t)}{\mathrm{d}\varphi(t)}\right)^{-1}\right]\end{aligned}$$

$$
\begin{aligned}
&< (\psi\_\overline{\text{dens}}E+\varepsilon)\psi(r)\left(\frac{\mathrm{d}\psi(r)}{\mathrm{d}\varphi(r)}\right)^{-1} \\
&\quad + \int_{r_0}^{r_1} \psi\_m(E(t))\mathrm{d}\left[-\left(\frac{\mathrm{d}\psi(t)}{\mathrm{d}\varphi(t)}\right)^{-1}\right] \\
&\quad + \int_{r_1}^{r} (\psi\_\overline{\text{dens}}E+\varepsilon)\psi(t)\mathrm{d}\left[-\left(\frac{\mathrm{d}\psi(t)}{\mathrm{d}\varphi(t)}\right)^{-1}\right] \\
&= O(1)+(\psi\_\overline{\text{dens}}E+\varepsilon)\left[\psi(r)\left(\frac{\mathrm{d}\psi(r)}{\mathrm{d}\varphi(r)}\right)^{-1} - \psi(t)\left(\frac{\mathrm{d}\psi(t)}{\mathrm{d}\varphi(t)}\right)^{-1}\Bigg|_{r_1}^{r}\right. \\
&\quad \left. + \int_{r_1}^{r}\left(\frac{\mathrm{d}\psi(t)}{\mathrm{d}\varphi(t)}\right)^{-1}\mathrm{d}\psi(t)\right] \\
&= O(1)+(\psi\_\overline{\text{dens}}E+\varepsilon)(O(1)+\varphi(r)).
\end{aligned}
$$

Thus

$$\limsup_{r\to\infty}\frac{\varphi\_m(E(r))}{\varphi(r)} \leqslant \psi\_\overline{\text{dens}}E.$$

The remainder inequality follows from this by taking complements. □

Specially, from Lemma 1.1.4 we get

$$\underline{\text{dens}}E \leqslant \log\_\,\underline{\text{dens}}E \leqslant \log\_\,\overline{\text{dens}}E \leqslant \overline{\text{dens}}E,$$

for r is a convex function of $\log r$.

Generally, a monotone continuous function may be complicated in the sense of its regular behavior and such an irregular behavior may cause difficulties to our discussion. However, fortunately, after a small set is ignored, such a function possess some regularities which are sufficient in certain discussions. The following is a fundamental lemma of E. Borel.

Lemma 1.1.5. *Let $T(r)$ be a non-decreasing continuous function in $[r_0,+\infty)$ such that $T(r_0)\geqslant 1$. Then with possible exception of values of r in a set with measure at most 2, we have*

$$T\left(r+\frac{1}{T(r)}\right) < 2T(r).$$

The following is Lemma 10.1 of Edrei and Fuchs [7], a modified version of the Borel Lemma 1.1.5.

Lemma 1.1.6. *Let $\psi(r)$ and $\varphi(r)$ be two positive functions on the positive real axis. Assume that for $r\geqslant r_0\geqslant 0$, $\psi(r)$ is non-decreasing while $\varphi(r)$ is non-increasing and that for some $r_1(>r_0)$ and a given positive number c, $\psi(r_1)>r_0+c$. Set*

$$E=\{r\geqslant r_1 : \psi(r+\varphi(\psi(r)))\geqslant \psi(r)+c\}.$$

Then we have

$$m(E(a,A)) \leqslant \frac{1}{c}\int_{\psi(a)-c}^{\psi(A)} \varphi(t)\mathrm{d}t,$$

provided that $r_1 \leqslant a < A < +\infty$, where $E(a,A)$ stands for the intersection of E with the interval (a,A).

Proof. Under the assumption that $\psi(r_1) > r_0 + c$, it is easy to see that $\psi(r) - c > r_0$ and $\varphi(t)$ is non-increasing for $t \geqslant \psi(r) - c$ and $r \geqslant r_1$.

Assume conversely that Lemma 1.1.6 is false, that is,

$$m(E(a_0,A)) \geqslant \varepsilon + \frac{1}{c}\int_{\psi(a_0)-c}^{\psi(A)} \varphi(t)\mathrm{d}t, \tag{1.1.6}$$

for three fixed numbers $\varepsilon > 0$, a_0 and A with $r_1 \leqslant a_0 < A < \infty$.

Put

$$\lambda(x) = \inf_{r\in E(x,A)} \{r\}$$

and in view of the definition of the infimum we can find $b_1 \in E(a_0,A)$ with $\lambda(a_0) \leqslant b_1 < \lambda(a_0) + \frac{\varepsilon}{2}$. Set $a_1 = b_1 + \varphi(\psi(b_1))$ and since $b_1 \in E$,

$$\psi(a_1) \geqslant \psi(b_1) + c.$$

Next we want to get the similar estimate from below of $m(E(a_1,A))$ to (1.1.6). Notice that if $a_1 \leqslant A$, $m(E(a_1,A)) = m(E(a_0,A)) - m(E(a_0,a_1))$, and to the end we respectively estimate $m(E(a_0,A))$ and $m(E(a_0,a_1))$ as follows: as $\varphi(r)$ is non-increasing, we have

$$\begin{aligned} m(E(a_0,a_1)) &\leqslant a_1 - \lambda(a_0) = (a_1 - b_1) + (b_1 - \lambda(a_0)) \\ &\leqslant \varphi(\psi(b_1)) + \frac{\varepsilon}{2} \\ &\leqslant \frac{1}{c}\int_{\psi(b_1)-c}^{\psi(b_1)} \varphi(t)\mathrm{d}t + \frac{\varepsilon}{2} \end{aligned}$$

and in view of (1.1.6) and $A \geqslant b_1 \geqslant a_0$, we have

$$\begin{aligned} m(E(a_0,A)) &\geqslant \varepsilon + \frac{1}{c}\int_{\psi(a_0)-c}^{\psi(A)} \varphi(t)\mathrm{d}t \\ &= \varepsilon + \frac{1}{c}\int_{\psi(b_1)}^{\psi(A)} \varphi(t)\mathrm{d}t + \frac{1}{c}\int_{\psi(a_0)-c}^{\psi(b_1)} \varphi(t)\mathrm{d}t \\ &\geqslant \varepsilon + \frac{1}{c}\int_{\psi(b_1)}^{\psi(A)} \varphi(t)\mathrm{d}t + \frac{1}{c}\int_{\psi(b_1)-c}^{\psi(b_1)} \varphi(t)\mathrm{d}t \\ &\geqslant \frac{\varepsilon}{2} + \frac{1}{c}\int_{\psi(b_1)}^{\psi(A)} \varphi(t)\mathrm{d}t + m(E(a_0,a_1)). \end{aligned}$$

This implies that $E(a_1,A)$ is not empty and $a_1 < A$ so that,

$$m(E(a_1,A)) \geqslant \frac{\varepsilon}{2} + \frac{1}{c}\int_{\psi(a_1)-c}^{\psi(A)} \varphi(t)\mathrm{d}t > 0.$$

Starting from this inequality we may repeat our previous construction with a_0 replaced by a_1 and ε by $\varepsilon/2$ and thus such construction can be repeated infinitely to obtain a sequence of intervals $[b_k, a_k]$ such that

$$a_0 \leqslant b_1 < a_1 \leqslant b_2 < a_2 \leqslant \cdots < A$$

and $b_k \in E$. Since $\psi(r)$ is non-decreasing, we have

$$\psi(b_{k+1}) \geqslant \psi(a_k) > \psi(b_k) + c,$$

so that $\psi(A) \geqslant \psi(b_{k+1}) \geqslant \psi(b_1) + kc$. This is impossible and therefore Lemma 1.1.6 is proved. □

Corollary 1.1.1. *Under the same assumption as in Lemma 1.1.6, assume, in addition, that*

$$\int^{\infty} \varphi(t)\mathrm{d}t < \infty.$$

Then E has only finite measure. In particular, let $T(r)$ be a continuous non-decreasing function of r with $T(r) > 1$. Then for $\varepsilon > 0$

$$T(re^{\alpha(r)}) \leqslant \mathrm{e}^c T(r), \ \alpha(r) = \frac{1}{(\log T(r))^{1+\varepsilon}}$$

holds for all r possibly outside a set of finite logarithmic measure.

Proof. The first part is obvious and we provide proof for the latter part only.

Set

$$\psi(r) = \log T(\mathrm{e}^r), \ \ \varphi(r) = \frac{1}{r^{1+\varepsilon}}.$$

It is obvious that $\psi(r)$ and $\varphi(r)$ satisfy the assumption of the first part. Since

$$\psi(\log r + \varphi(\psi(\log r))) = \log T(re^{\alpha(r)}) \text{ and } \psi(\log r) + c = \log \mathrm{e}^c T(r),$$

the first part implies that

$$E = \{x = \log r : T(re^{\alpha(r)}) \geqslant \mathrm{e}^c T(r)\}$$

has finite measure and therefore $F = \{r : \log r \in E\}$ has finite logarithmic measure by the formula for integration by substitution. Thus, the latter part has been proved. □

In the theory of value distribution, we have to often avoid some exceptional sets from the situation we consider, whence the following result is useful in treating this case.

Lemma 1.1.7. *Let $\psi(r)$ and $\varphi(r)$ be non-decreasing positive functions. Assume that*

$$\psi(r) \leqslant \varphi(r)$$

for all r possibly outside a set E with $\overline{\text{dens}}E < 1$. Then for each k with $(1-\overline{\text{dens}}E)^{-1} < k < +\infty$, for all sufficiently large r we have

$$\psi(r) \leqslant \varphi(kr).$$

If E is of finite measure or of finite logarithmic measure, then for each $k > 1$ and all sufficiently large r the above inequality is true.

Proof. Suppose for some $(1-\overline{\text{dens}}E)^{-1} < k < +\infty$ there exists an unbounded sequence $\{r_n\}$ such that $\psi(r_n) > \varphi(kr_n)$. Set $F = \bigcup_{n=1}^{\infty}[r_n, kr_n]$. Then

$$\overline{\text{dens}}F \geqslant \limsup_{n\to\infty} \frac{1}{kr_n} m(F(kr_n)) \geqslant \limsup_{n\to\infty} \frac{1}{kr_n}(kr_n - r_n) = \frac{k-1}{k} > \overline{\text{dens}}E.$$

This asserts an existence of a $r \in F \setminus E$ and so for some n, $r_n \leqslant r \leqslant kr_n$. Therefore in view of the monotonicity of ψ and φ, we have

$$\psi(r_n) \leqslant \psi(r) \leqslant \varphi(r) \leqslant \varphi(kr_n).$$

This contradicts the hypothesis about r_n and Lemma 1.1.7 follows. □

We remark that from Lemma 1.1.7 it follows that if $\log\_\overline{\text{dens}}E < 1$, then for $k > (1-\log\_\overline{\text{dens}}E)^{-1}$ and all sufficiently large r we have

$$\psi(r) \leqslant \varphi(r^k).$$

The following is due to Hayman [9].

Lemma 1.1.8. *Let $T(r)$ be a non-negative, non-constant and non-decreasing continuous function for $r \geqslant a$ with the order λ and lower order μ. Given two real numbers C_1 and C_2 greater than 1, set*

$$G = G(C_1, C_2) = \{r :\ T(C_1 r) \geqslant C_2 T(r)\}.$$

Then

$$\overline{\log\text{dens}}G \leqslant \lambda \frac{\log C_1}{\log C_2} \text{ and } \underline{\log\text{dens}}G \leqslant \mu \frac{\log C_1}{\log C_2}.$$

Proof. Set $r_1 = \inf\{r \geqslant 1 : r \in G\}$. Suppose that r_n has been chosen. Take $r_{n+1} = \inf\{r \geqslant C_1 r_n :\ r \in G\}$, and thus we inductively obtain a sequence of positive numbers $\{r_n\}$ such that $G \subset \bigcup_{n=1}^{\infty}[r_n, C_1 r_n]$. For $r \geqslant r_1$ with $r \in G$, we have $r_q \leqslant r < C_1 r_q$ for some $q \geqslant 1$. This implies that

$$\log\_ m(G(r)) = \int_{G(r)} \frac{dt}{t} \leqslant \sum_{k=1}^{q} \int_{r_k}^{C_1 r_k} \frac{dt}{t} = q \log C_1.$$

where $G(r) = G \cap [1, r]$.

Now we want to estimate q. Generally it is easy to see that

$$T(r_{n+1}) \geqslant T(C_1 r_n) \geqslant C_2 T(r_n)$$

so that

$$T(r_n) \geqslant C_2^{n-1} T(r_1)$$

and therefore,

$$q \leqslant 1 + \frac{1}{\log C_2} \log \frac{T(r_q)}{T(r_1)} \leqslant 1 + \frac{1}{\log C_2} \log \frac{T(r)}{T(1)}.$$

This deduces that

$$\frac{\log_{-} m(G(r))}{\log r} \leqslant \frac{\log C_1}{\log r} + \frac{\log C_1}{\log C_2} \frac{\log T(r) - \log T(1)}{\log r},$$

from which the desired inequalities follows directly by letting $r \to \infty$. □

1.1.4 Quasi-invariance of Inequalities

We begin the subsection with quasi-invariance of inequality under differentiation, that is to say, establish the following, the first part of which was proved in Barry [1].

Lemma 1.1.9. *Let $\psi(r)$ be non-decreasing and $\varphi(r)$ non-constant, non-decreasing and convex for $r \geqslant a$. Assume that*

$$0 \leqslant \psi(r) \leqslant \varphi(r),\ r \notin W$$

for a subset W of $[a, \infty)$ with $\tau = \varphi\_\overline{\mathrm{dens}} W < 1$. Then for arbitrary $K > 1/(1-\tau)$, we have

$$\underline{\mathrm{dens}} E \geqslant \frac{K-1}{K} - \tau,\ E = \{r : \psi'(r) \leqslant K\varphi'(r)\}.$$

Further, if $\psi(r)$ is convex, for all sufficiently large r we have

$$\psi'(r) \leqslant K\varphi'(dr),\ d > \frac{K}{(1-\tau)K - 1} > 0. \tag{1.1.7}$$

Proof. From the convexity of $\varphi(r)$, it is easy to see that $\varphi(r) \to \infty$ as $r \to \infty$ and $\varphi'(r)$ is non-negative and monotone non-decreasing and $\varphi(r)$ is absolutely continuous. Set

$$F = \{r : \psi'(r) \geqslant K\varphi'(r)\}$$

and $r' = \sup\{x \in F \setminus W : x \leqslant r\}$ for $r \geqslant a$. Then, for $r > a$, we have

$$\begin{aligned}
\int_{F(r)} \varphi'(t)\mathrm{d}t &= \int_{(F\backslash W)(r)} \varphi'(t)\mathrm{d}t + \int_{W(r)} \varphi'(t)\mathrm{d}t \\
&= \int_{(F\backslash W)(r')} \varphi'(t)\mathrm{d}t + \int_{W(r)} \varphi'(t)\mathrm{d}t \\
&\leqslant K^{-1} \int_{F(r')} \psi'(t)\mathrm{d}t + \int_{W(r)} \varphi'(t)\mathrm{d}t \\
&\leqslant K^{-1} \int_{a}^{r'} \psi'(t)\mathrm{d}t + \int_{W(r)} \varphi'(t)\mathrm{d}t \\
&\leqslant K^{-1}(\psi(r') - \psi(a)) + \int_{W(r)} \varphi'(t)\mathrm{d}t \\
&\leqslant K^{-1}\varphi(r') - K^{-1}\psi(a) + \int_{W(r)} \varphi'(t)\mathrm{d}t \\
&\leqslant K^{-1}\varphi(r) - K^{-1}\psi(a) + \int_{W(r)} \varphi'(t)\mathrm{d}t
\end{aligned}$$

and, thus, $\varphi\_\overline{\mathrm{dens}}F \leqslant K^{-1} + \tau$ and in view of Lemma 1.1.4 we get $\overline{\mathrm{dens}}F \leqslant K^{-1} + \tau$ and equivalently $\underline{\mathrm{dens}}E \geqslant 1 - K^{-1} - \tau$.

(1.1.7) follows from application of Lemma 1.1.7, for $\psi'(r)$ is non-decreasing under the assumption of convexity of $\psi(r)$ and $(1 - \overline{\mathrm{dens}}F)^{-1} < (1 - K^{-1} - \tau)^{-1} = \frac{K}{(1-\tau)K-1}$. □

Hayman and Stewart [10], and Hayman and Rossi [11] investigated the case of any order derivatives. The following result was obtained in [10]: if $\psi(r)$ and $\varphi(r)$ and their derivatives up to $n-1$ order are non-negative, non-decreasing and convex for $r \geqslant a$, then from $0 \leqslant \psi(r) \leqslant \varphi(r)$ for all $r \geqslant a$, we have

$$\psi^{(n)}(r) \leqslant Kn! \left(\frac{\mathrm{e}}{n}\right)^n \varphi^{(n)}(r) \tag{1.1.8}$$

on a set E of r with positive lower density depending only on K, n and φ but not on ψ and furthermore, Hayman and Rossi [11] proved that $\underline{\mathrm{dens}}E \geqslant (\sqrt[n]{K} - 1)/(\sqrt[n]{K} - 1 + n)$. What we should emphasize is that in Hayman and Stewart's result, the inequality (1.1.8) holds on the above fixed set E for any function $\psi(r)$ satisfying those assumptions determined by a given function $\varphi(r)$. Naturally we ask whether the set E in Lemma 1.1.9 is independent of $\psi(r)$, which concerns a question posed in page 256 of [10].

Finally, we consider quasi-invariance of inequality under integration. Here are two non-negative, non-decreasing real functions $A(r)$ and $B(r)$. If for all r in $[r_0, +\infty)$ but outside a subset E, we have

$$A(r) \leqslant B(r), \tag{1.1.9}$$

then could we compare $\int_{r_0}^{r} A(t)\mathrm{d}t$ to $\int_{r_0}^{r} B(t)\mathrm{d}t$? This is an important question in the value distribution of meromorphic functions. In terms of (1.1.9), we have

$$\begin{aligned}\int_{r_0}^{r} A(t)\mathrm{d}t &= \int_{[r_0,r]\setminus E} A(t)\mathrm{d}t + \int_{E\cap[r_0,r]} A(t)\mathrm{d}t\\ &\leqslant \int_{[r_0,r]\setminus E} B(t)\mathrm{d}t + \int_{E\cap[r_0,r]} A(t)\mathrm{d}t\\ &\leqslant \int_{r_0}^{r} B(t)\mathrm{d}t + \int_{E\cap[r_0,r]} A(t)\mathrm{d}t.\end{aligned}$$

Obviously, we cannot directly control $\int_{E\cap[r_0,r]} A(t)\mathrm{d}t$ in terms of $\int_{r_0}^{r} B(t)\mathrm{d}t$, but we can hope to use $\int_{r_0}^{r} A(t)\mathrm{d}t$ to control it. The following result realizes this purpose, which is a generalization of Lemma 9 of Eremenko and Sodin [8] but the basic idea is due to them.

Lemma 1.1.10. *Let E be a measurable subset of $[r_0,+\infty)$ and $\varepsilon>0$ and let $\varphi(x)$ be a positive non-increasing function in $[r_0,+\infty)$ such that $\int^{\infty}\varphi(t)\mathrm{d}t=+\infty$. Then there exists a subset E^* of $[r_0,+\infty)$ with*

$$\int_{E^*(r)} \varphi(t)\mathrm{d}t \leqslant \frac{2}{\varepsilon}\int_{E(r)} \varphi(t)\mathrm{d}t \tag{1.1.10}$$

such that for any non-negative, non-decreasing function $\psi(x)$ and $r\notin E^*$ and any $\tau<r$, we have

$$\int_{E(\tau,r)} \psi(t)\mathrm{d}t < 2\varepsilon\int_{\tau}^{r} \psi(t)\mathrm{d}t. \tag{1.1.11}$$

Proof. Define

$$E^* = \left\{ r\geqslant r_0 : \exists x=x(r)<r \text{ such that } \int_{E(x,r)} \varphi(t)\mathrm{d}t \geqslant \varepsilon\int_{x(r)}^{2r-x(r)} \varphi(t)\mathrm{d}t \right\}.$$

It is obvious that s is the center point of the interval $(x(s),2s-x(s))$ and so for a fixed $r\geqslant r_0$, $\{(x(s),2s-x(s)) : s\in E^*(r)\}$ is a covering of $E^*(r)$. As $E^*(r)$ is a bounded, closed set, there exist finitely many intervals $\{(x(s_j),2s_j-x(s_j)) : 1\leqslant j\leqslant q\}$ to cover $E^*(r)$ and each point in $E^*(r)$ is covered at most two times. Thus, as $s_j\in E^*$, we have

$$\begin{aligned}\int_{E^*(r)} \varphi(t)\mathrm{d}t &\leqslant \sum_{j=1}^{q}\int_{x(s_j)}^{2s_j-x(s_j)} \varphi(t)\mathrm{d}t\\ &\leqslant \frac{1}{\varepsilon}\sum_{j=1}^{q}\int_{E(x(s_j),s_j)} \varphi(t)\mathrm{d}t\\ &\leqslant \frac{2}{\varepsilon}\int_{E\cap(\cup_{j=1}^{q}(x(s_j),s_j))} \varphi(t)\mathrm{d}t\\ &\leqslant \frac{2}{\varepsilon}\int_{E(r)} \varphi(t)\mathrm{d}t.\end{aligned}$$

Now let us prove (1.1.11). For $r\notin E^*$ and for all $r_0\leqslant t\leqslant r$, we set $\eta(t)=\int_{E(t,r)}\varphi(t)\mathrm{d}t$ and, then, have

$$\eta(t) < \varepsilon \int_t^{2r-t} \varphi(x)\mathrm{d}x.$$

Noting that for $t < r$, $\varphi(2r-t) \leqslant \varphi(t)$ and $\eta(t)$ is non-increasing, but $\frac{\psi(t)}{\varphi(t)}$ is non-decreasing, we have

$$\begin{aligned}
&2\varepsilon \int_\tau^r \psi(t)\mathrm{d}t - \int_{E[\tau,r]} \psi(t)\mathrm{d}t \\
&\geqslant \varepsilon \int_\tau^r \frac{\psi(t)}{\varphi(t)}(\varphi(2r-t)+\varphi(t))\mathrm{d}t - \int_{E[\tau,r]} \psi(t)\mathrm{d}t \\
&= \int_\tau^r \frac{\psi(t)}{\varphi(t)}\mathrm{d}\left(-\varepsilon \int_t^{2r-t} \varphi(x)dx\right) + \int_\tau^r \frac{\psi(t)}{\varphi(t)}\mathrm{d}\eta(t) \\
&= \int_\tau^r \frac{\psi(t)}{\varphi(t)}\mathrm{d}\left(\eta(t) - \varepsilon \int_t^{2r-t} \varphi(x)\mathrm{d}x\right) \\
&\geqslant \frac{\psi(t)}{\varphi(t)}\left(\eta(t) - \varepsilon \int_t^{2r-t} \varphi(x)\mathrm{d}x\right)\Bigg|_\tau^r - \int_\tau^r \left(\eta(t) - \varepsilon \int_t^{2r-t} \varphi(x)\mathrm{d}x\right)\mathrm{d}\frac{\psi(t)}{\varphi(t)} \\
&= \frac{\psi(\tau)}{\varphi(\tau)}\left(\varepsilon \int_\tau^{2r-\tau} \varphi(x)\mathrm{d}x - \eta(\tau)\right) + \int_\tau^r \left(\varepsilon \int_t^{2r-t} \varphi(x)\mathrm{d}x - \eta(t)\right)\mathrm{d}\frac{\psi(t)}{\varphi(t)} \\
&\geqslant 0.
\end{aligned}$$

This yields (1.1.11). □

We make a remark on Lemma 1.1.10. If $\int_E \varphi(t)\mathrm{d}t < +\infty$, then $\int_{E^*} \varphi(t)\mathrm{d}t < +\infty$ and hence if $\varphi(t) \equiv 1$ or $\varphi(t) = 1/t$, that is to say, E is of finite measure or of finite logarithmic measure, then so is E^* in turn. Further, we can take into account the density of E and E^* in view of (1.1.10). Set $\phi(t) = \int^t \varphi(x)\mathrm{d}x$. Then we have

$$\phi\_\overline{\underline{\mathrm{dens}}}E^* \leqslant \frac{2}{\varepsilon}\phi\_\overline{\underline{\mathrm{dens}}}E$$

so that when $\phi\_\overline{\underline{\mathrm{dens}}}E = 0$, we have $\phi\_\overline{\underline{\mathrm{dens}}}E^* = 0$. What we should stress is that E^* in Lemma 1.1.10 does not rely on $\psi(r)$.

Now let us turn to answer the question mentioned before Lemma 1.1.10. Assume (1.1.9) holds for all r outside a set E with the properties $\int_E \varphi(t)\mathrm{d}t < +\infty$ for a $\varphi(x)$ stated in Lemma 1.1.10. Take a sequence of positive numbers $\{\varepsilon_j\}$ such that $0 < \varepsilon_j \leqslant 1$ and $\varepsilon_j \to 0$ as $j \to \infty$. In view of Lemma 1.1.10, we have E_j^* for each ε_j such that $\int_{E_j^*} \varphi(t)\mathrm{d}t < +\infty$ and for $r \notin E_j^*$

$$\int_{E(\tau,r)} \psi(t)\mathrm{d}t \leqslant \varepsilon_j \int_\tau^r \psi(t)\mathrm{d}t \tag{1.1.12}$$

for any non-negative, non-decreasing function $\psi(x)$. There exist a sequence of positive numbers $\{r_j\}$ such that $r_{j-1} < r_j \to \infty$ and

$$\int_{E_j^*\cap[r_j,\infty)} \varphi(t)\mathrm{d}t < \frac{1}{2^j}.$$

Define

$$E^* = (E_1^* \cap [r_0, r_1]) \cup \bigcup_{j=1}^{\infty} E_j^* \cap [r_j, r_{j+1}]. \tag{1.1.13}$$

Then we have

$$\begin{aligned}\int_{E^*} \varphi(t)\mathrm{d}t &\leqslant \int_{E_1^*\cap[r_0,r_1]} \varphi(t)\mathrm{d}t + \sum_{j=1}^{\infty} \int_{E_j^*\cap[r_j,r_{j+1}]} \varphi(t)\mathrm{d}t \\ &< \int_{E_1^*\cap[r_0,r_1]} \varphi(t)\mathrm{d}t + 1 \\ &< +\infty.\end{aligned}$$

Now define a function $\varepsilon(r)$ by $\varepsilon(r) = \varepsilon_j$ for $r_j \leqslant r < r_{j+1}$ $(j = 1, 2, \cdots)$ and $\varepsilon(r) = \varepsilon_1$ for $r_0 \leqslant r < r_1$. Obviously, $\varepsilon(r) \to 0$ as $r \to \infty$. For $r \notin E^*$, we have $r_j \leqslant r < r_{j+1}$ for some $j \in \mathbb{N}$ but $r \notin E_j^*$ and thus (1.1.12) holds. Further, in terms of (1.1.9), we can get

$$\begin{aligned}\int_{\tau}^{r} A(t)\mathrm{d}t &= \int_{[\tau,r]\setminus E} A(t)\mathrm{d}t + \int_{E\cap[\tau,r]} A(t)\mathrm{d}t \\ &\leqslant \int_{[\tau,r]\setminus E} B(t)\mathrm{d}t + \varepsilon_j \int_{\tau}^{r} A(t)\mathrm{d}t\end{aligned}$$

so that

$$(1 - \varepsilon(r)) \int_{\tau}^{r} A(t)\mathrm{d}t \leqslant \int_{\tau}^{r} B(t)\mathrm{d}t. \tag{1.1.14}$$

Now we consider the case when $\phi\_\overline{\mathrm{dens}}E = 0$ and $\phi(r) \to \infty$ as $r \to \infty$. Then there exists a set E_j^* for each ε_j such that $\phi\_\overline{\mathrm{dens}}E_j^* = 0$ and for $r \notin E_j^*$ we have (1.1.12). Take a r_j by induction on j such that $r_j > r_{j-1}$ and for $r \geqslant r_j$, we have

$$\frac{1}{\phi(r)} \int_{E_j^*(r)} \varphi(t)\mathrm{d}t < \frac{\varepsilon_j}{2^j}$$

and $\frac{\phi(r_{j-1})}{\phi(r)} < \varepsilon_j$. Define E^* by (1.1.13). Then for $r \notin E^*$, (1.1.14) holds.

Below we prove $\phi\_\overline{\mathrm{dens}}E^* = 0$ for this case. For arbitrary $\varepsilon > 0$, there exists a $N \in \mathbb{N}$ such that for all $j > N$, $\varepsilon_j < \varepsilon$. For $r \geqslant r_N$, we have $r_M \leqslant r < r_{M+1}$ for some $M \in \mathbb{N}$ with $M \geqslant N$ and therefore

$$
\begin{aligned}
\frac{1}{\phi(r)}\int_{E^*(r)}\varphi(t)\mathrm{d}t &= \frac{1}{\phi(r)}\sum_{j=1}^{M-1}\int_{E_j^*[r_j,r_{j+1}])}\varphi(t)\mathrm{d}t \\
&\quad +\frac{1}{\phi(r)}\left(\int_{E_1^*(r_1)}\varphi(t)\mathrm{d}t+\int_{E_M^*[r_M,r]}\varphi(t)\mathrm{d}t\right) \\
&< \sum_{j=1}^{M-1}\frac{\phi(r_{j+1})}{\phi(r)}\frac{\varepsilon_j}{2^j}+\frac{\phi(r_1)}{\phi(r)}\frac{\varepsilon_1}{2}+\frac{\varepsilon_M}{2^M} \\
&< \sum_{j=1}^{M-2}\varepsilon_M\frac{\varepsilon_j}{2^j}+\frac{\varepsilon_{M-1}}{2^{M-1}}+\frac{1}{2}\varepsilon_M+\frac{\varepsilon_M}{2^M} \\
&< \varepsilon_M+\frac{\varepsilon_{M-1}}{2^{M-1}}+\frac{1}{2}\varepsilon_M+\frac{\varepsilon_M}{2^M}<3\varepsilon
\end{aligned}
$$

taking note that $\frac{\phi(r_{j+1})}{\phi(r)}\leqslant\frac{\phi(r_{M-1})}{\phi(r)}<\varepsilon_M$ for $1\leqslant j\leqslant M-2$. This implies $\phi\_\overline{\mathrm{dens}}E^*=0$.

For the case when $\phi\_\underline{\mathrm{dens}}E=0$, we can attain the corresponding result whose proof is left to the reader. Let us formulate the above result as a lemma stated as follows.

Lemma 1.1.11. *Let E and $\varphi(x)$ be given as in Lemma 1.1.10. Then there exists a set E^* such that if (1.1.9) holds for $r\notin E$, we have (1.1.14) for $r\notin E^*$ with properties that:*

(1) *if $\int_E\varphi(t)\mathrm{d}t<+\infty$, then $\int_{E^*}\varphi(t)\mathrm{d}t<+\infty$;*

(2) *if $\phi\_\overline{\mathrm{dens}}E=0$ ($\phi\_\underline{\mathrm{dens}}E=0$, respectively), then $\phi\_\overline{\mathrm{dens}}E^*=0$ ($\phi\_\underline{\mathrm{dens}}E^*=0$), where $\phi(t)=\int^t\varphi(x)\mathrm{d}x$.*

1.2 Integral Formula and Integral Inequalities

For completeness and in order to bring the reader convenience in their readings, this section recall the Green formula and collect several integral inequalities. They are useful in the sequel and certain proofs will be provided taking into account that they are not easy to find or not well-known in the general literatures.

1.2.1 The Green Formula for Functions with Two Real Variables

Various characteristics, except the Ahlfors-Shimizu's, of a meromorphic function, we introduce in the next chapter, stem from the Green formula for functions with two real variables.

Let U be a domain in $\mathbb{C}$ surrounded by finitely many piecewise differentiable simple curves and let $X(x,y)$ and $Y(x,y)$ be two continuous differentiable functions in the closure of U. Then we have the Green formula

$$\iint_U \left(\frac{\partial Y}{\partial x} - \frac{\partial X}{\partial y}\right) \mathrm{d}\sigma = \int_{\partial U} X\mathrm{d}x + Y\mathrm{d}y$$

where $\mathrm{d}\sigma$ is the area element. We mean by $\mathrm{d}s$ the arc element, and by $\boldsymbol{n}$ the inner normal of ∂U with respect to U, and by Δ the Laplacian.

Assume further that $X(x,y)$ and $Y(x,y)$ are the second order continuous differentiable functions in the closure of U. In view of the Green formula, we have the following

$$\begin{aligned}\int_{\partial U} Y\frac{\partial X}{\partial \boldsymbol{n}}\mathrm{d}s &= \int_{\partial U} Y\left(\frac{\partial X}{\partial x}\cos\alpha + \frac{\partial X}{\partial y}\cos\beta\right)\mathrm{d}s\\ &= -\int_{\partial U}\left(Y\frac{\partial X}{\partial x}\mathrm{d}y - Y\frac{\partial X}{\partial y}\mathrm{d}x\right)\\ &= -\iint_U Y\Delta X\mathrm{d}\sigma - \iint_U \left(\frac{\partial X}{\partial x}\frac{\partial Y}{\partial x} + \frac{\partial X}{\partial y}\frac{\partial Y}{\partial y}\right)\mathrm{d}\sigma,\end{aligned}$$

where $\boldsymbol{n} = (\cos\alpha, \cos\beta)$. Thus

$$\iint_U (X\Delta Y - Y\Delta X)\mathrm{d}\sigma = \int_{\partial U}\left(Y\frac{\partial X}{\partial \boldsymbol{n}} - X\frac{\partial Y}{\partial \boldsymbol{n}}\right)\mathrm{d}s. \tag{1.2.1}$$

This formula is known as the second Green formula. We have two special formulae:

If $X(x,y)$ and $Y(x,y)$ are harmonic in U, that is, $\Delta X = 0 = \Delta Y$, then

$$\int_{\partial U}\left(X\frac{\partial Y}{\partial \boldsymbol{n}} - Y\frac{\partial X}{\partial \boldsymbol{n}}\right)\mathrm{d}s = 0 \tag{1.2.2}$$

and

$$\int_{\partial U}\frac{\partial X}{\partial \boldsymbol{n}}\mathrm{d}s = 0. \tag{1.2.3}$$

Furthermore, if U is doubly connected and Γ is the outer boundary and γ the inner boundary, then

$$\oint_\Gamma \left(X\frac{\partial Y}{\partial \boldsymbol{n}} - Y\frac{\partial X}{\partial \boldsymbol{n}}\right)\mathrm{d}s = \oint_\gamma \left(X\frac{\partial Y}{\partial \boldsymbol{n}} - Y\frac{\partial X}{\partial \boldsymbol{n}}\right)\mathrm{d}s. \tag{1.2.4}$$

These formulae will be used often in the next chapter.

1.2.2 Several Integral Inequalities

Let $(X, \mathscr{A}, \mu)$ be an arbitrary measure space. For a positive real number p, $L^p(X, \mathscr{A}, \mu)$ is the set of all real-valued $\mathscr{A}$-measurable function f defined μ-a.e. on X such that $\int_X |f(x)|^p \mathrm{d}\mu(x)$ exists and is finite. We write L^p for $L^p(X, \mathscr{A}, \mu)$ where confusion seems impossible. Define for $f \in L^p$

$$||f||_p = \left(\int_X |f(x)|^p \mathrm{d}\mu(x) \right)^{1/p}.$$

Hölder Inequality *For $f \in L^p$ and $g \in L^q$ with $p > 1$ and $\frac{1}{p} + \frac{1}{q} = 1$, we have*

$$\left| \int_X fg \mathrm{d}\mu \right| \leqslant \int_X |fg| \mathrm{d}\mu \leqslant ||f||_p ||g||_q. \tag{1.2.5}$$

For $p = q = 2$, the inequality (1.2.5) is called Schwartz inequality.

Minkowski Inequality *For $f,\ g \in L^p$ with $1 \leqslant p < \infty$, we have*

$$||f + g||_p \leqslant ||f||_p + ||g||_p.$$

The following lemma is very intuitive.

Lemma 1.2.1. *Let $\psi(x)$ be even, real and integrable in $(-a,a)$ and non-increasing in $(0,a)$ (with $\psi(0) = +\infty$ allowed). Assume that E is a measurable subset of $(-a,a)$ with $\mathrm{meas} E = 2b$. Then*

$$\int_E \psi(x)\mathrm{d}x \leqslant \int_{-b}^{b} \psi(x)\mathrm{d}x.$$

Proof. Set $E_1 = E \cap (-b,b)$, $E_2 = E \setminus E_1$ and $E_3 = (-b,b) \setminus E_1$. It is easy to see that for all $x \in E_2$ and $y \in E_3$ we have $\psi(x) \leqslant \psi(b) \leqslant \psi(y)$. Therefore, by noting $\mathrm{meas} E_2 = 2b - \mathrm{meas} E_1 = \mathrm{meas} E_3$ we deduce

$$\int_{E_2} \psi(x)\mathrm{d}x \leqslant \psi(b)\mathrm{meas} E_2 = \psi(b)\mathrm{meas} E_3 \leqslant \int_{E_3} \psi(x)\mathrm{d}x$$

and adding $\int_{E_1} \psi(x)\mathrm{d}x$ to the both sides implies the desired inequality. □

As an application of Lemma 1.2.1 we establish the following:

Lemma 1.2.2. *Let E be a measurable set of $[-\pi,\pi)$ with $\mathrm{meas} E = \delta \leqslant \pi$ and $a \in \mathbb{C}$. Then for $r \leqslant R$ we have*

$$\int_E \log \frac{R}{|r\mathrm{e}^{\mathrm{i}\theta} - a|} \mathrm{d}\theta \leqslant \delta \left(\log \frac{\pi R}{\delta r} + 1 \right) < \frac{R + 2r}{r} \delta \left(1 + \log^+ \frac{1}{\delta} \right).$$

Proof. In view of Lemma 1.2.1 and writing $a = |a|\mathrm{e}^{\mathrm{i}\phi}$ we have the estimation

$$\begin{aligned}
\int_E \log \frac{R}{|r\mathrm{e}^{\mathrm{i}\theta} - a|} \mathrm{d}\theta &= \int_{E-\phi} \log \frac{R}{|r\mathrm{e}^{\mathrm{i}\theta} - |a||} \mathrm{d}\theta \\
&\leqslant \int_{-\delta/2}^{\delta/2} \log \frac{R}{|r\mathrm{e}^{\mathrm{i}\theta} - |a||} \mathrm{d}\theta \\
&\leqslant \int_{-\delta/2}^{\delta/2} \log \frac{R}{r|\sin\theta|} \mathrm{d}\theta
\end{aligned}$$

$$\leqslant 2\int_0^{\delta/2} \log \frac{\pi R}{2r\theta} \mathrm{d}\theta$$
$$= \delta \left(\log \frac{\pi R}{\delta r} + 1 \right),$$

where $E - \phi = \{\theta - \phi : \theta \in E\}$. Thus, the first inequality we intend to prove follows. Since

$$1 < \log \frac{\pi R}{r} + 1 < \frac{R + 2r}{r},$$

this easily deduces the second desired inequality. □

Finally, we take into account the Jensen's general inequality for a convex function in the sense of integral.

Lemma 1.2.3. *Let $\psi(x)$ be a convex function in an interval I. For two integrable functions $f(x)$ and $g(x)$ in an interval $[a,b]$ such that $f([a,b]) \subset I$, $g(x) \geqslant 0$ and $A = \int_a^b g(x)\mathrm{d}x \neq 0$, then we have*

$$\psi\left(\frac{1}{A}\int_a^b f(x)g(x)\mathrm{d}x\right) \leqslant \frac{1}{A}\int_a^b \psi(f(x))g(x)\mathrm{d}x.$$

The inequality for the case $g(x) \equiv 1$ is known as Jensen's inequality.

Proof. Set

$$m = \frac{1}{A}\int_a^b f(x)g(x)\mathrm{d}x.$$

Since $\psi(x)$ is convex, we can find a real number α such that for all $x \in [a,b]$,

$$\psi(f(x)) \geqslant \alpha(f(x) - m) + \psi(m).$$

Therefore

$$\int_a^b \psi(f(x))g(x)\mathrm{d}x \geqslant \alpha \int_a^b (f(x) - m)g(x)\mathrm{d}x + \psi(m)\int_a^b g(x)\mathrm{d}x = A\psi(m),$$

from which the desired inequality follows. □

For the sake of application in the sequel, we consider the special case, that is, $\psi(x) = -\log x$ is convex in $(0,\infty)$ and $f(x)$ and $g(x)$ are both non-negative. Set $f^{\wedge}(x) = \max\{f(x), 1\}$. Then

$$\frac{1}{A}\int_a^b [\log^+ f(x)]g(x)\mathrm{d}x = \frac{1}{A}\int_a^b [\log f^{\wedge}(x)]g(x)\mathrm{d}x$$
$$\leqslant \log\left(\frac{1}{A}\int_a^b f^{\wedge}(x)g(x)\mathrm{d}x\right)$$

$$\leqslant \log\left(\frac{1}{A}\int_a^b (f(x)+1)g(x)\mathrm{d}x\right)$$
$$\leqslant \log^+\left(\frac{1}{A}\int_a^b f(x)g(x)\mathrm{d}x\right)+\log 2,$$

that is,

$$\frac{1}{A}\int_a^b [\log^+ f(x)]g(x)\mathrm{d}x \leqslant \log^+\left(\frac{1}{A}\int_a^b f(x)g(x)\mathrm{d}x\right)+\log 2. \tag{1.2.6}$$

References

1. Barry, P. D., The minimum modulus of small integral and subharmonic functions, Proc. London Math. Soc., (3)**12**(1962), 445-495.
2. Chuang, C. T., Sur la comparaison de la croissance d'une fonction méromorphe et de celle de sa dérivée, Bull. Sci. Math., **75**(1951), 171-190.
3. Chuang, C. T., On the Pólya peak, Science in China, **1**(1980), 9-18 (in Chinese).
4. Chuang, C. T., Singular directions of a meromorphic function, Science Press of China, 1982 (in Chinese).
5. Drasin, D. and Shea, D. F., Pólya peaks and the oscillation of positive functions, Proc. Amer. Math. Soc., (2)**34**(1972), 403-411.
6. Edrei, A., Sums of deficiencies of meromorphic functions, J. Analyse Math., I. **14**(1965), 79-107; II. **19**(1967), 53-74.
7. Edrei, A. and Fuchs, W. H. J., Bounds for the number of deficient values of certain classes of meromorphic functions, Proc. London Math. Soc., **12**(1962), 315-344.
8. Eremenko A. E. and Sodin, M. L., The value distribution of meromorphic functions and meromorphic curves from the point of view of potential theory, St. Petersburg Math. J., (1) **3**(1992), 109-136.
9. Hayman, W. K., On the characteristic of functions meromorphic in the plane and of their integrals, Proc. London Math. Soc., (3)**14A**(1965), 93-128.
10. Hayman, W. K. and Stewart, F. M., Real inequlities with application to function theory, Proc. Cambridge Philos. Soc., **50**(1954), 250-260.
11. Hayman, W. K. and Rossi, J. F., Characteristic, maximum modulus, and value distribution, Trans. Amer. Math. Soc., **284**(1984),651-664.
12. Yang, L., New researches on value distribution, Science Press of China, 1982.

Chapter 2
Characteristics of a Meromorphic Function

Jianhua Zheng
Department of Mathematical Sciences, Tsinghua University, Beijing 100084, P. R. China
jzheng@math.tsinghua.edu.cn

Abstract: We characterize meromorphic functions in terms of points at which they assume some values. The purpose is realized by using their characteristics. In this chapter, we introduce the Nevanlinna's characteristic in a domain (especially in a disk centered at the origin), the Nevanlinna's characteristic in an angle and the Tsuji's characteristic in terms of the generalized Poisson formula, Carleman formula and Levin formula respectively. These formulae are derived from the second Green formula. Similarly, the introduction of the Ahlfors-Shimizu characteristic originates from the second Green formula from the point of view of analysis. We exhibit the first and second fundamental theorems for every type of characteristics and the estimates of error terms, especially that of corresponding error terms to the Nevanlinna's characteristic in an angle. The relationship among various characteristics and among the integrated counting functions are found out. These relationship make us to produce new results and applications. We compare the characteristics of meromorphic functions and their derivatives. Next in terms of the Ahlfors-Shimizu's characteristic for an angle, we make a careful discussion of value distribution of functions meromorphic in an angle, especially theorems of the Borel-type. Then we discuss deficiency and deficient values which includes an introduction to Baernstein's spread relation along with related results. Finally, we establish a series of unique theorems of meromorphic functions in an angle in terms of Tsuji's characteristic. This is a new topic.

Key words: Nevanlinna characteristic, Tsuji characteristic, Ahlfors-Shimizu characteristic, Angular domain, Unique theorem

The main object to study in this book is transcendental meromorphic functions on the complex plane or in an angular domain. A meromorphic function on the complex plane is transcendental if it has ∞ as its unique essential singular point, in other word, it is not a rational function, and equivalently it can assume infinitely often on the complex plane all but at most two values on the extended complex plane. This can be characterized by its characteristic on the complex plane. By a transcendental meromorphic function, we mean a transcendental meromorphic function on the complex plane. Similarly, we shall define transcendental meromorphic function in an angular domain in terms of the corresponding characteristics in the angular domain. Actually, in study of value distribution of meromorphic functions, one of important topics is to characterize meromorphic functions in terms of points at which they assume some values. Then the characteristic functions of a meromorphic function play a crucial role in such discussions. This is realized by several fundamental theorems, that is, the characteristic can be controlled by means of the integrated counting functions of the number of points of several values assumed. In this chapter, we shall introduce characteristics for several different domains in an analogue approach originated from the Green formulae.

2.1 Nevanlinna's Characteristic in a Domain

We confine our discussion in the complex plane. By $\mathbb{C}$ we denote the complex plane and by $\widehat{\mathbb{C}}$ the extended complex plane. Let D be a domain on $\mathbb{C}$ surrounded by finitely many piecewise analytic curves. Then for any $a \in D$, there exists the Green function, denoted by $G_D(z,a)$, for D with singularity at $a \in D$ which is uniquely determined by the following conditions:

(1) $G_D(z,a)$ is harmonic in $D \setminus \{a\}$;

(2) in a neighborhood of a, $G_D(z,a) = \log \frac{1}{|z-a|} + \omega(z,a)$ for some function $\omega(z,a)$ harmonic in D;

(3) $G_D(z,a) \equiv 0$ on the boundary of D.

By Γ we denote the boundary of D and $\mathbf{n}$ the inner normal of Γ with respect to D. Since for $z \in D$, $G_D(z,a) > 0$ and for $z \in \Gamma$, $G_D(z,a) = 0$, from the definition of directional derivative it follows that the directional derivative of $G_D(z,a)$ on Γ in the inner normal is non-negative, that is, $\frac{\partial G}{\partial \mathbf{n}} \geqslant 0$ $(G = G_D(z,a))$. From the Green formula, in view of the Green function, we can establish the following formula, which is an extension of the Poisson formula for a disk (For a generalization of the formula, the reader is referred to Theorem 1.1 of [11]).

Lemma 2.1.1. *Let $u(z)$ be a harmonic function in D and have the second order continuous differentiation on ∂D except at most finitely many points $\{a_k\}_{k=1}^{q}$. Assume that in a neighborhood of a_k, we have*

$$u(z) = d_k \log|z - a_k| + u_k(z)$$

for some second order continuous differentiable $u_k(z)$. Then

$$u(z) = \frac{1}{2\pi}\int_{\partial D} u(\zeta)\frac{\partial G_D(\zeta,z)}{\partial \mathbf{n}}\mathrm{d}s.$$

Proof. Given arbitrarily a point $z \in D$, choose a $\varepsilon > 0$ such that $\{\zeta : |\zeta - z| \leqslant \varepsilon\} \subset D$ and in this disk we can write

$$G_D(\zeta,z)=\log\frac{1}{|\zeta-z|}+\omega(\zeta,z)$$

where $\omega(\zeta,z)$ is harmonic in ζ in D. Set $\Gamma_\varepsilon=\{\zeta:|\zeta-z|=\varepsilon\}$. Taking a sufficiently small $\delta>0$, put $D_\delta=D\setminus\cup_{k=1}^{q}B(a_k,\delta)$. Noting that $G_D(\zeta,z)\equiv 0,\ z\in\partial D$ and in view of (1.2.2), (1.2.3) and (1.2.4), we have

$$\begin{aligned}\int_{\partial D}u(\zeta)\frac{\partial G_D(\zeta,z)}{\partial n}\mathrm{d}s&=\int_{\partial D}\left(u(\zeta)\frac{\partial G_D(\zeta,z)}{\partial n}-G_D(\zeta,z)\frac{\partial u}{\partial n}\right)\mathrm{d}s\\&=\lim_{\delta\to 0}\int_{\partial D_\delta}\left(u(\zeta)\frac{\partial G_D(\zeta,z)}{\partial n}-G_D(\zeta,z)\frac{\partial u}{\partial n}\right)\mathrm{d}s\\&=\int_{\Gamma_\varepsilon}\left(u(\zeta)\frac{\partial G_D(\zeta,z)}{\partial n}-G_D(\zeta,z)\frac{\partial u}{\partial n}\right)\mathrm{d}s\\&=\int_{\Gamma_\varepsilon}\left(-u(\zeta)\frac{\partial \log|\zeta-z|}{\partial n}-\log\frac{1}{|\zeta-z|}\frac{\partial u}{\partial n}\right)\mathrm{d}s\\&=-\int_{\Gamma_\varepsilon}u(\zeta)\frac{\partial\log|\zeta-z|}{\partial n}\mathrm{d}s-\int_{\Gamma_\varepsilon}\log\frac{1}{\varepsilon}\frac{\partial u}{\partial n}\mathrm{d}s\\&=\int_{\Gamma_\varepsilon}u(\zeta)\frac{1}{|\zeta-z|}\mathrm{d}s=\frac{1}{\varepsilon}\int_{\Gamma_\varepsilon}u(\zeta)\mathrm{d}s\\&=\int_0^{2\pi}u(z+\varepsilon e^{i\theta})\mathrm{d}\theta=2\pi u(z).\end{aligned}$$

This yields the desired formula. □

In particular, given $u(z)$ being a constant, we deduce

$$\frac{1}{2\pi}\int_{\partial D}\frac{\partial G_D(\zeta,z)}{\partial n}\mathrm{d}s=1 \tag{2.1.1}$$

for all $z\in D$. From Lemma 2.1.1, we deduce the following, which is our starting point to introduce the characteristic of a meromorphic function on a domain.

Theorem 2.1.1. *Let $f(z)$ be a meromorphic function on $\overline{D}$. Then for arbitrary $z\in D$ such that $f(z)\neq 0,\infty$, we have*

$$\begin{aligned}\log|f(z)|=&\frac{1}{2\pi}\int_\Gamma\log|f(\zeta)|\frac{\partial G_D(\zeta,z)}{\partial n}\mathrm{d}s\\&-\sum_{a_m\in D}G_D(a_m,z)+\sum_{b_n\in D}G_D(b_n,z),\end{aligned}\tag{2.1.2}$$

where $\Gamma=\partial D$, and a_m is a zero of $f(z)$ and b_n a pole of $f(z)$, and a_m and b_n appear often in (2.1.2) according to their multiplicities.

Proof. Set

$$u(z)=\log|f(z)|+\sum_{a_m\in D}G_D(a_m,z)-\sum_{b_n\in D}G_D(b_n,z).$$

It is easy to see that $u(z)$ satisfies the conditions of Lemma 2.1.1. Application of Lemma 2.1.1 to the $u(z)$ implies the desired result by noticing that $G_D(a_m,\zeta)=0$ and $G_D(b_n,\zeta)=0$ on Γ. □

We can also consider the case when $f(z)=0$ or ∞. In this case, we use Theorem 2.1.1 to the function $\log|f(w)|-mG_D(w,z)$ to obtain the formula (2.1.2) replacing $\log|f(z)|$ in the left side of (2.1.2) with

$$\lim_{w\to z}(\log|f(w)|-mG_D(w,z))=\log|c(z)|-m\omega_D(z,z),$$

where $c(z)$ is the coefficient of the first term in Laurent series of $f(w)$ centered at z, and when $f(z)=0$, m is negative multiplicity of zero of $f(w)$ at z; when $f(z)=\infty$, m is multiplicity of pole of $f(w)$ at z .

The formula (2.1.2) with D being a disk is known as the Poisson-Jensen formula. Let us introduce several notations according to the formula (2.1.2). Define

$$N(D,a,f)=\sum_{b_n\in D}G_D(b_n,a)+n(0,a,f)\omega_D(a,a), \tag{2.1.3}$$

where a is a point in D, and b_n a pole of $f(z)$ appearing often according to their multiplicities, and $n(0,a,f)$ is the multiplicity of pole of $f(z)$ at a; $\overline{N}(D,a,f)$ is the sum in (2.1.3) counting all distinct b_n in D and with $n(0,a,f)$ replaced by 1 when $f(a)=\infty$;

$$m(D,a,f)=\frac{1}{2\pi}\int_\Gamma \log^+|f(\zeta)|\frac{\partial G_D(\zeta,a)}{\partial n}ds. \tag{2.1.4}$$

Define

$$T(D,a,f)=m(D,a,f)+N(D,a,f), \tag{2.1.5}$$

which is called Nevanlinna Characteristic of $f(z)$ with the center at a for D. From the formula (2.1.2) and the equality $\log x=\log^+ x-\log^+\frac{1}{x}$, $x>0$, it immediately follows that for $a\in D$ such that $f(a)\neq 0,\infty$, we have

$$T(D,a,f)=T\left(D,a,\frac{1}{f}\right)+\log|f(a)|. \tag{2.1.6}$$

In view of the remark following the proof of Theorem 2.1.1, for $f(a)=0$ or ∞, (2.1.6) holds for $f(a)$ replaced by the coefficient of the first term of the Laurent series of $f(z)$ at a.

We have to stress that $\omega_D(a,a)$ may not be non-negative and so $N(D,a,f)$ may be negative. However, for $D\subset U$, we have $G_D(a,z)\leqslant G_U(a,z)$, $\omega_D(a,a)\leqslant\omega_U(a,a)$ and so $N(D,a,f)\leqslant N(U,a,f)$. It is easy to prove the following basic inequalities: for p functions f_j $(j=1,2,\cdots,p)$ meromorphic on $\overline{D}$, we have

$$m(D,a,\sum_{j=1}^p f_j)\leqslant\sum_{j=1}^p m(D,a,f_j)+\log p,$$

$$m(D,a,\prod_{j=1}^{p} f_j) \leqslant \sum_{j=1}^{p} m(D,a,f_j)$$

and when $f_j(a) \neq \infty (1 \leqslant j \leqslant p-1)$,

$$N(D,a,\sum_{j=1}^{p} f_j) \leqslant \sum_{j=1}^{p} N(D,a,f_j),$$

$$N(D,a,\prod_{j=1}^{p} f_j) \leqslant \sum_{j=1}^{p} N(D,a,f_j),$$

and, therefore, when $f_j(a) \neq \infty (1 \leqslant j \leqslant p-1)$,

$$T(D,a,\sum_{j=1}^{p} f_j) \leqslant \sum_{j=1}^{p} T(D,a,f_j) + \log p,$$

$$T(D,a,\prod_{j=1}^{p} f_j) \leqslant \sum_{j=1}^{p} T(D,a,f_j).$$

Now we can establish the Nevanlinna first fundamental theorem.

Theorem 2.1.2. *Let $f(z)$ be a meromorphic function on $\overline{D}$. Then for a fixed complex number b and arbitrary $a \in D$ such that $f(a) \neq b, \infty$, we have*

$$T\left(D,a,\frac{1}{f-b}\right) = T(D,a,f) - \log|f(a)-b| + \varepsilon(b,D), \tag{2.1.7}$$

where $|\varepsilon(b,D)| \leqslant \log^+|b| + \log 2$.

Proof. Using the formula (2.1.6) implies that

$$\begin{aligned} T\left(D,a,\frac{1}{f-b}\right) &= T(D,a,f-b) - \log|f(a)-b| \\ &\leqslant T(D,a,f) + T(D,a,b) + \log 2 - \log|f(a)-b| \\ &= T(D,a,f) + \log^+|b| + \log 2 - \log|f(a)-b|; \end{aligned} \tag{2.1.8}$$

where the equality $T(D,a,b) = \log^+|b|$ follows from (2.1.5) and (2.1.1). By the same argument as above, we can deduce

$$T(D,a,f) \leqslant T\left(D,a,\frac{1}{f-b}\right) + \log^+|b| + \log 2 + \log|f(a)-b|.$$

Set

$$\varepsilon(b,D) = T\left(D,a,\frac{1}{f-b}\right) - T(D,a,f) + \log|f(a)-b|.$$

Then the equality (2.1.7) holds for $\varepsilon(b,D)$ with the desired property. □

We extend Theorem 2.1.2 to prove the following result.

Theorem 2.1.3. *Let $f(z)$ be a meromorphic function on $\overline{D}$ and $R(z) = \frac{P(z)}{Q(z)}$ a rational function with degree $d = \max\{\deg P, \deg Q\}$. Then for $f(a) \neq 0, b_j$ and ∞, we have*

$$T(D,a,R(f)) = dT(D,a,f) - \sum_{j=1}^{q} n_j \log|f(a) - b_j| + \upsilon(R,D), \tag{2.1.9}$$

where n_j is the multiplicity of zero b_j of $Q(z)$, q the number of distinct poles of $R(z)$ and $\upsilon(R,D)$ is a bounded quantity independent of $f(z)$ and a.

Proof. First of all, we discuss the case when $R(z)$ is a non-constant polynomial. We can write $R(z) = c\prod_{j=1}^{d}(z-a_j)$, and so we have

$$\begin{aligned} T(D,a,R(f)) &\leqslant \sum_{j=1}^{d} T(D,a,f-a_j) + \log^+|c| \\ &\leqslant dT(D,a,f) + d\log 2 + \sum_{j=1}^{d} \log^+|a_j| + \log^+|c|. \end{aligned} \tag{2.1.10}$$

Now we consider the case when $R(z)$ is a rational function and can write $R(z) = P_1(z) + R_1(z)$ with a polynomial $P_1(z)$ and a proper rational function $R_1(z)$, and so we have the form $R_1(z) = \sum_{j=1}^{q} Q_j\left(\frac{1}{z-b_j}\right)$, where $Q_j(w)$ is a polynomial in w with degree $n_j = \deg Q_j$, b_j a pole of $R(z)$ and q the number of distinct poles of $R(z)$. It is obvious that $\deg R = \deg P_1 + \sum_{j=1}^{q} \deg Q_j$. Then applying the above result (2.1.10) about polynomial yields that

$$\begin{aligned} T(D,a,R(f)) &\leqslant T(D,a,P_1(f)) + T(D,a,R_1(f)) + \log 2 \\ &\leqslant \deg P_1 T(D,a,f) + \sum_{j=1}^{q} T\left(D,a,Q_j\left(\frac{1}{f-b_j}\right)\right) + O(1) \\ &\leqslant \deg P_1 T(D,a,f) + \sum_{j=1}^{q} \deg Q_j T\left(D,a,\frac{1}{f-b_j}\right) + O(1) \\ &\leqslant dT(D,a,f) - \sum_{j=1}^{q} \deg Q_j(\log|f(a)-b_j| - \varepsilon(b_j,D)) + O(1), \end{aligned}$$

where $O(1)$ is independent of $f(z)$ and a.

On the other hand, we want to establish the reversal of the inequality. There exists a $K > 1$ such that for $|z| > K$, $|R(z)| > c|z|^p$, $p = \deg P - \deg Q$ (If $R(z)$ is a proper rational function, this case does not occur). Write $Q(w) = \alpha\prod_{j=1}^{q}(w-b_j)^{n_j}$

with $\sum_{j=1}^{q} n_j = \deg Q$. Then for $|w-b_j|<\delta,\ \delta=\frac{1}{2}\min\{|b_i-b_j| : 1\leqslant i\neq j\leqslant q\}$, we have $K_j>0$ such that $|R(w)|>\frac{K_j}{|w-b_j|^{n_j}}$. Set $E=\{\zeta\in\Gamma : |f(\zeta)|>K\}$ and $F_j=\{\zeta\in\Gamma : |f(\zeta)-b_j|<\delta\}$ $(j=1,2,\cdots,q)$.We can assume that $E\cap F_j=\varnothing$ for each j. Then

$$\begin{aligned}
m(D,a,R(f)) &\geqslant \frac{1}{2\pi}\int_E \log^+|R(f)(\zeta)|\frac{\partial G_D(\zeta,a)}{\partial \boldsymbol{n}}\mathrm{d}s \\
&\quad + \sum_{j=1}^{q}\frac{1}{2\pi}\int_{F_j}\log^+|R(f)(\zeta)|\frac{\partial G_D(\zeta,a)}{\partial \boldsymbol{n}}\mathrm{d}s \\
&\geqslant \frac{1}{2\pi}\int_E \log^+\{c|f(\zeta)|^p\}\frac{\partial G_D(\zeta,a)}{\partial \boldsymbol{n}}\mathrm{d}s \\
&\quad + \sum_{j=1}^{q}\frac{1}{2\pi}\int_{F_j}\log^+\frac{K_j}{|f(\zeta)-b_j|^{n_j}}\frac{\partial G_D(\zeta,a)}{\partial \boldsymbol{n}}\mathrm{d}s \\
&\geqslant \frac{p}{2\pi}\int_E \log|f(\zeta)|\frac{\partial G_D(\zeta,a)}{\partial \boldsymbol{n}}\mathrm{d}s-\log^+\frac{1}{c} \\
&\quad + \sum_{j=1}^{q}\frac{n_j}{2\pi}\int_{F_j}\log\frac{1}{|f(\zeta)-b_j|}\frac{\partial G_D(\zeta,a)}{\partial \boldsymbol{n}}\mathrm{d}s-\sum_{j=1}^{q}\log^+\frac{1}{K_j} \\
&\geqslant \frac{p}{2\pi}\int_E \log\frac{|f(\zeta)|}{K}\frac{\partial G_D(\zeta,a)}{\partial \boldsymbol{n}}\mathrm{d}s+\frac{p\log K}{2\pi}\int_E\frac{\partial G_D(\zeta,a)}{\partial \boldsymbol{n}}\mathrm{d}s-\log^+\frac{1}{c} \\
&\quad + \sum_{j=1}^{q}\left(\frac{n_j}{2\pi}\int_{F_j}\log\frac{\delta}{|f(\zeta)-b_j|}\frac{\partial G_D(\zeta,a)}{\partial \boldsymbol{n}}\mathrm{d}s-n_j\log^+\delta-\log^+\frac{1}{K_j}\right) \\
&\geqslant pm\left(D,a,\frac{f}{K}\right)+\sum_{j=1}^{q}n_jm\left(D,a,\frac{\delta}{f-b_j}\right)-\sum_{j=1}^{q}n_j\log^+\delta+O(1) \\
&\geqslant pm(D,a,f)-p\log^+K+\sum_{j=1}^{q}n_jm\left(D,a,\frac{1}{f-b_j}\right) \\
&\quad -\sum_{j=1}^{q}n_j\log^+\frac{1}{\delta}-\sum_{j=1}^{q}n_j\log^+\delta+O(1) \\
&= pm(D,a,f)+\sum_{j=1}^{q}n_jm\left(D,a,\frac{1}{f-b_j}\right) \\
&\quad -\sum_{j=1}^{q}n_j\left(\log^+\frac{1}{\delta}+\log^+\delta\right)+O(1).
\end{aligned} \tag{2.1.11}$$

It is obvious that

$$\begin{aligned} N(D,a,R(f)) &= N(D,a,P_1(f)) + N\left(D,a,\frac{1}{Q(f)}\right) \\ &= pN(D,a,f) + \sum_{j=1}^{q} n_j N\left(D,a,\frac{1}{f-b_j}\right). \end{aligned} \tag{2.1.12}$$

From Theorem 2.1.2, combination of (2.1.11) with (2.1.12) yields that

$$\begin{aligned} T(D,a,R(f)) &\geqslant pT(D,a,f) + \sum_{j=1}^{q} n_j T\left(D,a,\frac{1}{f-b_j}\right) + O(1) \\ &= dT(D,a,f) - \sum_{j=1}^{q} n_j \log|f(a)-b_j| + O(1). \end{aligned}$$

Thus we complete the proof of Theorem 2.1.3. □

The following result is called the Nevanlinna second fundamental theorem with the center at a for D.

Theorem 2.1.4. *Let $f(z)$ be a meromorphic function on $\overline{D}$ and a_j $(j=1,2,\cdots,q)$ be q distinct finite complex numbers and $\delta = \min_{1\leqslant i\neq j\leqslant q} |a_i - a_j|$. Then for $a \in D$ with $f(a) \neq 0, a_j$ and ∞, we have*

$$\begin{aligned} (q-1)T(D,a,f) \leqslant N(D,a,f) &+ \sum_{j=1}^{q} N\left(D,a,\frac{1}{f-a_j}\right) \\ &-N_1(D,a,f) + S(D,a,f), \end{aligned} \tag{2.1.13}$$

where

$$\begin{aligned} S(D,a,f) = m\left(D,a,\frac{f'}{f}\right) &+ \sum_{j=1}^{q} m\left(D,a,\frac{f'}{f-a_j}\right) \\ &+q\left(\log^+\frac{2q}{\delta} + \log^+\frac{\delta}{2q} + \log 2\right) + \log q - \log|f'(a)| \\ &+\sum_{j=1}^{q}\left(\log|f(a)-a_j| + \varepsilon(a_j,D)\right) \end{aligned} \tag{2.1.14}$$

and

$$N_1(D,a,f) = 2N(D,a,f) - N(D,a,f') + N\left(D,a,\frac{1}{f'}\right).$$

Proof. Set

$$F(z) = \sum_{j=1}^{q} \frac{1}{f(z)-a_j}.$$

From Theorem 2.1.2, we have following estimation

$$
\begin{aligned}
m(D,a,F) &\leqslant m\left(D,a,\frac{1}{f'}\right)+m(D,a,f'F) \\
&= T(D,a,f')-N\left(D,a,\frac{1}{f'}\right)-\log|f'(a)|+m(D,a,f'F) \\
&= N(D,a,f)+\overline{N}(D,a,f)+m(D,a,f') \\
&\quad -N\left(D,a,\frac{1}{f'}\right)-\log|f'(a)|+m(D,a,f'F) \\
&\leqslant T(D,a,f)+(\overline{N}(D,a,f)-N\left(D,a,\frac{1}{f'}\right))-\log|f'(a)| \\
&\quad +m\left(D,a,\frac{f'}{f}\right)+\sum_{j=1}^{q} m\left(D,a,\frac{f'}{f-a_j}\right)+\log q. \qquad (2.1.15)
\end{aligned}
$$

On the other hand, from (2.1.11) we want to estimate $m(D,a,F)$ from below. For z with $|z-a_j|<\frac{\delta}{2q}$ and for $i\neq j$, we have

$$
|z-a_i| \geqslant |a_i-a_j|-|z-a_j| \geqslant \delta-\frac{\delta}{2q} > (2q-1)|z-a_j|
$$

so that

$$
\left|\sum_{j=1}^{q}\frac{1}{z-a_j}\right| = \frac{1}{|z-a_j|}\left(1-\sum_{i\neq j}\frac{|z-a_j|}{|z-a_i|}\right) > \frac{q}{2q-1}\frac{1}{|z-a_j|}.
$$

In view of (2.1.11), we obtain

$$
m(D,a,F) \geqslant \sum_{j=1}^{q} m\left(D,a,\frac{1}{f-a_j}\right)-\sum_{j=1}^{q}\log^{+}\frac{1}{K_j}-q(\log^{+}\frac{2q}{\delta}+\log^{+}\frac{\delta}{2q}),
$$

where $K_j=\frac{q}{2q-1}$. Thus the inequality (2.1.13) is shown by combining the above inequality with (2.1.15) and then by using Theorem 2.1.2. □

It is an important step to take the derivatives of meromorphic functions into account, as H. Milloux did. We can also establish the following Milloux inequality, whose proof will be omitted.

Theorem 2.1.5. *Let $f(z)$ be a meromorphic function on $\overline{D}$. Then for $a\in D$ with $f(a)\neq 0,\infty$ and $f^{(n)}(a)\neq 1$ and $f^{(n+1)}(a)\neq 0$, we have*

$$
\begin{aligned}
T(D,a,f) &\leqslant \overline{N}(D,a,f)+N\left(D,a,\frac{1}{f}\right)+N\left(D,a,\frac{1}{f^{(n)}-1}\right) \\
&\quad -N\left(D,a,\frac{1}{f^{(n+1)}}\right)+S(D,a,f), \qquad (2.1.16)
\end{aligned}
$$

where

$$S(D,a,f) = m\left(D,a,\frac{f^{(n)}}{f}\right) + m\left(D,a,\frac{f^{(n+1)}}{f}\right) + m\left(D,a,\frac{f^{(n+1)}}{f^{(n)}-1}\right) + \log\left|\frac{f(a)(f^{(n)}(a)-1)}{f^{(n+1)}(a)}\right| + \log 2. \tag{2.1.17}$$

Now we come to consider the monotone increasing property of $T(D,a,f)$ with respect to the domain D in the inclusion sense, which is indeed confirmed by the following result.

Theorem 2.1.6. *Let $f(z)$ be a meromorphic function on $\overline{D}$. Then*

$$T(D,a,f) = \frac{1}{2\pi}\int_0^{2\pi} N\left(D,a,\frac{1}{f-e^{i\theta}}\right)d\theta + \log^+|f(a)|,$$

where $\log^+|f(a)|$ will be replaced by $\log|c(a)|$ when $f(a)=\infty$, and $c(a)$ is the coefficient of first term of the Laurent series of $f(z)$ at a.

Proof. For any complex number w, applying (2.1.2) to $f(z)=z-w$ in the unit disk $\{z:|z|<1\}$ yields that

$$\frac{1}{2\pi}\int_0^{2\pi}\log|w-e^{i\theta}|d\theta = \log^+|w|.$$

First of all we assume $f(a)\neq\infty$. From the formula (2.1.6), we have

$$\begin{aligned} N\left(D,a,\frac{1}{f-e^{i\theta}}\right) &= \frac{1}{2\pi}\int_\Gamma \log|f(\zeta)-e^{i\theta}|\frac{\partial G_D(\zeta,a)}{\partial n}ds \\ &\quad + N(D,a,f) - \log|f(a)-e^{i\theta}|. \end{aligned}$$

Integrating in θ both sides of the above equality yields, from the Fubini Theorem, that

$$\begin{aligned} \frac{1}{2\pi}\int_0^{2\pi} N\left(D,a,\frac{1}{f-e^{i\theta}}\right)d\theta &= \frac{1}{2\pi}\int_\Gamma \frac{1}{2\pi}\int_0^{2\pi}\log|f(\zeta)-e^{i\theta}|d\theta\frac{\partial G_D(\zeta,a)}{\partial n}ds \\ &\quad + N(D,a,f) - \log^+|f(a)| \\ &= \frac{1}{2\pi}\int_\Gamma \log^+|f(\zeta)|\frac{\partial G_D(\zeta,a)}{\partial n}ds \\ &\quad + N(D,a,f) - \log^+|f(a)| \\ &= T(D,a,f) - \log^+|f(a)|. \end{aligned}$$

Next, we consider the case when $f(a)=\infty$. Then from the above result it follows that

$$\begin{aligned}
T\left(D,a,\frac{1}{f}\right) &= \frac{1}{2\pi}\int_0^{2\pi} N\left(D,a,\frac{1}{1/f-\mathrm{e}^{\mathrm{i}\theta}}\right)\mathrm{d}\theta \\
&= \frac{1}{2\pi}\int_0^{2\pi} N\left(D,a,\frac{1}{f\mathrm{e}^{\mathrm{i}\theta}-1}\right)\mathrm{d}\theta \\
&= \frac{1}{2\pi}\int_0^{2\pi} N\left(D,a,\frac{1}{f-\mathrm{e}^{\mathrm{i}\theta}}\right)\mathrm{d}\theta
\end{aligned}$$

and

$$T\left(D,a,\frac{1}{f}\right) = T(D,a,f) - \log|c(a)|.$$

This completes the proof of Theorem 2.1.6. □

According to the increase of the Green function with respect to the domain in the inclusion sense, we have for $D \subset U$, $N(D,a,f) \leqslant N(U,a,f)$ and, therefore, from Theorem 2.1.6, it follows that $T(D,a,f) \leqslant T(U,a,f)$.

If $f(a) \neq \infty$, then $N(D,a,f)$ and $T(D,a,f)$ is non-negative, while for the case $f(a) = \infty$, $T(D,a,f)$ may be negative. Consider the function $f(z) = 1/(2z^p)$, and in view of Theorem 2.1.6, we have $T(D,0,f) = -\log 2$ for $D = \{z : |z| < 1\}$ as $N(D,0,1/(f-\mathrm{e}^{\mathrm{i}\theta})) = 0$; It is obvious that when $\omega_D(a,a) \geqslant 0$, $N(D,a,f) \geqslant 0$ and so $T(D,a,f) \geqslant 0$.

In what follows, we consider some special domains. First of all, we need to calculate the Green function for a simply connected domain. Let D be simply connected. There exists the Riemann mapping $\phi_a(z) : D \to \{w : |w| < 1\}$ such that $\phi_a(a) = 0$. Then it is easy to see that

$$G_D(z,a) = -\log|\phi_a(z)|.$$

Along the boundary ∂D of D, we have

$$\frac{\phi_a(z)'}{\phi_a(z)}\mathrm{d}z = \mathrm{i}\frac{\partial}{\partial \boldsymbol{n}} G_D(z,a)\mathrm{d}s \tag{2.1.18}$$

by the Cauchy-Riemann condition which says that for two orthogonal directions $\boldsymbol{s}$ and $\boldsymbol{n}$ such that $\boldsymbol{s}$ becomes $\boldsymbol{n}$ after $\boldsymbol{s}$ is rotated $\pi/2$ anticlockwise, we have

$$\frac{\partial u}{\partial \boldsymbol{s}} = \frac{\partial v}{\partial \boldsymbol{n}} \text{ and } \frac{\partial u}{\partial \boldsymbol{n}} = -\frac{\partial v}{\partial \boldsymbol{s}}$$

if $u(z) + \mathrm{i}v(z)$ is analytic. Indeed, (2.1.18) follows from the following calculation

$$\begin{aligned}
\mathrm{d}\log\phi_a(z) &= \frac{\partial}{\partial \boldsymbol{s}}\log\phi_a(z)\mathrm{d}s = \mathrm{i}\frac{\partial}{\partial \boldsymbol{s}}\arg\phi_a(z)\mathrm{d}s \\
&= -\mathrm{i}\frac{\partial}{\partial \boldsymbol{n}}\log|\phi_a(z)|\mathrm{d}s = \mathrm{i}\frac{\partial G_D(z,a)}{\partial \boldsymbol{n}}\mathrm{d}s.
\end{aligned}$$

In this way, we can obtain the Green functions for some special domains.

(1) For $D = \{z : |z| < R\}$ and $|a| < R$, we have

$$\phi_a(z) = \frac{R(z-a)}{R^2 - \bar{a}z}, \quad G_D(z,a) = \log\left|\frac{R^2-\bar{a}z}{R(z-a)}\right|$$

and

$$\frac{\partial}{\partial \boldsymbol{n}} G_D(\zeta,z)\mathrm{d}s = \frac{R^2-|z|^2}{|R\mathrm{e}^{\mathrm{i}\theta}-z|^2}\mathrm{d}\theta = \mathrm{Re}\frac{R\mathrm{e}^{\mathrm{i}\theta}+z}{R\mathrm{e}^{\mathrm{i}\theta}-z}\mathrm{d}\theta$$

where $\zeta = R\mathrm{e}^{\mathrm{i}\theta}$. $\frac{R^2-|z|^2}{|R\mathrm{e}^{\mathrm{i}\theta}-z|^2}$ is the Poisson kernel.

(2) For $D = \{z : |z| < R,\ \mathrm{Im}z > 0\}$ and $a \in D$, we have

$$\phi_a(z) = \frac{R(z-a)}{R^2-\bar{a}z}\,\frac{R^2-az}{R(z-\bar{a})}, \quad G_D(z,a) = \log\left|\frac{R^2-\bar{a}z}{R(z-a)}\,\frac{R(z-\bar{a})}{R^2-az}\right|.$$

Then on ∂D,

$$\begin{aligned}
\frac{\partial G_D}{\partial \boldsymbol{n}}(z,a)\mathrm{d}s &= -\mathrm{i}\left(\log\frac{R(z-a)}{R^2-\bar{a}z}\,\frac{R^2-az}{R(z-\bar{a})}\right)'\mathrm{d}z\\
&= -\mathrm{i}\left(\frac{1}{z-a} - \frac{1}{z-\bar{a}} - \frac{a}{R^2-az} + \frac{\bar{a}}{R^2-\bar{a}z}\right)\mathrm{d}z\\
&= -\mathrm{i}(a-\bar{a})\left[\frac{1}{(z-a)(z-\bar{a})} - \frac{R^2}{(R^2-az)(R^2-\bar{a}z)}\right]\mathrm{d}z\\
&= -\mathrm{i}\left(\frac{R^2-|a|^2}{(z-a)(R^2-\bar{a}z)} - \frac{R^2-|a|^2}{(z-\bar{a})(R^2-az)}\right)\mathrm{d}z. \qquad (2.1.19)
\end{aligned}$$

Thus on the upper half circle $\{\zeta = R\mathrm{e}^{\mathrm{i}\theta} : 0 < \theta < \pi\}$, for $z \in D$ we have

$$\frac{\partial G_D}{\partial \boldsymbol{n}}(\zeta,z)\mathrm{d}s = \left(\frac{R^2-|z|^2}{|\zeta-z|^2} - \frac{R^2-|z|^2}{|\zeta-\bar{z}|^2}\right)\mathrm{d}\theta \qquad (2.1.20)$$

and on the segment $\{\zeta = t : -R < t < R\}$ and $z = r\mathrm{e}^{\mathrm{i}\phi}$,

$$\frac{\partial G_D}{\partial \boldsymbol{n}}(\zeta,z)\mathrm{d}s = 2\left(\frac{r\sin\phi}{|z-t|^2} - \frac{R^2 r\sin\phi}{|R^2-zt|^2}\right)\mathrm{d}t. \qquad (2.1.21)$$

When D involved is a disk $\{z : |z-a| < r\}$, we briefly write $m(r,a,f)$, $N(r,a,f)$ and $T(r,a,f)$ for $m(D,a,f)$, $N(D,a,f)$ and $T(D,a,f)$, and if $a = 0$, then we write $m(r,f)$, $N(r,f)$ and $T(r,f)$ for $m(r,0,f)$, $N(r,0,f)$ and $T(r,0,f)$. Thus

$$m(r,f) = \frac{1}{2\pi}\int_0^{2\pi}\log^+|f(r\mathrm{e}^{\mathrm{i}\theta})|\mathrm{d}\theta$$

and

$$\begin{aligned} N(r,f) &= \sum_{b_n \in D} \log \frac{r}{|b_n|} + n(0,f)\log r \\ &= \int_0^r \log \frac{r}{t}\, \mathrm{d}(n(t,f)-n(0,f)) + n(0,f)\log r \\ &= \int_0^r \frac{n(t,f)-n(0,f)}{t}\mathrm{d}t + n(0,f)\log r, \end{aligned} \tag{2.1.22}$$

by noticing that $\omega_D(0,0) = \log r$, where $n(r,f)$ denotes the number of poles of f in $\{z : |z| < r\}$, and the Nevanlinna characteristic with center at the origin for the disk $D = \{z : |z| < r\}$ is

$$T(r,f) = m(r,f) + N(r,f).$$

We want to stress that Theorems 2.1.2~2.1.6 still hold for $m(r,a,f)$, $N(r,a,f)$ and $T(r,a,f)$ without restriction imposed on the primitive value of $f(z)$ at a, as long as we consider the coefficient of the first terms of the Laurent series of suitable functions at a.

Throughout this book, for an unbounded subset X of the complex plane, we denote by $n(r,X,f=a)$ and $\overline{n}(r,X,f=a)$ the number of, respectively, the roots repeated according to their multiplicities and distinct roots of $f(z) = a$, $a \in \widehat{\mathbb{C}}$ in $X \cap \{z : |z| < r\}$ and define $N(r,X,f=a)$ and $\overline{N}(r,X,f=a)$ in the same way as in (2.1.22). We shall use breviate notation $N(r,f=a)$ for $N(r,\mathbb{C},f=a)$ and then $N\left(r,\frac{1}{f-a}\right) = N(r,f=a)$ and $\overline{N}\left(r,\frac{1}{f-a}\right) = \overline{N}(r,f=a)$.

There exist relations of some delicacy between $N(D,a,f)$ and the number, denoted by $n(D,f)$, of poles of $f(z)$ in D, and those among other pairs of $n(*)$ and $N(*)$ for example, $n(r,X,f=a)$ and $N(r,X,f=a)$.

By means of the basic properties of the Green function, we compare $N(D,a,f)$ with $n(D,f)$. Let $R_D(a) = \sup\{|\zeta - a| : \zeta \in \Gamma\}$, $\Gamma = \partial D$. Since $G_D(z,a) + \log|z-a| - \log R_D(a)$ is harmonic in D and non-positive on Γ, it follows from the basic property of harmonic function that $G_D(z,a) + \log|z-a| - \log R_D(a)$ is negative in D, that is,

$$G_D(z,a) < \log \frac{R_D(a)}{|z-a|}. \tag{2.1.23}$$

Thus

$$\begin{aligned} N(D,a,f) &= \sum_{b_n \in D} G_D(b_n,a) \\ &\leqslant \sum_{b_n \in D} \log \frac{R}{|b_n - a|} \\ &= \int_0^R \frac{n_D(t,a,f)}{t}\mathrm{d}t, \ R = R_D(a), \end{aligned} \tag{2.1.24}$$

when $f(a) \neq \infty$, where b_n is a pole of $f(z)$ and $n_D(t,a,f)$ is the number of poles of $f(z)$ in $D \cap \{z : |z-a| < t\}$. Let $r_D(a) = \mathrm{dist}(a,\Gamma)$. By the same argument as above, we have

$$G_D(z,a) > \log \frac{r_D(a)}{|z-a|}. \tag{2.1.25}$$

Thus

$$\begin{aligned} N(D,a,f) &= \sum_{b_n \in D} G_D(b_n,a) \\ &\geqslant \sum_{\substack{b_n \in D \\ |b_n - a| < r}} \log \frac{r}{|b_n - a|} \\ &= \int_0^r \frac{n_D(t,a,f)}{t} \mathrm{d}t, \; r = r_D(a), f(a) \neq \infty. \end{aligned} \tag{2.1.26}$$

Important is the choice of a in the above estimation of $N(D,a,f)$ from below, which is related to the structure of D.

However, the following Boutroux-Cartan Theorem is often used in estimation of $N(D,a,f)$ in terms of $n(D,f)$.

Lemma 2.1.2. *Let a_j, $j = 1,2,\cdots,n$, be n complex numbers. Then the set of the point z satisfying*

$$\prod_{j=1}^{n} |z - a_j| < h^n$$

can be contained in several disks, denoted by (γ), the total sum of whose diameters does not exceed e*h.*

We shall call (γ) Boutroux-Cartan exceptional disks for these n complex numbers and h. Lemma 2.1.2 holds for the chordal metric, either. The chordal distance of two points a and b on the extended complex plane, denoted by $|a,b|$, is

$$|a,b| = \frac{|a-b|}{\sqrt{1+|a|^2}\sqrt{1+|b|^2}}, \text{ when } a,b \neq \infty, \text{ and } |a,\infty| = \frac{1}{\sqrt{1+|a|^2}}.$$

From Lemma 2.1.2 we can immediately obtain the following basic inequality, which is often used in the sequel. For n complex numbers a_j $(j = 1,2,\cdots,n)$ in D and for $1 \leqslant q \leqslant n$, in view of Lemma 2.1.2, we have

$$\prod_{j=1}^{q} \frac{R}{|a_j - a|} \leqslant R^q \hat{R}^{n-q} \prod_{j=1}^{n} \frac{1}{|a_j - a|} \leqslant R^q \hat{R}^{n-q} \left(\frac{1}{h}\right)^n = \left(\frac{R}{h}\right)^n \left(\frac{\hat{R}}{R}\right)^{n-q},$$

and therefore

$$\sum_{j=1}^{q} \log \frac{R}{|a_j - a|} \leqslant n \log \frac{R}{h} + (n-q) \log \frac{\hat{R}}{R}, \tag{2.1.27}$$

where $\hat{R} = \max\{|a_j - a| : q+1 \leqslant j \leqslant n\}$ and a is chosen outside the Boutroux-Cartan exceptional disks for these n complex numbers and $h < R$. Thus from (2.1.24) and (2.1.27), we have

$$N(D,a,f) \leqslant n(D,f) \log \frac{R}{h}.$$

On the other hand, we always have the inequality

$$\begin{aligned} N(r,X,f=a) &= \int_0^r \frac{n(t,X,f=a)-n(0,X,f=a)}{t}\mathrm{d}t + n(0,X,f=a)\log r \\ &\geqslant \int_1^r \frac{n(t,X,f=a)}{t}\mathrm{d}t \geqslant n(dr,X,f=a)\log\frac{1}{d},\ 0<d<1 \end{aligned}$$

for $1 < dr$ and

$$N(r,X,f=a) = \int_1^r \frac{n(t,X,f=a)}{t}\mathrm{d}t + O(1) \leqslant n(r,X,f=a)\log r + O(1).$$

Next we establish the estimations of $\log|f(z)|$ in a disk or a sector in terms of Theorem 2.1.1, because from (2.1.2) it follows that for $z \in D$ with $f(z) \neq \infty$, we have

$$\log|f(z)| \leqslant m(D,z,f) + N(D,z,f)$$

and since the quantity in the right side is non-negative, we have

$$\log^+|f(z)| \leqslant m(D,z,f) + N(D,z,f). \tag{2.1.28}$$

Lemma 2.1.3. *Let $f(z)$ be a meromorphic function on $\{z : |z| \leqslant R\}$. By (γ) we mean Boutroux-Cartan exceptional disks for the poles of $f(z)$ in $\{z : |z| < \tilde{R}\}$, $\tilde{R} < R$ and h. Then for $z \notin (\gamma)$ with $|z| = r < \tilde{R}$, we have*

$$\log^+|f(z)| \leqslant \left(\frac{\tilde{R}+r}{\tilde{R}-r} + \left(\log\frac{R}{\tilde{R}}\right)^{-1}\log\frac{\tilde{R}}{h}\right)T(R,f). \tag{2.1.29}$$

If $f(z)$ is analytic, then for each z with $|z| = r < R$, we have

$$\log^+|f(z)| \leqslant \frac{R+r}{R-r}T(R,f).$$

Proof. Set $D = \{z : |z| < \tilde{R}\}$. Then

$$\begin{aligned} m(D,z,f) &= \frac{1}{2\pi}\int_{\partial D}\log^+|f(\zeta)|\frac{\partial G_D(\zeta,z)}{\partial n}\mathrm{d}s \\ &= \frac{1}{2\pi}\int_0^{2\pi}\log^+|f(\tilde{R}\mathrm{e}^{\mathrm{i}\theta})|\frac{\tilde{R}^2-r^2}{|\tilde{R}\mathrm{e}^{\mathrm{i}\theta}-z|^2}\mathrm{d}\theta \\ &\leqslant \frac{\tilde{R}+r}{\tilde{R}-r}m(\tilde{R},f) \end{aligned}$$

and

$$N(D,z,f) \leqslant n(\tilde{R},f)\log\frac{\tilde{R}}{h} \leqslant \left(\log\frac{R}{\tilde{R}}\right)^{-1}\log\frac{\tilde{R}}{h}N(R,f).$$

In view of (2.1.28), combining the two above inequalities immediately yields (2.1.29). □

It is natural to consider analogy of Lemma 2.1.3 for an angular domain, which is basic in the investigation of argument distribution of a meromorphic function.

Lemma 2.1.4. *Let $f(z)$ be a meromorphic function on $\{z : |z| \leqslant R$ and $\operatorname{Im} z \geqslant 0\}$. By (γ) we mean Boutroux-Cartan exceptional disks for the poles of $f(z)$ and h. Then for $z \notin (\gamma)$ with $|z| = r < R$ and $\delta < \phi = \arg z < \pi - \delta$, $0 < \delta < \pi/2$, we have*

$$\begin{aligned}\log^+|f(z)| \leqslant & \frac{1}{\pi r \sin\phi}\frac{R+r}{R-r}\int_{-R}^{R}\left(1-\left(\frac{t}{R}\right)^2\right)\log^+|f(t)|\mathrm{d}t \\ & + \frac{\sin\phi}{\pi(1-\cos\delta)}\frac{R+r}{R-r}\int_0^{\pi}\log^+|f(Re^{\mathrm{i}\theta})|\sin\theta \mathrm{d}\theta \\ & + n(D_R, f)\log\frac{2R}{h}, \end{aligned} \tag{2.1.30}$$

where $n(D_R, f)$ is the number of poles of $f(z)$ counting the multiplicities in $D_R = \{z : |z| < R$ and $0 < \arg z < \pi\}$. If $f(z)$ is analytic, then for any z we have (2.1.30) without the final term $n(D_R, f)\log\frac{2R}{h}$.

Proof. In view of (2.1.2), (2.1.19), (2.1.20) and (2.1.21) we easily get the following formula

$$\begin{aligned}\log|f(z)| = & \frac{z-\bar{z}}{2\pi\mathrm{i}}\int_{-R}^{R}\log|f(t)|\left\{\frac{1}{|z-t|^2} - \frac{R^2}{|R^2-zt|^2}\right\}\mathrm{d}t \\ & + \frac{R^2-|z|^2}{2\pi}\int_0^{\pi}\log|f(Re^{\mathrm{i}\theta})|\left\{\frac{1}{|Re^{\mathrm{i}\theta}-z|^2} - \frac{1}{|Re^{-\mathrm{i}\theta}-z|^2}\right\}\mathrm{d}\theta \\ & - \sum_{\substack{|a_m|<R \\ \operatorname{Im} a_m>0}}\log\left|\frac{R^2-\bar{a}_m z}{R(z-a_m)}\frac{R(z-\bar{a}_m)}{R^2-a_m z}\right| \\ & + \sum_{\substack{|b_m|<R \\ \operatorname{Im} b_m>0}}\log\left|\frac{R^2-\bar{b}_m z}{R(z-b_m)}\frac{R(z-\bar{b}_m)}{R^2-b_m z}\right|. \end{aligned} \tag{2.1.31}$$

A straightforward calculation derives the following basic equalities and inequalities for $z = re^{\mathrm{i}\phi} \in D_R$, $\delta < \phi < \pi - \delta$,

$$|R^2 - tz|^2 - R^2|t-z|^2 = (R^2 - t^2)(R^2 - |z|^2), \tag{2.1.32}$$

$$|Re^{-\mathrm{i}\theta} - z|^2 - |Re^{\mathrm{i}\theta} - z|^2 = (\bar{z} - z)(Re^{\mathrm{i}\theta} - Re^{-\mathrm{i}\theta}) = 4Rr\sin\theta\sin\phi, \tag{2.1.33}$$

$$\frac{R^2-r^2}{|R^2-zt|^2} \leqslant \frac{R+r}{R^2(R-r)}, \quad |z-t| \geqslant r\sin\phi,$$

and, $\delta < \theta + \phi < 2\pi - \delta$,

$$\begin{aligned}\frac{2Rr}{|Re^{-i\theta}-z|^2} &= \frac{2Rr}{R^2+r^2-2Rr\cos(\theta+\phi)}\\ &\leqslant \frac{2Rr}{R^2+r^2-2Rr\cos\delta}\\ &\leqslant \frac{1}{1-\cos\delta}.\end{aligned}$$

We therefore have

$$\begin{aligned}\log^+|f(z)| &\leqslant \frac{r\sin\phi}{\pi}\int_{-R}^{R}\log^+|f(t)|\frac{(R^2-t^2)(R^2-r^2)}{|z-t|^2|R^2-zt|^2}\mathrm{d}t\\ &\quad+\frac{R^2-r^2}{2\pi}\int_0^{\pi}\log^+|f(Re^{i\theta})|\frac{4Rr\sin\theta\sin\phi}{|Re^{i\theta}-z|^2|Re^{-i\theta}-z|^2}\mathrm{d}\theta\\ &\quad+\sum_{\substack{|b_m|<R\\ \mathrm{Im}b_m>0}}\log\left|\frac{R^2-\bar{b}_m z}{R(z-b_m)}\frac{R(z-\bar{b}_m)}{R^2-b_m z}\right| \qquad (2.1.34)\\ &\leqslant \frac{1}{\pi r\sin\phi}\frac{R+r}{R-r}\int_{-R}^{R}\frac{R^2-t^2}{R^2}\log^+|f(t)|\mathrm{d}t\\ &\quad+\frac{\sin\phi}{\pi(1-\cos\delta)}\frac{R+r}{R-r}\int_0^{\pi}\log^+|f(Re^{i\theta})|\sin\theta\mathrm{d}\theta\\ &\quad+\sum_{\substack{|b_m|<R\\ \mathrm{Im}b_m>0}}\log\left|\frac{R^2-\bar{b}_m z}{R(z-b_m)}\right|.\\ &\leqslant \frac{1}{\pi r\sin\phi}\frac{R+r}{R-r}\int_{-R}^{R}\left(1-\left(\frac{t}{R}\right)^2\right)\log^+|f(t)|\mathrm{d}t\\ &\quad+\frac{\sin\phi}{\pi(1-\cos\delta)}\frac{R+r}{R-r}\int_0^{\pi}\log^+|f(Re^{i\theta})|\sin\theta\mathrm{d}\theta\\ &\quad+\sum_{\substack{|b_m|<R\\ \mathrm{Im}b_m>0}}\log\frac{2R}{|z-b_m|}.\end{aligned}$$

This immediately implies the desired result. □

The following lemma will be often used in the sequel, which is Lemma C of [8].

Lemma 2.1.5. *Let $f(z)$ be a meromorphic function on $\mathbb{C}$. With each $r(>0)$ we associate a measurable set $I(r)$ (of values of θ) of measure* $\mathrm{mes}I(r)\leqslant\pi$. *Then for $1\leqslant r<R$, we have*

$$\int_{I(r)}\log^+|f(re^{i\theta})|\mathrm{d}\theta \leqslant \frac{14R}{R-r}T(R,f)\mathrm{mes}I(r)\left[1+\log^+\frac{1}{\mathrm{mes}I(r)}\right]. \qquad (2.1.35)$$

Proof. Take a $\tilde{R}$ with $r<\tilde{R}<R$ and consider $D=\{z:|z|\leqslant\tilde{R}\}$. For $z=re^{i\theta}$ with $f(z)\neq\infty$, we have

$$m(D,z,f) \leqslant \frac{\tilde{R}+r}{\tilde{R}-r} m(\tilde{R},f)$$

and

$$N(D,z,f) = \sum_{b_n \in D} G_D(z,b_n) = \sum_{b_n \in D} \log \left| \frac{\tilde{R}^2 - \overline{b_n} z}{\tilde{R}(z-b_n)} \right| \leqslant \sum_{b_n \in D} \log \frac{\tilde{R}+r}{|z-b_n|}.$$

In view of (2.1.28) and the above inequalities, letting $\delta = \text{mes} I(r) \leqslant \pi$ and then applying Lemma 1.2.2, we have

$$\begin{aligned}
\int_{I(r)} \log^+ |f(re^{i\theta})| d\theta &\leqslant \frac{\tilde{R}+r}{\tilde{R}-r} m(\tilde{R},f)\delta + \sum_{b_n \in D} \int_{I(r)} \log \frac{\tilde{R}+r}{|re^{i\theta}-b_n|} d\theta \\
&\leqslant \frac{\tilde{R}+r}{\tilde{R}-r} m(\tilde{R},f)\delta + \frac{\tilde{R}+3r}{r} n(\tilde{R},f)\delta \left(1+\log^+ \frac{1}{\delta}\right) \\
&\leqslant \left(\frac{\tilde{R}+r}{\tilde{R}-r} + \frac{R(\tilde{R}+3r)}{r(R-\tilde{R})} \right) T(R,f)\delta \left(1+\log^+ \frac{1}{\delta}\right),
\end{aligned}$$

where the final inequality results from the following inequality

$$n(\tilde{R},f) \leqslant \left(\log \frac{R}{\tilde{R}}\right)^{-1} N(R,f) \leqslant \frac{R}{R-\tilde{R}} N(R,f).$$

Now let $\tilde{R} = \min\{\frac{R+r}{2}, 2r\}$. If $\tilde{R} = \frac{R+r}{2} \leqslant 2r$ and $\frac{R}{r} \leqslant 3$, then

$$\frac{\tilde{R}+r}{\tilde{R}-r} + \frac{R(\tilde{R}+3r)}{r(R-\tilde{R})} = \frac{R+3r+\frac{R}{r}R+7R}{R-r} \leqslant \frac{14R}{R-r};$$

If $\tilde{R} = 2r \leqslant \frac{R+r}{2}$, then $\frac{2}{R-r} \geqslant \frac{1}{R-2r}$ and

$$\frac{\tilde{R}+r}{\tilde{R}-r} + \frac{R(\tilde{R}+3r)}{r(R-\tilde{R})} \leqslant 3 + \frac{5R}{R-2r} \leqslant \frac{13R}{R-r}.$$

Thus the above inequalities immediately produce the desired inequality (2.1.35). □

For the case of entire functions, Hayman and Rossi [18] obtained the following result, which is produced from a combination of Theorem 3 and Theorem 5 of [18]: Let $f(z)$ be an entire function on $\mathbb{C}$. For any $\varepsilon > 0$, there exists a set F with $\overline{\text{dens}}F < \varepsilon$ such that for any measurable subset I of $[0,2\pi)$, we have

$$\int_I \log^+ \frac{1}{|f(re^{i\theta})-a|} d\theta \leqslant K(\log M(r,f)) \text{mes} I \log^+ \frac{4\pi}{\text{mes} I}, \quad r \notin F,$$

where K is a constant only depending on ε and a. This inequality is essentially a more precise estimation than (2.1.35), as in view of Lemma 2.1.3 we have $\log M(r,f) \leqslant \frac{R+r}{R-r} T(R,f)$.

The following is a transfiguration of the Nevanlinna second fundamental theorem for a disk, whose proof can be obtained from the proof of Theorem 3.3, that is, essentially Lemma 6.4, of Yang [36].

Lemma 2.1.6. *Let $f(z)$ be a meromorphic function in $\{z : |z| < R\}$ $(0 < R < +\infty)$ and let*

$$N = n(R, f = a) + n(R, f = b) + n(R, f = c)$$

for three distinct a, b and c in $\hat{\mathbb{C}}$. If $0 \notin (\gamma)$, (γ) is the set of Boutroux-Cartan exceptional disks corresponding to these N a, b, c-value points and poles and $h \leqslant \frac{R-r}{32e}$, then

$$T(r, f) < C\frac{R}{R-r}\left(N \log^{+}\frac{2R}{h} + \log\frac{R}{R-r}\right) + \log^{+}|f(0)|,$$

where C is a constant only depending on a, b and c.

Now it is turn of estimation of the number of valued-points. Let us present a modified format of the basic theorem of Valiron's [32] [33], which is important in discussion of value distribution of meromorphic functions and whose proof will be completed in view of Lemma 2.1.6.

Lemma 2.1.7. *Let $f(z)$ and $g(z)$ be both meromorphic functions in $\{z : |z| < R\}$ $(0 < R < +\infty)$ and g is allowed to be a value in $\widehat{\mathbb{C}}$. Set*

$$N = n(R, f = a) + n(R, f = b) + n(R, f = c) \text{ and } p = n(R, g)$$

for three distinct a, b and c in $\hat{\mathbb{C}}$. Then for $0 < r < R$, we have

$$\begin{aligned} n(r, f = g) < C\frac{R^2}{(R-r)^2}\Bigg((N+p)\log^{+}\frac{2R}{h} + \log\frac{1}{|f(z_0), g(z_0)|} \\ + \log\frac{R}{R-r} + (m(\tau, z_0, g) - \log^{+}|g(z_0)|)\Bigg), \end{aligned} \qquad (2.1.36)$$

for each $z_0 \in \{z : |z| < \frac{R-r}{5}\} \setminus ((\gamma) \cup (\gamma)')$, where (γ) is the set of Boutroux-Cartan exceptional disks for h and those N a, b, c-value points in $\{z : |z| < R\}$ and $(\gamma)'$ for the poles of $g(z)$, $\tau = r + 2(R-r)/5$ and C is a constant only depending on a, b and c. If $g \equiv \infty$, then we have

$$n(r, f = \infty) < C\frac{R^2}{(R-r)^2}\left(N \log^{+}\frac{2R}{h} + \log\frac{R}{R-r} + \log^{+}|f(z_0)|\right).$$

Proof. Assume that $g \not\equiv \infty$. Obviously we may assume $f(z_0) \neq g(z_0)$. A routine calculation yields the inclusion relation

$$\{z : |z| < r\} \subset \{z : |z - z_0| < r + (R-r)/5\} \subset \{z : |z - z_0| < \tau\} \subset \{z : |z| < R\},$$

$\tau = r + 2(R-r)/5$. It is easily seen from $f(z_0) \neq g(z_0)$ that

$$\begin{aligned} n(r, f = g) &\leqslant n\left(r + \frac{R-r}{5}, z_0, f = g\right) \\ &\leqslant \frac{5R}{R-r} N\left(\tau, z_0, \frac{1}{f-g}\right) \\ &\leqslant \frac{5R}{R-r} T\left(\tau, z_0, \frac{1}{f-g}\right) \\ &= \frac{5R}{R-r}\left(T(\tau, z_0, f-g) + \log \frac{1}{|f(z_0) - g(z_0)|}\right). \end{aligned}$$

Below we estimate the characteristic T-function in the above brace in view of Lemma 2.1.6. Then from $g(z_0) \neq \infty$, we have

$$\begin{aligned} T(\tau, z_0, f-g) &\leqslant T(\tau, z_0, f) + T(\tau, z_0, g) + \log 2 \\ &\leqslant \frac{CR}{R-r}\left((N+p)\log^+ \frac{2R}{h} + \log \frac{R}{R-r}\right) + \log^+ |f(z_0)| \\ &\quad + m(\tau, z_0, g) + \log 2. \end{aligned}$$

Noting that

$$\log^+ |f(z_0)| + \log^+ |g(z_0)| + \log \frac{1}{|f(z_0) - g(z_0)|} \leqslant \log \frac{1}{|f(z_0), g(z_0)|},$$

we get (2.1.36). The same argument as above yields the desired result for the case when $g \equiv \infty$. Thus we complete the proof of Lemma 2.1.7. □

Applying Lemmas 2.1.7 and 2.1.6, we establish the following result in [41], which will be often used in the sequel and is of independent significance.

Theorem 2.1.7. *Let $f(z)$ be a meromorphic function in $\{z : |z| < R\}$ $(0 < R < +\infty)$ and let*

$$N = n(R, f = 0) + n(R, f = 1) + n(R, f = \infty).$$

(γ) is the set of Boutroux-Cartan exceptional disks corresponding to these N zeros, one-value points and poles and $h = \frac{R}{K}, K > 32e$. Given $a \in \hat{\mathbb{C}}$, $(\gamma)_a$ is the set of Boutroux-Cartan exceptional disks corresponding to a-value points in $\{z : |z| \leqslant \frac{4R}{5}\}$ and h. Then for any $z_0 \notin (\gamma)$ and $z_1 \notin (\gamma)_a$ with $|z_0| < \frac{R}{5}$ and $|z_1| < \frac{R}{5}$, we have

$$\log^+ \frac{1}{|f(z_1) - a|} \leqslant C_{K,a}\left\{N + 1 + \log^+ \frac{1}{|f(z_0) - a|}\right\}, \tag{2.1.37}$$

where if $a = \infty$, $\log^+ \frac{1}{|f(*) - a|}$ is replaced by $\log^+ |f(*)|$, and

$$\log^+ \frac{1}{|f(z_1), a|} \leqslant C_{K,a}\left\{N + 1 + \log^+ \frac{1}{|f(z_0), a|}\right\}, \tag{2.1.38}$$

where $C_{K,a}$ is a constant only depending on K and a.

Proof. Let $F(z)=\frac{1}{f(z)-a}$ for $a\in\mathbb{C}$; $F(z)=f(z)$ for $a=\infty$. From Lemma 2.1.7 it follows that

$$n(\frac{4R}{5},F=\infty)<C_1\{N+1+\log^+|F(z_0)|\},$$

where C_1 is a constant only depending on a and K. It is easy to see that $\{z:|z-z_0|<\frac{4R}{5}\}\subset\{z:|z|<R\}$. Since $z_0\notin(\gamma)$ and the number of $0,1,\infty$-value points of $f(z)$ in $\{z:|z-z_0|<\frac{4R}{5}\}$ does not exceed N, from Lemma 2.1.6 we have

$$T(\frac{3R}{5},z_0,F)<C_2\{N+1\}+\log^+|F(z_0)|.$$

It is clear that $|z_1-z_0|<\frac{2R}{5}$ and $\{z:|z-z_0|<\frac{3R}{5}\}\subset\{z:|z|<\frac{4R}{5}\}$. Let c_j $(j=1,2,\cdots,p;\ p=n(\frac{4R}{5},F=\infty))$ be a-value points of $f(z)$ in $\{z:|z|<\frac{4R}{5}\}$ and c_j $(j=1,2,\cdots,p_1)$ are all points of $\{c_j\}_{j=1}^{p}$ in $\{z:|z-z_0|<\frac{3R}{5}\}$. From Lemma 2.1.2, we have

$$\prod_{j=1}^{p}|z_1-c_j|>h^p.$$

It is obvious that $R^{p-p_1}\prod_{j=1}^{p_1}|z_1-c_j|>h^p$ so that

$$K^p=\left(\frac{R}{h}\right)^p\geqslant\prod_{j=1}^{p_1}\frac{R}{|z_1-c_j|}. \tag{2.1.39}$$

By using the Poisson-Jensen Formula, that is, (2.1.2) with $D=\{z:|z-z_0|<3R/5\}$ and (2.1.39), we have

$$\begin{aligned}
\log^+|F(z_1)| &\leqslant \frac{1}{2\pi}\int_{\partial D}\log^+|F(\zeta)|\frac{\partial G_D(\zeta,z_1)}{\partial n}\mathrm{d}s+\sum_{c_j\in D}G_D(c_j,z_1)\\
&\leqslant \frac{\frac{3R}{5}+\frac{2R}{5}}{\frac{3R}{5}-\frac{2R}{5}}m\left(\frac{3R}{5},z_0,F\right)+\sum_{j=1}^{p_1}\log\left|\frac{\left(\frac{3R}{5}\right)^2-\overline{c_j-z_0}(z_1-z_0)}{\frac{3R}{5}(z_1-c_j)}\right|\\
&\leqslant 5m\left(\frac{3R}{5},z_0,F\right)+\sum_{j=1}^{p_1}\log\frac{R}{|z_1-c_j|}\\
&\leqslant 5m\left(\frac{3R}{5},z_0,F\right)+p\log K\\
&\leqslant C_3\left\{T\left(\frac{3R}{5},z_0,F\right)+n\left(\frac{4R}{5},F\right)\right\}\\
&\leqslant C_4\{N+1+\log^+|F(z_0)|\}.
\end{aligned}$$

This is the inequality (2.1.37). (2.1.38) follows from (2.1.37) by noting that

$$\frac{1}{|b,a|}\leqslant\sqrt{1+|a|^2}\left(1+\frac{1+|a|}{|a-b|}\right)\quad\text{and}\quad\frac{1}{|b,\infty|}\leqslant 1+|b|$$

and

$$\frac{1}{|a-b|} \leqslant \frac{1}{|a,b|}$$

for $a \neq b$.

Theorem 2.1.7 follows. □

What we should emphasize is that (2.1.37) and (2.1.38) hold for all z_1 without any exception provided that $f(z) \neq a$ on $\{z : |z| \leqslant R\}$.

Finally we take into account the second Nevanlinna's fundamental theorem for small functions as targets. A meromorphic function $a(z)$ is called small with respect to another meromorphic function $f(z)$ provided that $T(r,a) = o(T(r,f))$ as $r \notin E \to \infty$ for a set E of finite measure and if no exceptional set E is considered, then $a(z)$ is called absolutely small. Nevanlinna proposed whether the second fundamental theorem for complex numbers could be extended to that for small functions as targets. Recently, Yamanoi [34] completely solved the Nevanlinna's question by proving the following.

Theorem 2.1.8. *Let $f(z)$ be a transcendental meromorphic function and $a_j(z)$ $(j = 1,2,\cdots,q)$ be q distinct meromorphic functions small with respect to $f(z)$. Then for $\varepsilon > 0$,*

$$(q-2-\varepsilon)T(r,f) \leqslant \sum_{j=1}^{q} \overline{N}(r,a_j,f) + o(T(r,f)), \tag{2.1.40}$$

as $r \notin E \to \infty$ for a set E of finite measure.

Actually, the ε in the inequality (2.1.40) can be removed. Here we prove it. In view of Theorem 2.1.8, there exist a $r_1 > 0$ and a set E_1 such that $\int_{E_1(r_1,+\infty)} \mathrm{d}t < \frac{1}{2}$, and

$$(q-2)T(r,f) \leqslant \sum_{j=1}^{q} \overline{N}(r,a_j,f) + \frac{1}{2}T(r,f),\ r \notin E_1.$$

Then we can find a sequence of positive numbers $\{r_n\}$ with $r_n < r_{n+1} \to \infty$ and a sequence of sets $\{E_n\}$ with $\int_{E_n(r_n,+\infty)} \mathrm{d}t < \frac{1}{2^n}$ such that

$$(q-2)T(r,f) \leqslant \sum_{j=1}^{q} \overline{N}(r,a_j,f) + \frac{1}{n}T(r,f),\ r \notin E_n.$$

Set $E = \cup_{n=1}^{\infty} E_n(r_n, r_{n+1}) \cup [0, r_1)$ and define $\varepsilon(r)$ by $\varepsilon(r) = \frac{1}{n}$ for $r_n \leqslant r < r_{n+1}$. Obviously,

$$\int_E \mathrm{d}t = \sum_{n=1}^{\infty} \int_{E_n(r_n,r_{n+1})} \mathrm{d}t + \int_0^{r_1} \mathrm{d}t < \sum_{n=1}^{\infty} \frac{1}{2^n} + r_1 = 1 + r_1$$

and $\varepsilon(r) \to 0$ as $r \to \infty$. Then we have for $r \notin E$

$$(q-2)T(r,f) \leqslant \sum_{j=1}^{q} \overline{N}(r,a_j,f) + \varepsilon(r)T(r,f).$$

Namely,

$$(q-2)T(r,f) \leqslant \sum_{j=1}^{q} \overline{N}(r,a_j,f) + o(T(r,f)), \text{ as } r \notin E \to \infty.$$

Chi-Tai Chuang [4] was the first one to make progress in study of the Nevanlinna problem and confirmed this problem for the case of entire functions without the bar over the letter N in (2.1.40), and, we should mention that he introduced the Wronskian determinant into the investigation of value distribution of meromorphic functions. This problem without the bar over the letter N in (2.1.40) was solved by Frank and Weissenborn [10] for rational functions as targets and by, independently, Osgood [28] and Steinmetz [30] for a general small functions. The methods of Frank-Weissenborn and Steinmetz's to solve the problem is a continuation of Chuang's method.

2.2 Nevanlinna's Characteristic in an Angle

The Nevanlinna's characteristic of a meromorphic function in an angle stems from the following Carleman formula, which is similar to the formula (2.1.2). For a function $f(z)$ meromorphic in the half ring $\Omega_{0,\pi}(R,R_0) = \{z : R_0 \leqslant |z| \leqslant R, \mathrm{Im} z \geqslant 0\}$, we have

$$\begin{aligned}
&\sum_m \left(\frac{1}{|a_m|} - \frac{|a_m|}{R^2}\right) \sin\phi_m - \sum_n \left(\frac{1}{|b_n|} - \frac{|b_n|}{R^2}\right) \sin\theta_n \\
&= \frac{1}{2\pi} \int_{R_0}^{R} \left(\frac{1}{t^2} - \frac{1}{R^2}\right) (\log|f(t)| + \log|f(-t)|) \mathrm{d}t \\
&\quad + \frac{1}{\pi R} \int_0^{\pi} \log|f(R\mathrm{e}^{\mathrm{i}\theta})| \sin\theta \mathrm{d}\theta + Q(R,R_0,f),
\end{aligned} \tag{2.2.1}$$

where $a_m = |a_m|\mathrm{e}^{\mathrm{i}\phi_m}$ are the zeros of $f(z)$ and $b_n = |b_n|\mathrm{e}^{\mathrm{i}\theta_n}$ are the poles of $f(z)$ on $\Omega_{0,\pi}(R,R_0)$, and $\sum\limits_m$ is the sum taken over all the zeros and $\sum\limits_n$ is that over all the poles of $f(z)$ on $\Omega_{0,\pi}(R,R_0)$, and

$$\begin{aligned}
Q(R,R_0,f) = &-\frac{1}{2\pi} \int_0^{\pi} \left\{ \left(\frac{1}{R_0^2} + \frac{1}{R^2}\right) \log|f(R_0\mathrm{e}^{\mathrm{i}\theta})| \right. \\
&\left. + \left(\frac{1}{R_0} + \frac{R_0}{R^2}\right) \frac{\partial}{\partial R_0} \log|f(R_0\mathrm{e}^{\mathrm{i}\theta})| \right\} R_0 \sin\theta \mathrm{d}\theta = O(1),
\end{aligned}$$

as $R \to \infty$.

The Carleman formula (2.2.1) can be also derived directly from the second Green formula. In fact, let $\Omega(\varepsilon)$ be the domain obtained by removing disks of sufficiently small radius $\varepsilon > 0$ with centers at a_m and b_n from $\Omega_{0,\pi}(R,R_0)$. Then

$$\int_{\partial\Omega(\varepsilon)} \left(u\frac{\partial v}{\partial n} - v\frac{\partial u}{\partial n} \right) \mathrm{d}s = 0, \tag{2.2.2}$$

for $u(z) = \log|f(z)|$ and $v(z) = -\mathrm{Im}\left(\frac{1}{z} + \frac{z}{R^2}\right)$ by noting that these two functions are harmonic in $\overline{\Omega}(\varepsilon)$. Obviously, we have

$$v(z) = 0, \ \frac{\partial v}{\partial n} = \frac{2}{R^2}\sin\phi, \ z = Re^{\mathrm{i}\phi}$$

on the half circle $\{z : |z| = R, \mathrm{Im}z \geqslant 0\}$, and

$$v(z) = 0, \ \frac{\partial v}{\partial n} = \frac{1}{t^2} - \frac{1}{R^2}$$

on the interval $\{z = \pm t : R_0 < t < R\}$, and for a zero or pole a of $f(z)$, we can write in a neighborhood of a

$$u(z) = p\log|z-a| + \varphi(z)$$

for some integer p and harmonic $\varphi(z)$, thus

$$\begin{aligned}
\int_{|z-a|=\varepsilon} \left(u\frac{\partial v}{\partial n} - v\frac{\partial u}{\partial n} \right) \mathrm{d}s &= p\int_{|z-a|=\varepsilon} \left(\log|z-a|\frac{\partial v}{\partial n} - v\frac{\partial \log|z-a|}{\partial n} \right) \mathrm{d}s \\
&= -p\int_{|z-a|=\varepsilon} v\frac{\partial \log|z-a|}{\partial n}\mathrm{d}s \\
&= 2p\pi v(a) \\
&= 2\pi p\left(\frac{1}{|a|} - \frac{|a|}{R^2}\right)\sin\varphi, \varphi = \arg a.
\end{aligned}$$

Therefore, (2.2.1) follows from (2.2.2).

For the sake of simplicity, throughout this book, we denote by $\Omega(\alpha,\beta)$ the angle $\{z : \alpha < \arg z < \beta\}$ and by $\overline{\Omega}(\alpha,\beta)$ the closed angle, and set

$$\Omega(\alpha,\beta;R,R_0) = \Omega(\alpha,\beta) \cap \{z : R_0 < |z| < R\}$$

$$\Omega(\alpha,\beta;R) = \Omega(\alpha,\beta) \cap \{z : 1 < |z| < R\}.$$

Let $f(z)$ be a meromorphic function on the angle $\overline{\Omega}(\alpha,\beta) = \{z : \ \alpha \leqslant \arg z \leqslant \beta\}$, where $0 < \beta - \alpha \leqslant 2\pi$. Following Nevanlinna (see [11]), define

$$\begin{aligned}
A_{\alpha,\beta}(r,f) &= \frac{\omega}{\pi}\int_1^r \left(\frac{1}{t^\omega} - \frac{t^\omega}{r^{2\omega}} \right) \{\log^+|f(te^{\mathrm{i}\alpha})| + \log^+|f(te^{\mathrm{i}\beta})|\}\frac{\mathrm{d}t}{t}, \\
B_{\alpha,\beta}(r,f) &= \frac{2\omega}{\pi r^\omega}\int_\alpha^\beta \log^+|f(re^{\mathrm{i}\theta})|\sin\omega(\theta-\alpha)\mathrm{d}\theta, \qquad (2.2.3) \\
C_{\alpha,\beta}(r,f) &= 2\sum_{1<|b_n|<r} \left(\frac{1}{|b_n|^\omega} - \frac{|b_n|^\omega}{r^{2\omega}} \right) \sin\omega(\theta_n - \alpha),
\end{aligned}$$

where $\omega = \frac{\pi}{\beta-\alpha}$ and $b_n = |b_n|\mathrm{e}^{\mathrm{i}\theta_n}$ are the poles of $f(z)$ on $\overline{\Omega}(\alpha,\beta)$ appeared according to their multiplicities. And define $\overline{C}_{\alpha,\beta}(r,f)$ in the same form of $C_{\alpha,\beta}(r,f)$ for distinct poles b_n of $f(z)$, that is, ignoring their multiplicities. $C_{\alpha,\beta}(r,f)$ (resp. $\overline{C}_{\alpha,\beta}(r,f)$) is called the angular (precise) integrated counting function of the poles of $f(z)$ on $\overline{\Omega}(\alpha,\beta)$. For $a \in \mathbb{C}$, we write $C_{\alpha,\beta}(r,f=a)$ for $C_{\alpha,\beta}\left(r,\frac{1}{f-a}\right)$; $C_{\alpha,\beta}(r,f=\infty)$ for $C_{\alpha,\beta}(r,f)$ sometimes in the sequel. Furthermore, we can give an integral expression of $C_{\alpha,\beta}(r,f)$ and $\overline{C}_{\alpha,\beta}(r,f)$. Set

$$c_{\alpha,\beta}(r,f) = \sum_{1<|b_n|<r} \sin(\omega(\theta_n - \alpha)).$$

Then

$$\begin{aligned} C_{\alpha,\beta}(r,f) &= 2\int_1^r \left(\frac{1}{t^\omega} - \frac{t^\omega}{r^{2\omega}}\right) \mathrm{d}c_{\alpha,\beta}(t,f) \\ &= 2\omega \int_1^r c_{\alpha,\beta}(t,f)\left(\frac{1}{t^\omega} + \frac{t^\omega}{r^{2\omega}}\right)\frac{\mathrm{d}t}{t}. \end{aligned} \tag{2.2.4}$$

The Nevanlinna's angular characteristic is defined as follows:

$$S_{\alpha,\beta}(r,f) = A_{\alpha,\beta}(r,f) + B_{\alpha,\beta}(r,f) + C_{\alpha,\beta}(r,f).$$

It is easy to see that all the inequalities listed before Theorem 2.1.2 are available for $S_{\alpha,\beta}$, $(A+B)_{\alpha,\beta}$ and $C_{\alpha,\beta}$. For instance, we have

$$S_{\alpha,\beta}(r,\sum_{j=1}^p f_j) \leqslant \sum_{j=1}^p S_{\alpha,\beta}(r,f_j) + 3\log p$$

and

$$S_{\alpha,\beta}(r,\prod_{j=1}^p f_j) \leqslant \sum_{j=1}^p S_{\alpha,\beta}(r,f_j).$$

In view of the transformation $w = (\mathrm{e}^{-\mathrm{i}\alpha}z)^\omega$ which maps $\overline{\Omega}(\alpha,\beta)$ onto $\overline{\Omega}(0,\pi)$, and setting $F(w) = f(\mathrm{e}^{\mathrm{i}\alpha}w^{1/\omega}) = f(z)$, we have

$$\begin{aligned} A_{0,\pi}(r^\omega,F) &= \frac{1}{\pi}\int_1^{r^\omega}\left(\frac{1}{t} - \frac{t}{r^{2\omega}}\right)\{\log^+|F(t)| + \log^+|F(-t)|\}\frac{\mathrm{d}t}{t} \\ &= \frac{1}{\pi}\int_1^{r^\omega}\left(\frac{1}{t} - \frac{t}{r^{2\omega}}\right)\{\log^+|f(t^{1/\omega}\mathrm{e}^{\mathrm{i}\alpha})| + \log^+|f(t^{1/\omega}\mathrm{e}^{\mathrm{i}\beta})|\}\frac{\mathrm{d}t}{t} \\ &= \frac{\omega}{\pi}\int_1^r\left(\frac{1}{t^\omega} - \frac{t^\omega}{r^{2\omega}}\right)\{\log^+|f(t\mathrm{e}^{\mathrm{i}\alpha})| + \log^+|f(t\mathrm{e}^{\mathrm{i}\beta})|\}\frac{\mathrm{d}t}{t} \\ &= A_{\alpha,\beta}(r,f), \end{aligned}$$

so that

$$A_{\alpha,\beta}(r,f) = A_{0,\pi}(r^{\omega},F).$$

The same equality still holds for $B_{\alpha,\beta}$ and $C_{\alpha,\beta}$ as well so that

$$S_{\alpha,\beta}(r,f) = S_{0,\pi}(r^{\omega},F).$$

Using the Carleman formula (2.2.1) to the function $F(w)$, we can obtain

$$S_{\alpha,\beta}(r,f) = S_{0,\pi}(r^{\omega},F) = S_{0,\pi}\left(r^{\omega},\frac{1}{F}\right) + O(1) = S_{\alpha,\beta}\left(r,\frac{1}{f}\right) + O(1)$$

and further the Nevanlinna first fundamental theorem for the angular domain, that is,

$$S_{\alpha,\beta}(r,f) = S_{\alpha,\beta}\left(r,\frac{1}{f-a}\right) + O(1) \tag{2.2.5}$$

for a complex number $a \in \mathbb{C}$. Therefore, we want to emphasize that Theorems 2.1.2, 2.1.3, 2.1.4 and 2.1.5 still hold for $S_{\alpha,\beta}$, $(A+B)_{\alpha,\beta}$ and $C_{\alpha,\beta}$ in the place of, respectively, T, m and N there, with error terms denoted by $R_{\alpha,\beta}(r,f)$ replacing $S(D,a,f)$, by applying the same methods which produce the inequalities there and (2.2.5). For instance, we have the Nevanlinna second fundamental theorems, the Milloux inequality and the Hayman inequality for the angular domain $\Omega(\alpha,\beta)$:

$$(q-2)S_{\alpha,\beta}(r,f) \leqslant \sum_{j=1}^{q} \overline{C}_{\alpha,\beta}\left(r,\frac{1}{f-a_j}\right) + R_{\alpha,\beta}(r,f), \tag{2.2.6}$$

for q distinct points $a_j \in \hat{\mathbb{C}}$, where

$$R_{\alpha,\beta}(r,f) = (A+B)_{\alpha,\beta}\left(r,\frac{f'}{f}\right) + \sum_{j=1}^{q}(A+B)_{\alpha,\beta}\left(r,\frac{f'}{f-a_j}\right) + O(1), \tag{2.2.7}$$

and

$$\begin{aligned} S_{\alpha,\beta}(r,f) \leqslant{}& \overline{C}_{\alpha,\beta}(r,f) + C_{\alpha,\beta}(r,f=0) + C_{\alpha,\beta}(r,f^{(k)}=1) \\ & - C_{\alpha,\beta}(r,f^{(k+1)}=0) + R_{\alpha,\beta}(r,f), \end{aligned} \tag{2.2.8}$$

where

$$\begin{aligned} R_{\alpha,\beta}(r,f) ={}& (A+B)_{\alpha,\beta}\left(r,\frac{f^{(k)}}{f}\right) + (A+B)_{\alpha,\beta}\left(r,\frac{f^{(k+1)}}{f}\right) \\ & + (A+B)_{\alpha,\beta}\left(r,\frac{f^{(k+1)}}{f^{(k)}-1}\right) + O(1), \end{aligned} \tag{2.2.9}$$

and

$$S_{\alpha,\beta}(r,f) \leqslant \left(2+\frac{1}{k}\right) C_{\alpha,\beta}\left(r,\frac{1}{f}\right) + \left(2+\frac{2}{k}\right) \overline{C}_{\alpha,\beta}(r,f^{(k)}=1)$$
$$+R_{\alpha,\beta}(r,f),\ k>0, \tag{2.2.10}$$

where

$$R_{\alpha,\beta}(r,f) = \left(2+\frac{2}{k}\right)(A+B)_{\alpha,\beta}\left(r,\frac{f^{(k+1)}}{f^{(k)}-1}\right) + \left(2+\frac{1}{k}\right)(A+B)_{\alpha,\beta}\left(r,\frac{f^{(k+1)}}{f}\right)$$
$$+\left(2+\frac{1}{k}\right)(A+B)_{\alpha,\beta}\left(r,\frac{f^{(k)}}{f}\right) + O(1). \tag{2.2.11}$$

Throughout this book, $R_{\alpha,\beta}(r,f)$ is called error term associated with the Nevanlinna characteristic for the angle $\Omega(\alpha,\beta)$, and it may not be the same at each occurrence.

By $f^{\#}(z)$ we mean the sphere derivative of $f(z)$, that is, for $f(z)\neq\infty$

$$f^{\#}(z) = \lim_{\zeta\to z}\frac{|f(\zeta),f(z)|}{|\zeta-z|} = \frac{|f'(z)|}{1+|f(z)|^2}$$

and for $f(z)=\infty$, $f^{\#}(z)=\lim_{\zeta\to z} f^{\#}(\zeta)$. It is easy to derive in view of simple calculation

$$\Delta\log(1+|f(z)|^2) = \frac{4|f'(z)|^2}{(1+|f(z)|^2)^2} = 4(f^{\#}(z))^2,$$

where Δ is the Laplacian operator. Set

$$\dot{S}_{\alpha,\beta}(r,f) = \frac{1}{\pi}\int_1^r\int_\alpha^\beta\left(\frac{1}{t^\omega}-\frac{t^\omega}{r^{2\omega}}\right)(f^{\#}(te^{i\theta}))^2\sin\omega(\theta-\alpha)\,t\mathrm{d}t\mathrm{d}\theta.$$

It is obvious that $\dot{S}_{\alpha,\beta}(r,f)$ is increasing with respect to r. Set

$$D_{\alpha,\beta}(r) = \int_1^r\int_\alpha^\beta (f^{\#}(te^{i\theta}))^2\sin\omega(\theta-\alpha)\,t\mathrm{d}t\mathrm{d}\theta.$$

By means of the formula for integration by parts, we immediately have

$$\dot{S}_{\alpha,\beta}(r,f) = \frac{1}{\pi}\int_1^r\left(\frac{1}{t^\omega}-\frac{t^\omega}{r^{2\omega}}\right)\mathrm{d}D_{\alpha,\beta}(t)$$
$$= \frac{\omega}{\pi}\int_1^r\left(\frac{1}{t^{\omega+1}}+\frac{t^{\omega-1}}{r^{2\omega}}\right)D_{\alpha,\beta}(t)\mathrm{d}t. \tag{2.2.12}$$

From $F(w)=f(e^{i\alpha}w^{1/\omega})$ on $\overline{\Omega}(0,\pi)$, we have

$$F^{\#}(te^{i\theta}) = \omega^{-1}t^{1/\omega-1}f^{\#}(t^{1/\omega}e^{i(\alpha+\theta/\omega)}),$$

and then

$$\dot{S}_{0,\pi}(r^\omega,F) = \frac{1}{\pi}\int_1^{r^\omega}\int_0^\pi \left(\frac{1}{t}-\frac{t}{r^{2\omega}}\right) F^\#(te^{i\theta})^2 \sin\theta \, t\mathrm{d}t\mathrm{d}\theta$$
$$= \frac{1}{\pi}\int_1^{r^\omega}\int_0^\pi \left(\frac{1}{t}-\frac{t}{r^{2\omega}}\right) f^\#(t^{1/\omega}e^{i(\alpha+\theta/\omega)})^2\omega^{-2}(t^{1/\omega-1})^2 \sin\theta \, t\mathrm{d}t\mathrm{d}\theta$$
$$= \frac{1}{\pi}\int_1^{r}\int_\alpha^\beta \left(\frac{1}{x^\omega}-\frac{x^\omega}{r^{2\omega}}\right) f^\#(xe^{i\phi})^2 \sin\omega(\phi-\alpha) \, x\mathrm{d}x\mathrm{d}\phi,\ x=t^{\frac{1}{\omega}},\ \phi=\alpha+\frac{\theta}{\omega}$$
$$= \dot{S}_{\alpha,\beta}(r,f).$$

There is a close relation between $\dot{S}_{\alpha,\beta}(r,f)$ and $S_{\alpha,\beta}(r,f)$. We can obtain

$$S_{0,\pi}(r,f) = \dot{S}_{0,\pi}(r,f) + O(1)$$

by employing the second Green formula to the functions $u(z) = \frac{1}{2}\log(1+|f(z)|^2)$ and $v(z) = -\mathrm{Im}\left(\frac{1}{z}+\frac{z}{R^2}\right)$, and by noting

$$|\log^+|f(z)| - \frac{1}{2}\log[1+|f(z)|^2]| \leqslant \frac{1}{2}\log 2.$$

The reader is referred to Chapter 3 of [11] for the detail implication. Therefore we have

Lemma 2.2.1. *Let $f(z)$ be a meromorphic function on $\overline{\Omega}(\alpha,\beta)$. Then we have the following*

$$S_{\alpha,\beta}(r,f) = \dot{S}_{\alpha,\beta}(r,f) + O(1),$$

and for $\delta > 0$, we have

$$\dot{S}_{\alpha,\beta}(r,f) \geqslant \frac{\omega}{\hat{\omega}}\dot{S}_{\alpha+\delta,\beta-\delta}(r,f), \tag{2.2.13}$$

where $\hat{\omega} = \frac{\pi}{\beta-\alpha-2\delta}$.

Proof. The first equality follows from $S_{\alpha,\beta}(r,f) = S_{0,\pi}(r^\omega,F)$ and $\dot{S}_{\alpha,\beta}(r,f) = \dot{S}_{0,\pi}(r^\omega,F)$, where $F(w) = f(e^{i\alpha}w^{1/\omega})$. Below we prove the inequality (2.2.13). When $\alpha+\delta \leqslant \theta \leqslant \frac{\alpha+\beta}{2}$, we have $(\beta-\alpha-2\delta)(\theta-\alpha) > (\beta-\alpha)(\theta-\alpha-\delta)$ and so $\hat{\omega}(\theta-\alpha-\delta) < \omega(\theta-\alpha) \leqslant \frac{\pi}{2}$; When $\beta-\delta \geqslant \theta \geqslant \frac{\alpha+\beta}{2}$, we obtain $\frac{\pi}{2} \leqslant \omega(\theta-\alpha) < \hat{\omega}(\theta-\alpha-\delta) \leqslant \pi$; Thus we always have $\sin\omega(\theta-\alpha) > \sin\hat{\omega}(\theta-\alpha-\delta)$, and therefore $D_{\alpha,\beta}(r) \geqslant D_{\alpha+\delta,\beta-\delta}(r)$. Consider the function

$$h(x) = \frac{1}{t^{x+1}} + \frac{t^{x-1}}{r^{2x}},\ x > 0 \text{ and } 1 \leqslant t \leqslant r.$$

It is easy to see that it is decreasing, and so $h(\omega) > h(\hat{\omega})$ for $\hat{\omega} > \omega > 0$ and for $1 \leqslant t \leqslant r$. (2.2.13) follows immediately from the representation (2.2.12) of $\dot{S}_{\alpha,\beta}(r,f)$ and the above facts. □

An important application of Lemma 2.2.1 is to show that $S_{\alpha,\beta}(r,f)$ is increasing up to a bounded quantity as $\dot{S}_{\alpha,\beta}(r,f)$ is increasing.

Certainly, it is important and necessary to determine relations between $C_{\alpha,\beta}(r,f)$ and $N(r,\Omega,f)$, which will be helpful in characterizing meromorphic functions in an angle in terms of the number of points of some values.

Lemma 2.2.2. *Let $f(z)$ be a meromorphic function on $\overline{\Omega}(\alpha,\beta)$. Then the following inequalities hold:*

$$C_{\alpha,\beta}(r,f) \leqslant 4\omega\frac{N(r)}{r^{\omega}} + 2\omega^2\int_1^r \frac{N(t)}{t^{\omega+1}}\mathrm{d}t \tag{2.2.14}$$

and

$$C_{\alpha,\beta}(r,f) \geqslant 2\omega\sin(\omega\delta)\frac{N_0(r)}{r^{\omega}} + 2\omega^2\sin(\omega\delta)\int_1^r \frac{N_0(t)}{t^{\omega+1}}\mathrm{d}t, \tag{2.2.15}$$

where $N(t) = N(t,\Omega,f) = \int_1^t \frac{n(t,\Omega,f)}{t}\mathrm{d}t$, $n(t,\Omega,f)$ is the number of poles of $f(z)$ in $\Omega\cap\{z : 1 < |z| \leqslant t\}$, and $N_0(t) = N(t,\Omega_\delta,f) = \int_1^t \frac{n(t,\Omega_\delta,f)}{t}\mathrm{d}t$, and $\Omega_\delta = \Omega(\alpha+\delta,\beta-\delta)$. (2.2.14) and (2.2.15) still hold for $\overline{C}$ and $\overline{N}$ in the place of C and N.

Proof. For the sake of simplicity, we write $n(t)$ for $n(t,\Omega,f)$ and dropping the subscripts of $C_{\alpha,\beta}(r,f)$, we instead use $C(r,f)$. From the representation of $C(r,f)$, it follows, by means of the formula for integration by parts, that

$$\begin{aligned}
C(r,f) &\leqslant 2\sum_{1<|b_n|<r}\left(\frac{1}{|b_n|^{\omega}} - \frac{|b_n|^{\omega}}{r^{2\omega}}\right)\\
&= 2\int_1^r \left(\frac{1}{t^{\omega}} - \frac{t^{\omega}}{r^{2\omega}}\right)\mathrm{d}n(t)\\
&= 2\omega\int_1^r n(t)\left(\frac{1}{t^{\omega+1}} + \frac{t^{\omega-1}}{r^{2\omega}}\right)\mathrm{d}t, \quad \text{noting } n(1)=0\\
&= 2\omega\int_1^r \left(\frac{1}{t^{\omega}} + \frac{t^{\omega}}{r^{2\omega}}\right)\mathrm{d}N(t)\\
&\leqslant 4\omega\frac{N(r)}{r^{\omega}} + 2\omega^2\int_1^r \frac{N(t)}{t^{\omega+1}}\mathrm{d}t,
\end{aligned}$$

this is (2.2.14), and

$$\begin{aligned}
C(r,f) &\geqslant 2\sin(\omega\delta)\sum_{b_n\in\Omega_\delta}\left(\frac{1}{|b_n|^{\omega}} - \frac{|b_n|^{\omega}}{r^{2\omega}}\right)\\
&= 2\sin(\omega\delta)\int_1^r \left(\frac{1}{t^{\omega}} - \frac{t^{\omega}}{r^{2\omega}}\right)\mathrm{d}n_0(t)
\end{aligned}$$

$$
\begin{aligned}
&= 2\omega\sin(\omega\delta)\int_1^r n_0(t)\left(\frac{1}{t^{\omega+1}}+\frac{t^{\omega-1}}{r^{2\omega}}\right)\mathrm{d}t\\
&\geqslant 2\omega\sin(\omega\delta)\int_1^r \frac{1}{t^{\omega}}\mathrm{d}N_0(t)\\
&= 2\omega\sin(\omega\delta)\frac{N_0(r)}{r^{\omega}}+2\omega^2\sin(\omega\delta)\int_1^r\frac{N_0(t)}{t^{\omega+1}}\mathrm{d}t,
\end{aligned}
$$

which is (2.2.15). □

Now we estimate $\log^+|f(z)|$ in terms of the Nevanlinna characteristic in an angle.

Theorem 2.2.1. *Let $f(z)$ be a meromorphic function on $\overline{\Omega}=\overline{\Omega}(\alpha,\beta;R,0)=\{z: \alpha\leqslant\arg z\leqslant\beta, |z|\leqslant R\}$ with $R\geqslant 1$. By (γ) we mean Boutroux-Cartan exceptional disks for the poles of $f(z)$ and h. Then for $z=re^{i\phi}\in\Omega(\alpha+\delta,\beta-\delta;R_0,0)\setminus(\gamma)$ with $1<r<R_0<R$, we have*

$$
\begin{aligned}
\log^+|f(z)| \leqslant K_{\delta,\alpha,\beta}R_0^{\tilde{\omega}}&\left[\frac{R_0^{\tilde{\omega}}+r^{\tilde{\omega}}}{R_0^{\tilde{\omega}}-r^{\tilde{\omega}}}+\left(\log\frac{2R^{\tilde{\omega}}}{h^{\tilde{\omega}}}\right)\left(\log\frac{R}{R_0}\right)^{-1}\frac{1}{2\omega\sin(\delta/2)}\right]\\
&\times S_{\alpha,\beta}(R,f)+O\left(\frac{R_0{}^{\tilde{\omega}}+r^{\tilde{\omega}}}{R_0{}^{\tilde{\omega}}-r^{\tilde{\omega}}}\right) \qquad (2.2.16)
\end{aligned}
$$

for a positive constant $K_{\delta,\alpha,\beta}$ and $\tilde{\omega}=\frac{\pi}{\beta-\alpha-\delta}$.

Proof. First of all, assume that $\alpha=0$ and $\beta=\pi$ and $|z|=r<R$ and $\delta<\arg z<\pi-\delta$. We shall use (2.1.34) in this case. Now we estimate $\frac{t^2}{|z-t|^2}$. Letting $z=x+iy$, set

$$
h(t)=\frac{t^2}{(x-t)^2+y^2},\ t\geqslant 0.
$$

It is easy to see that if $x\leqslant 0$, then $h(t)\leqslant 1$. For $x>0$, a straightforward calculation implies that $h(t)$ assumes its maximum value at $t_0=x^{-1}(x^2+y^2)$ and therefore

$$
h(t)\leqslant h(t_0)=\frac{x^{-2}(x^2+y^2)^2}{x^{-2}y^4+y^2}=1+\left(\frac{x}{y}\right)^2\leqslant(\sin\delta)^{-2}.
$$

In view of (2.1.34), by combining the representations of $A(R,f)$ and $B(R,f)$, we immediately have

$$
\begin{aligned}
\log^+|f(z)|\leqslant{}&\frac{r\sin\phi}{\pi}(\sin\delta)^{-2}\frac{R+r}{R-r}\int_1^R\frac{R^2-t^2}{t^2R^2}[\log^+|f(t)|+\log^+|f(-t)|]\mathrm{d}t\\
&+\frac{1}{\pi r\sin\phi}\frac{R+r}{R-r}\int_{-1}^1\log^+|f(t)|\mathrm{d}t\\
&+\frac{R\sin\phi}{2(1-\cos\delta)}\frac{R+r}{R-r}B(R,f)+n(R,\Omega,f)\log\frac{2R}{h}
\end{aligned}
$$

$$\leqslant r\sin\phi(\sin\delta)^{-2}\frac{R+r}{R-r}A(R,f)+O(\frac{R+r}{R-r})$$
$$+\frac{R\sin\phi}{2(1-\cos\delta)}\frac{R+r}{R-r}B(R,f)+n(R,\Omega,f)\log\frac{2R}{h}.$$

Now let us consider general case. Letting $w=(e^{-i\alpha}z)^{\omega}$, we shall use the above result to function $f(z)=f(e^{i\alpha}w^{1/\omega})=F(w)$ in the upper half disk $\{w:0\leqslant\arg w\leqslant\pi,|w|\leqslant R^{\omega}\}$ to obtain the desired inequality (2.2.16) by noting that $A_{0,\pi}(R^{\omega},F)=A_{\alpha,\beta}(R,f)$ and $B_{0,\pi}(R^{\omega},F)=B_{\alpha,\beta}(R,f)$.

By (γ) we mean Boutroux-Cartan exceptional disks for the poles of $F(w)$ and h^{ω}. Then for $z=re^{i\phi}\in\Omega(\alpha+\delta,\beta-\delta;R)$ such that $w=(e^{-i\alpha}z)^{\omega}\notin(\gamma)$, we have

$$\log^{+}|f(z)|\leqslant r^{\omega}\sin\omega(\phi-\alpha)(\sin(\omega\delta))^{-2}\frac{R^{\omega}+r^{\omega}}{R^{\omega}-r^{\omega}}A_{\alpha,\beta}(R,f)$$
$$+O(\frac{R^{\omega}+r^{\omega}}{R^{\omega}-r^{\omega}})+\frac{R^{\omega}\sin\omega(\phi-\alpha)}{2(1-\cos(\omega\delta))}\frac{R^{\omega}+r^{\omega}}{R^{\omega}-r^{\omega}}B_{\alpha,\beta}(R,f)$$
$$+n(R,\Omega,f)\log\frac{2R^{\omega}}{h^{\omega}}, \tag{2.2.17}$$

where $n(R,\Omega,f)$ is the number of poles of $f(z)$ in $\Omega\cap\{z:|z|<R\}$.

Now set $\Omega_{\delta/2}=\Omega(\alpha+\delta/2,\beta-\delta/2;R_0)$ for $R_0<R$ and write $\tilde{\omega}=\frac{\pi}{\beta-\alpha-\delta}$ and so $\tilde{\omega}>\omega$. Employing the above result we have obtained (2.2.17) for the angular domain $\Omega_{\delta/2}$, and noting that

$$n(R_0,\Omega_{\delta/2},f)\leqslant\left(\log\frac{R}{R_0}\right)^{-1}N(R,\Omega_{\delta/2},f)\leqslant\left(\log\frac{R}{R_0}\right)^{-1}\frac{R^{\omega}C_{\alpha,\beta}(R,f)}{2\omega\sin(\omega\delta/2)},$$

where the inequality (2.2.15) has been used, therefore we have

$$\log^{+}|f(z)|\leqslant K_{\delta}R_0{}^{\tilde{\omega}}\frac{R_0{}^{\tilde{\omega}}+r^{\tilde{\omega}}}{R_0{}^{\tilde{\omega}}-r^{\tilde{\omega}}}S_{\alpha+\delta/2,\beta-\delta/2}(R,f)+O(\frac{R_0{}^{\tilde{\omega}}+r^{\tilde{\omega}}}{R_0{}^{\tilde{\omega}}-r^{\tilde{\omega}}})$$
$$+n(R_0,\Omega_{\delta/2},f)\log\frac{2R^{\tilde{\omega}}}{h^{\tilde{\omega}}}$$
$$\leqslant K_{\delta}R_0{}^{\tilde{\omega}}\frac{R_0{}^{\tilde{\omega}}+r^{\tilde{\omega}}}{R_0{}^{\tilde{\omega}}-r^{\tilde{\omega}}}S_{\alpha+\delta/2,\beta-\delta/2}(R,f)+O(\frac{R_0{}^{\tilde{\omega}}+r^{\tilde{\omega}}}{R_0{}^{\tilde{\omega}}-r^{\tilde{\omega}}})$$
$$+\left(\log\frac{2R^{\tilde{\omega}}}{h^{\tilde{\omega}}}\right)\left(\log\frac{R}{R_0}\right)^{-1}\frac{R^{\omega}C_{\alpha,\beta}(R,f)}{2\omega\sin(\omega\delta/2)}, \tag{2.2.18}$$

for a positive constant K_{δ} only depending on δ. Thus (2.2.16) immediately follows. □

There is an excellent estimate for $\log^{+}|f(z)|$ due to Goldberg and Ostrovskii, which is Theorem 6.3.3 in [11] and whose proof is complicated as mentioned in [11].

Theorem 2.2.2. *Let $f(z)$ be a meromorphic function on $\overline{\Omega}(\alpha,\beta)$ and let $\kappa(r)$ be a continuous increasing function in $(0,+\infty)$, $U(r) = r(1+\kappa(r)^{-1})$. Then for any ε with $0<\varepsilon<1$, there exists a set E_ε on $(0,+\infty)$ with $\mathrm{dens}(E_\varepsilon) < \varepsilon$ such that for $z\in\Omega$ with $r=|z|\notin E_\varepsilon$, we have*

$$\log^+|f(z)| \leqslant c\kappa(r)^3 r^\omega\{S_{\alpha,\beta}(U(r),f)+1\}, \tag{2.2.19}$$

where c is a positive constant.

We remark that Theorem 2.2.2 is proved in [11] for the case when $\alpha=0$ and $\beta=\pi$, but the general case follows directly from this special case. Actually, we can find a continuous increasing function $\widehat{k}(r)$ such that $1+\widehat{\kappa}(r^\omega)^{-1} = (1+\kappa(r)^{-1})^\omega$ and noticing for $x>0$ and $\omega\geqslant 1$, $(1+x)^\omega \geqslant 1+\omega x$, we have $\widehat{\kappa}(r^\omega)\leqslant \frac{1}{\omega}\kappa(r)$; noticing for $0<x\leqslant c$ and $0<\omega<1$, $(1+x)^\omega\geqslant 1+\omega(1+c)^{\omega-1}x$, we have $\widehat{\kappa}(r^\omega)\leqslant\frac{1}{\omega}(1+k(1)^{-1})^{1-\omega}\kappa(r)$, $r\geqslant 1$. Then applying the result for $\overline{\Omega}(0,\pi)$, $\widehat{\kappa}(r)$ and $F(w)=f(z)$ with $w=(\mathrm{e}^{-\mathrm{i}\alpha}z)^\omega$ yields

$$\begin{aligned}
\log^+|f(z)| &= \log^+|F(w)| \leqslant c\widehat{\kappa}(r^\omega)^3 r^\omega(S_{0,\pi}(\widehat{U}(r^\omega),F)+1)\\
&\leqslant c\omega^{-3}(1+k(1)^{-1})^{3(1+\omega)}\kappa(r)^3 r^\omega(S_{\alpha,\beta}(\widehat{U}(r^\omega)^{1/\omega},f)+1)\\
&= c\omega^{-3}(1+k(1)^{-1})^{3(1+\omega)}\kappa(r)^3 r^\omega(S_{\alpha,\beta}(U(r),f)+1),\ r^\omega\notin\widehat{E}_\varepsilon,
\end{aligned}$$

where $\widehat{U}(r) = r(1+\widehat{\kappa}(r)^{-1})$ and $\widehat{E}_\varepsilon$ is a set with $\mathrm{dens}\widehat{E}_\varepsilon<\varepsilon$. Set $E_\varepsilon = \{r: r^\omega\in\widehat{E}_\varepsilon\}$. Since for $1\leqslant x\leqslant r^\omega$, $r^{-1}x^{1/\omega-1} = r^{-\omega}(rx^{-1/\omega})^{\omega-1}\leqslant r^{-\omega}$ when $\omega\geqslant 1$; $r^{-1}x^{1/\omega-1}\leqslant r^{-1}r^{1-\omega}\leqslant r^{-\omega}$ when $0<\omega<1$, therefore we have

$$\frac{1}{r}\int_{E_\varepsilon(1,r)}\mathrm{d}t = \frac{1}{r}\int_{\widehat{E}_\varepsilon(1,r^\omega)}\frac{1}{\omega}x^{1/\omega-1}\mathrm{d}x \leqslant \frac{1}{\omega r^\omega}\int_{\widehat{E}_\varepsilon(1,r^\omega)}\mathrm{d}t$$

so that we have $\mathrm{dens}(E_\varepsilon)<\omega^{-1}\varepsilon$. Theorem 2.2.2 follows.

Set

$$m_{\alpha,\beta}(r,f) = \frac{1}{2\pi}\int_\alpha^\beta \log^+|f(r\mathrm{e}^{\mathrm{i}\theta})|\mathrm{d}\theta$$

for $0\leqslant\alpha<\beta\leqslant 2\pi$. The following is a consequence of Theorem 2.2.2, which is Theorem 6.2.3 of [11].

Lemma 2.2.3. *Let $f(z)$ be a meromorphic function on $\overline{\Omega}=\overline{\Omega}(\alpha,\beta)$. Given $d>1$ and $\varepsilon>0$, we have*

$$m_{\alpha,\beta}(r,f)\leqslant Kr^\omega\{S_{\alpha,\beta}(r,f)+1\}^d$$

and

$$m_{\alpha,\beta}(r,f)\leqslant Kr^\omega\{S_{\alpha,\beta}(dr,f)+1\},\ r\notin E,$$

$\omega = \frac{\pi}{\beta-\alpha}$ *and* $\mathrm{dens}(E)<\varepsilon$.

Ostrovskii characterized the meromorphic function in an angular domain when its corresponding Nevanlinna characteristic is bounded. This is the following, which is Theorem 6.2.7 of [11].

Theorem 2.2.3. *Let $f(z)$ be a meromorphic function on $\overline{\Omega}=\overline{\Omega}(\alpha,\beta)$. If $S_{\alpha,\beta}(r,f)=O(1)$, then*

$$\log|f(re^{i\phi})| = r^{\omega}c\sin(\omega(\phi-\alpha))+o(r^{\omega})$$

uniformly holds for $\alpha\leqslant\phi\leqslant\beta$ as $r\notin F\to\infty$, where F is a set of finite logarithmic measure.

We obtain a consequence of Theorem 2.2.3.

Corollary 2.2.1. *Let $f(z)$ be a meromorphic function on $\overline{\Omega}=\overline{\Omega}(\alpha,\beta)$. Assume that for three distinct points a_ν $(\nu=1,2,3)$ in $\widehat{\mathbb{C}}$*

$$\sum_{\nu=1}^{3}\overline{N}(r,\Omega,f=a_\nu)=O(r^{\omega}(\log r)^{-\tau})$$

for some $\tau>1$ and $R_{\alpha,\beta}(r,f)=O(1)$. Then the result of Theorem 2.2.3 is true.

Proof. In view of (2.2.14) under the assumption of Corollary 2.2.1 we have

$$\sum_{\nu=1}^{3}\overline{C}_{\alpha,\beta}(r,f=a_\nu)=O(1),$$

and then from (2.2.6) it follows that $S_{\alpha,\beta}(r,f)=O(1)$. The condition of Theorem 2.2.3 is satisfied, and so Corollary 2.2.1 follows. □

We have the following consequence of Theorem 2.2.2.

Corollary 2.2.2. *Let $f(z)$ be an analytic function on $\overline{\Omega}(\alpha,\beta)$ with $0<\alpha<\beta<2\pi$. Then we have*

$$\log M(r,\Omega,f)\leqslant Kr^{\omega}\{S_{\alpha,\beta}(2r,f)+1\}, \tag{2.2.20}$$

where $\log M(r,\Omega,f)=\max\{|f(te^{i\theta})|:\alpha\leqslant\theta\leqslant\beta,1\leqslant t\leqslant r\}$ and K is a positive constant. On the other hand, we have

$$S_{\alpha,\beta}(r,f)\leqslant\frac{2\omega}{\pi}\int_1^r\frac{\log^+M(t,\Omega,f)}{t^{\omega+1}}dt+\frac{4}{\pi}\frac{\log^+M(r,\Omega,f)}{r^{\omega}}. \tag{2.2.21}$$

In view of Lemma 1.1.7 and the non-decreasing property of $S_{\alpha,\beta}(r,f)$ up to a constant, we can deduce (2.2.20) for all r. The inequality (2.2.21) directly follows from the definitions of $A_{\alpha,\beta}(r,f)$ and $B_{\alpha,\beta}(r,f)$.

2.3 Tsuji's Characteristic

The Tsuji's Characteristic of a meromorphic function in an angle stems from the Levin formula: Let $f(z)$ be a meromorphic function in the half plane $\mathrm{Im}\, z > 0$. Then for $0 < r < R$, we have

$$\sum_{r<|a_n|<R\sin\alpha_n}\left(\frac{\sin\alpha_n}{|a_n|}-\frac{1}{R}\right)-\sum_{r<|b_m|<R\sin\beta_m}\left(\frac{\sin\beta_m}{|b_m|}-\frac{1}{R}\right)$$

$$=\frac{1}{2\pi}\int_{\arcsin(r/R)}^{\pi-\arcsin(r/R)}\log|f(Re^{i\theta}\sin\theta)|\frac{d\theta}{R\sin^2\theta}+O(1), \tag{2.3.1}$$

where $a_n = |a_n|e^{i\alpha_n}$ are zeros and $b_m = |b_m|e^{i\beta_m}$ are poles of $f(z)$ in $\{z : |z-\frac{1}{2}Ri| \leqslant \frac{1}{2}R\}\setminus\{z : |z-\frac{1}{2}ri| < \frac{1}{2}r\}$, appearing according to their multiplicities.

The Levin formula (2.3.1) can be derived from the second Green formula, that is, (1.2.1), for $u(z) = \log|f(z)|$ and $v(z) = -\mathrm{Im}\left(\frac{1}{z}+\frac{i}{R}\right)$.

Consider the following domains: for any pair of real numbers α and β in $[0, 2\pi)$ with $0 < \beta - \alpha \leqslant 2\pi$,

$$\Xi(\alpha,\beta;r) = \{z = te^{i\theta} : \alpha < \theta < \beta, 1 < t \leqslant r(\sin(\omega(\theta-\alpha)))^{1/\omega}\},$$

$\omega = \frac{\pi}{\beta-\alpha}$. A straightforward calculation implies that for each $0 < \varepsilon < \pi - \frac{\beta-\alpha}{2}$, we have the inclusions

$$\Xi(\alpha,\beta;r) \subset \Omega(\alpha,\beta;r) \subset \Xi(\alpha-\varepsilon,\beta+\varepsilon;\sigma r) \tag{2.3.2}$$

where $\sigma = \left(\sin\frac{\pi\varepsilon}{\beta-\alpha+2\varepsilon}\right)^{-\frac{\beta-\alpha+2\varepsilon}{\pi}} > 1$.

Now let us introduce the Tsuji characteristic as follows. Assume that $f(z)$ is a meromorphic function in an angular domain $\Omega(\alpha,\beta)$. Define

$$\mathfrak{m}_{\alpha,\beta}(r,f) = \frac{1}{2\pi}\int_{\arcsin(r^{-\omega})}^{\pi-\arcsin(r^{-\omega})}\log^+\left|f(re^{i(\alpha+\omega^{-1}\theta)}\sin^{\omega^{-1}}\theta)\right|\frac{1}{r^{\omega}\sin^2\theta}d\theta,$$

$$\mathfrak{N}_{\alpha,\beta}(r,f) = \sum_{1<|b_n|<r(\sin(\omega(\beta_n-\alpha)))^{\omega^{-1}}}\left(\frac{\sin\omega(\beta_n-\alpha)}{|b_n|^{\omega}}-\frac{1}{r^{\omega}}\right),$$

where b_n are the poles of $f(z)$ in $\Xi(\alpha,\beta;r)$ appearing often according to their multiplicities and then Tsuji characteristic of f is

$$\mathfrak{T}_{\alpha,\beta}(r,f) = \mathfrak{m}_{\alpha,\beta}(r,f)+\mathfrak{N}_{\alpha,\beta}(r,f). \tag{2.3.3}$$

We denote by $\mathfrak{n}_{\alpha,\beta}(r,f)$ the number of poles of $f(z)$ in $\Xi(\alpha,\beta;r)$, and then

$$\mathfrak{N}_{\alpha,\beta}(r,f) = \int_1^r\left(\frac{1}{t^{\omega}}-\frac{1}{r^{\omega}}\right)d\mathfrak{n}_{\alpha,\beta}(t,f) = \omega\int_1^r\frac{\mathfrak{n}_{\alpha,\beta}(t,f)}{t^{\omega+1}}dt. \tag{2.3.4}$$

This is because for each b_n, we can write $|b_n| = t(\sin(\omega(\beta_n - \alpha)))^{\omega^{-1}}$ for some definite t with $1 < t \leqslant r$.

It is obvious that $\Xi(0,\pi;r) = \{z : |z - \frac{1}{2}ri| \leqslant \frac{1}{2}r\} \setminus \{z : |z| < 1\}$ and $\Xi(\alpha,\beta;r)$ is exactly image of $\Xi(0,\pi;r^\omega)$ under the transformation $w = e^{i\alpha}z^{1/\omega}$, which is taken to be the main branch, because for $z = \rho e^{i\theta}$ and $w = \sigma e^{i\phi}$, under the transformation, we have $\rho = \sigma^\omega$ and $\theta = \omega(\phi - \alpha)$, and the curve $\rho = r^\omega \sin\theta$, that is, the boundary of $\Xi(0,\pi;r^\omega)$, is mapped onto $\sigma = r(\sin(\omega(\phi - \alpha)))^{1/\omega}$, the boundary of $\Xi(\alpha,\beta;r)$. Set $F(z) = f(e^{i\alpha}z^{1/\omega}) = f(w)$. It is easy to see that $\mathfrak{m}_{0,\pi}(r^\omega, F) = \mathfrak{m}_{\alpha,\beta}(r,f)$ by using $F(r^\omega e^{i\theta}\sin\theta) = f(e^{i\alpha}(r^\omega e^{i\theta}\sin\theta)^{1/\omega}) = f(re^{i(\alpha+\theta/\omega)}\sin^{1/\omega}\theta)$. For a pole b_n of $f(w)$, $(e^{-i\alpha}b_n)^\omega$ is a pole of $F(z)$ and thus $\mathfrak{N}_{0,\pi}(r^\omega, F) = \mathfrak{N}_{\alpha,\beta}(r,f)$. This shows that

$$\mathfrak{T}_{0,\pi}(r^\omega, F) = \mathfrak{T}_{\alpha,\beta}(r,f).$$

In view of the Levin formula (2.3.1), we have

$$\frac{1}{2\pi}\int_{\arcsin(r^{-\omega})}^{\pi-\arcsin(r^{-\omega})} \frac{d\theta}{r^\omega \sin^2\theta} = O(1)$$

and further

$$\mathfrak{T}_{\alpha,\beta}\left(r, \frac{1}{f-a}\right) = \mathfrak{T}_{\alpha,\beta}(r,f) + O(1)$$

for $a \in \mathbb{C}$. By means of the same method as in Sections 2.1 and 2.2, we also have the following fundamental inequalities

$$(q-2)\mathfrak{T}_{\alpha,\beta}(r,f) \leqslant \sum_{j=1}^{q} \overline{\mathfrak{N}}_{\alpha,\beta}\left(r, \frac{1}{f-a_j}\right) + Q_{\alpha,\beta}(r,f) \tag{2.3.5}$$

for q distinct points $a_j \in \hat{\mathbb{C}}$, where

$$Q_{\alpha,\beta}(r,f) = \mathfrak{m}_{\alpha,\beta}\left(r, \frac{f'}{f}\right) + \sum_{j=1}^{q} \mathfrak{m}_{\alpha,\beta}\left(r, \frac{f'}{f-a_j}\right) + O(1), \tag{2.3.6}$$

which is named as the Tsuji second fundamental theorem, and

$$\begin{aligned}\mathfrak{T}_{\alpha,\beta}(r,f) \leqslant{}& \overline{\mathfrak{N}}_{\alpha,\beta}(r,f) + \mathfrak{N}_{\alpha,\beta}(r, f=0) + \mathfrak{N}_{\alpha,\beta}(r, f^{(k)} = 1)\\ &- \mathfrak{N}_{\alpha,\beta}(r, f^{(k+1)} = 0) + Q_{\alpha,\beta}(r,f),\end{aligned} \tag{2.3.7}$$

where

$$\begin{aligned}Q_{\alpha,\beta}(r,f) ={}& \mathfrak{m}_{\alpha,\beta}\left(r, \frac{f^{(k)}}{f}\right) + \mathfrak{m}_{\alpha,\beta}\left(r, \frac{f^{(k+1)}}{f}\right)\\ &+ \mathfrak{m}_{\alpha,\beta}\left(r, \frac{f^{(k+1)}}{f^{(k)}-1}\right) + O(1),\end{aligned} \tag{2.3.8}$$

and

$$\mathfrak{T}_{\alpha,\beta}(r,f) \leqslant \left(2+\frac{1}{k}\right)\mathfrak{N}_{\alpha,\beta}(r,f)+\left(2+\frac{2}{k}\right)\overline{\mathfrak{N}}_{\alpha,\beta}(r,f^{(k)}=1)$$
$$+Q_{\alpha,\beta}(r,f),\ k>0, \tag{2.3.9}$$

where

$$Q_{\alpha,\beta}(r,f) = \left(2+\frac{2}{k}\right)\mathfrak{m}_{\alpha,\beta}\left(r,\frac{f^{(k+1)}}{f^{(k)}-1}\right)$$
$$+\left(2+\frac{1}{k}\right)\left[\mathfrak{m}_{\alpha,\beta}\left(r,\frac{f^{(k+1)}}{f}\right)+\mathfrak{m}_{\alpha,\beta}\left(r,\frac{f^{(k)}}{f}\right)\right]+O(1). \tag{2.3.10}$$

Throughout this book, $Q_{\alpha,\beta}(r,f)$ is called error term associated with the Tsuji characteristic for the angle $\Omega(\alpha,\beta)$, and it may not be the same at each occurrence.

Assume that $f(z)$ and $a(z)$ are two meromorphic functions in $\Omega(\alpha,\beta)$. Then $a(z)$ is called a small function with respect to $f(z)$ in $\Omega(\alpha,\beta)$ (in the sense of Tsuji characteristic) if $\mathfrak{T}_{\alpha,\beta}(r,a)=o(\mathfrak{T}_{\alpha,\beta}(r,f))$ as $r\notin E\to\infty$ for a set E with finite measure. Then we have the following theorem of Valiron-Mohon'ko type for the Tsuji characteristic.

Theorem 2.3.1. *Let $f(z)$ be a meromorphic function in $\Omega(\alpha,\beta)$. Then for all irreducible rational function $R(z,f)$ in f with coefficients meromorphic and small with respect to f in $\Omega(\alpha,\beta)$, we have*

$$\mathfrak{T}_{\alpha,\beta}(r,R(z,f)) = d\mathfrak{T}_{\alpha,\beta}(r,f)+Q_{\alpha,\beta}(r,f), \tag{2.3.11}$$

where $Q_{\alpha,\beta}(r,f)=o(\mathfrak{T}_{\alpha,\beta}(r,f))$ for $r\notin E$, E is a set with finite measure and d is the degree of $R(z,f)$ in f.

The proof of Theorem 2.3.1 can be completed by the method in the proof of Theorem 2.1.3 and the estimation of the error term is obtained using below Lemma 2.5.4.

In the Tsuji second fundamental theorem, could we consider small functions in the place of the constant targets? This is a natural question. We do not know whether the method of Yamanoi [34] is available to this question and however, fortunately the method of Chuang [4], Frank-Weissenborn [10] and Steinmetz [30] is available in such a generalization of the Tsuji second fundamental theorem concerning small functions as targets.

Theorem 2.3.2. *Let $f(z)$ be a meromorphic function in $\Omega(\alpha,\beta)$ and assume that $a_j(z)$ $(j=1,2,\cdots,q;q\geqslant 3)$ are distinct small functions with respect to $f(z)$. Then for any positive number ε, we have*

$$(q-2-\varepsilon)\mathfrak{T}_{\alpha,\beta}(r,f) \leqslant \sum_{j=1}^{q}\mathfrak{N}_{\alpha,\beta}\left(r,\frac{1}{f-a_j}\right)+Q_{\alpha,\beta}(r,f), \tag{2.3.12}$$

where

$$Q_{\alpha,\beta}(r,f) = O\left(\sum_{k=1}^{q} \mathfrak{m}_{\alpha,\beta}\left(r,\frac{f^{(k)}}{f}\right) + \sum_{j=1}^{q}\sum_{k=1}^{q} \mathfrak{m}_{\alpha,\beta}\left(r,\frac{(f-a_j)^{(k)}}{f-a_j}\right)\right)$$
$$+o(\mathfrak{T}_{\alpha,\beta}(r,f)) + O(\log r),\ r \notin E,$$

where E is a set of finite measure.

In view of below Lemma 2.5.4 and the argument following Theorem 2.1.8, we can actually establish (2.3.12) without ε and with $Q_{\alpha,\beta}(r,f)$ replaced by $o(\mathfrak{T}_{\alpha,\beta}(r,f)) + O(\log r)$.

In order to prove Theorem 2.3.2 we need a result which can be proved by using the method of Frank and Weissenborn [10]. Let $a_1(z),\cdots,a_p(z)$ be meromorphic in $\Omega(\alpha,\beta)$. The following is the Wronskian determinant of $a_1(z),\cdots,a_p(z)$

$$W(a_1(z),\cdots,a_p(z)) = \begin{vmatrix} a_1 & a_2 & \cdots & a_p \\ a_1' & a_2' & \cdots & a_p' \\ \vdots & \vdots & & \vdots \\ a_1^{(p-1)} & a_2^{(p-1)} & \cdots & a_p^{(p-1)} \end{vmatrix}.$$

Lemma 2.3.1. *Let $f(z)$ and $a_j(z)$ $(j=1,2,\cdots,p; p\geqslant 3)$ be as in Theorem 2.3.2. Set $W(f) = W(a_1,\cdots,a_p,f)$. If $a_j(z)$ $(j=1,2,\cdots,p; p\geqslant 3)$ are linearly independent, then for $\varepsilon > 0$*

$$p\overline{\mathfrak{N}}_{\alpha,\beta}(r,f) \leqslant \mathfrak{N}_{\alpha,\beta}\left(r,\frac{1}{W(f)}\right) + (1+\varepsilon)\mathfrak{N}_{\alpha,\beta}(r,f) + Q_{\alpha,\beta}(r,f). \qquad (2.3.13)$$

Now we can prove Theorem 2.3.2 in the way of Chuang [4] in view of Theorem 2.3.1 and Lemma 2.3.1(Compare Steinmetz [30]).

Proof of Theorem 2.3.2. Assume without any loss of generalities that $\{a_1,\cdots,a_p\}$ is a maximum linearly independent subset of $a_j(z)$ $(j=1,2,\cdots,q)$ and then $p\leqslant q$ and each a_j $(j=1,2,\cdots,q)$ can be linearly expressed in terms of a_j $(j=1,2,\cdots,p)$. Set $W(f) = W(a_1,\cdots,a_p,f)$. Then

$$W(f) = b_p f^{(p)} + b_{p-1} f^{(p-1)} + \cdots + b_1 f' + b_0 f$$

where b_j $(j=1,2,\cdots,p)$ are small functions with respect to f, so that

$$\mathfrak{N}_{\alpha,\beta}(r,W(f)) = p\overline{\mathfrak{N}}_{\alpha,\beta}(r,f) + \mathfrak{N}_{\alpha,\beta}(r,f) + Q_{\alpha,\beta}(r,f)$$

and

$$\mathfrak{m}_{\alpha,\beta}(r,W(f)) \leqslant \mathfrak{m}_{\alpha,\beta}(r,f) + \mathfrak{m}_{\alpha,\beta}\left(r,\frac{W(f)}{f}\right) = \mathfrak{m}_{\alpha,\beta}(r,f) + Q_{\alpha,\beta}(r,f).$$

Thus we obtain

$$\mathfrak{T}_{\alpha,\beta}(r,W(f)) \leqslant p\overline{\mathfrak{N}}_{\alpha,\beta}(r,f)+\mathfrak{T}_{\alpha,\beta}(r,f)+Q_{\alpha,\beta}(r,f). \tag{2.3.14}$$

It is easy to see that for each $j=1,2,\cdots,q$, we have $W(f-a_j)=W(f)$ so that

$$\mathfrak{m}_{\alpha,\beta}\left(r,\frac{W(f)}{f-a_j}\right)=\mathfrak{m}_{\alpha,\beta}\left(r,\frac{W(f-a_j)}{f-a_j}\right)=Q_{\alpha,\beta}(r,f).$$

Set

$$F(z)=\sum_{j=1}^{q}\frac{1}{f(z)-a_j(z)}.$$

In view of (2.3.14) and (2.3.13) we estimate

$$\begin{aligned}
\mathfrak{m}_{\alpha,\beta}(r,F) &\leqslant \mathfrak{m}_{\alpha,\beta}\left(r,\frac{1}{W(f)}\right)+\mathfrak{m}_{\alpha,\beta}(r,FW(f))\\
&\leqslant \mathfrak{T}_{\alpha,\beta}(r,W(f))-\mathfrak{N}_{\alpha,\beta}\left(r,\frac{1}{W(f)}\right)+Q_{\alpha,\beta}(r,f)\\
&\leqslant p\overline{\mathfrak{N}}_{\alpha,\beta}(r,f)+\mathfrak{T}_{\alpha,\beta}(r,f)-\mathfrak{N}_{\alpha,\beta}\left(r,\frac{1}{W(f)}\right)+Q_{\alpha,\beta}(r,f)\\
&\leqslant \mathfrak{T}_{\alpha,\beta}(r,f)+(1+\varepsilon)\mathfrak{N}_{\alpha,\beta}(r,f)+Q_{\alpha,\beta}(r,f).
\end{aligned}$$

By means of Theorem 2.3.1, we have

$$\begin{aligned}
q\mathfrak{T}_{\alpha,\beta}(r,f) &= \mathfrak{T}_{\alpha,\beta}(r,F)+Q_{\alpha,\beta}(r,f)\\
&\leqslant \sum_{j=1}^{q}\mathfrak{N}_{\alpha,\beta}\left(r,\frac{1}{f-a_j}\right)+\mathfrak{T}_{\alpha,\beta}(r,f)+(1+\varepsilon)\mathfrak{N}_{\alpha,\beta}(r,f)+Q_{\alpha,\beta}(r,f).
\end{aligned}$$

This immediately deduces (2.3.12). □

In what follows, let us make a further discussion of the Tsuji characteristic $\mathfrak{T}_{\alpha,\beta}(r,f)$. Set

$$\begin{aligned}
\dot{\mathfrak{T}}_{\alpha,\beta}(r,f) &= \frac{1}{\pi}\int\int_{\Xi(\alpha,\beta;r)}\left(\frac{\sin(\omega(\theta-\alpha))}{t^{\omega}}-\frac{1}{r^{\omega}}\right)(f^{\#}(te^{i\theta}))^2\,t\mathrm{d}t\mathrm{d}\theta\\
&= \frac{1}{\pi}\int_a^b \mathrm{d}\theta\int_1^{r(\sin\omega(\theta-\alpha))^{1/\omega}}\left(\frac{\sin\omega(\theta-\alpha)}{t^{\omega}}-\frac{1}{r^{\omega}}\right)(f^{\#}(te^{i\theta}))^2\,t\mathrm{d}t,
\end{aligned}$$

where $a=\alpha+\omega^{-1}\arcsin r^{-\omega}$ and $b=\beta-\omega^{-1}\arcsin r^{-\omega}$. It is obvious that $\dot{\mathfrak{T}}_{\alpha,\beta}(r,f)$ is increasing in r. Set

$$K(r,\theta)=\int_1^r (f^{\#}(te^{i\theta}))^2 t\mathrm{d}t.$$

By means of the formula for integration by parts, we immediately have

$$\dot{\mathfrak{T}}_{\alpha,\beta}(r,f) = \frac{1}{\pi}\int_a^b \mathrm{d}\theta \int_1^{r(\sin\omega(\theta-\alpha))^{1/\omega}} \left(\frac{\sin(\omega(\theta-\alpha))}{t^{\omega}} - \frac{1}{r^{\omega}}\right)\mathrm{d}K(t,\theta)$$
$$= \frac{\omega}{\pi}\int_a^b \int_1^{r(\sin\omega(\theta-\alpha))^{1/\omega}} \frac{\sin(\omega(\theta-\alpha))}{t^{\omega+1}} K(t,\theta)\mathrm{d}t\mathrm{d}\theta$$
$$= \frac{\omega}{\pi}\int\int_{\Xi(\alpha,\beta;r)} \frac{\sin\omega(\theta-\alpha)}{t^{\omega+1}} K(t,\theta)\mathrm{d}t\mathrm{d}\theta. \tag{2.3.15}$$

We have an analogy of Lemma 2.2.1 for $\dot{\mathfrak{T}}_{\alpha,\beta}(r,f)$ and $\mathfrak{T}_{\alpha,\beta}(r,f)$.

Lemma 2.3.2. *Assume that $f(z)$ is a meromorphic function in $\Omega(\alpha,\beta)$. Then*

$$\mathfrak{T}_{\alpha,\beta}(r,f) = \dot{\mathfrak{T}}_{\alpha,\beta}(r,f) + O(1)$$

and

$$\dot{\mathfrak{T}}_{\alpha,\beta}(r,f) \geqslant \frac{\omega}{\hat{\omega}}\dot{\mathfrak{T}}_{\alpha+\delta,\beta-\delta}(r,f),$$

for any $\delta > 0$ with $\alpha+\delta < \beta-\delta$, where $\hat{\omega} = \frac{\pi}{\beta-\alpha-2\delta}$.

Proof. A direct calculation implies that $\dot{\mathfrak{T}}_{\alpha,\beta}(r,f) = \dot{\mathfrak{T}}_{0,\pi}(r^{\omega},F)$ for $F(z) = f(\mathrm{e}^{\mathrm{i}\alpha}z^{1/\omega}) = f(w)$, for $w = \mathrm{e}^{\mathrm{i}\alpha}z^{1/\omega}$ maps conformally $\Xi(0,\pi;r^{\omega})$ onto $\Xi(\alpha,\beta;r)$. The first equality follows from the fact that $\mathfrak{T}_{0,\pi}(r^{\omega},F) = \dot{\mathfrak{T}}_{0,\pi}(r^{\omega},F) + O(1)$ which has been proved in [11], as mentioned in the paragraph before Lemma 2.2.1. Below we give out a proof of the second inequality. In view of (2.3.15) and noting $\hat{\omega} > \omega$, we have

$$\dot{\mathfrak{T}}_{\alpha,\beta}(r,f) \geqslant \frac{\omega}{\pi}\int\int_{\Xi(\alpha+\delta,\beta-\delta;r)} \frac{\sin\omega(\theta-\alpha)}{t^{\omega+1}} K(t,\theta)\mathrm{d}t\mathrm{d}\theta$$
$$\geqslant \frac{\omega}{\pi}\int\int_{\Xi(\alpha+\delta,\beta-\delta;r)} \frac{\sin\hat{\omega}(\theta-\alpha-\delta)}{t^{\hat{\omega}+1}} K(t,\theta)\mathrm{d}t\mathrm{d}\theta$$
$$= \frac{\omega}{\hat{\omega}}\dot{\mathfrak{T}}_{\alpha+\delta,\beta-\delta}(r,f).$$

This yields the desired inequality. □

Next we compare the Nevanlinna's characteristic for an angle to the Tsuji characteristic.

Theorem 2.3.3. *Assume that $f(z)$ is a meromorphic function in $\overline{\Omega}(\alpha,\beta)$. Then*

$$\dot{S}_{\alpha,\beta}(r,f) > \dot{\mathfrak{T}}_{\alpha,\beta}(r,f)$$

and

$$\dot{\mathfrak{T}}_{\alpha,\beta}(r,f) > \frac{2\omega}{\hat{\omega}}\dot{S}_{\alpha+\delta,\beta-\delta}(sr,f),$$

for any $\delta > 0$ with $\alpha+\delta < \beta-\delta$ and some $0 < s < 1$ which can be computed by means of (2.3.2).

Proof. In view of the definition of $\dot{S}_{\alpha,\beta}(r,f)$ and by using the inclusion (2.3.2), the first desired inequality follows from the following implication

$$\begin{aligned}\dot{S}_{\alpha,\beta}(r,f) &\geqslant \frac{1}{\pi}\int\int_{\Xi(\alpha,\beta;r)}\left(\frac{1}{t^\omega}-\frac{t^\omega}{r^{2\omega}}\right)(f^\sharp(te^{i\theta}))^2\ \sin\omega(\theta-\alpha)t\mathrm{d}t\mathrm{d}\theta\\ &> \dot{\mathfrak{T}}_{\alpha,\beta}(r,f),\end{aligned}$$

by noting the inequality

$$\left(\frac{1}{t^\omega}-\frac{t^\omega}{r^{2\omega}}\right)\sin\omega(\theta-\alpha) > \frac{\sin\omega(\theta-\alpha)}{t^\omega}-\frac{1}{r^\omega}.$$

We can produce the second desired inequality by means of the following steps

$$\begin{aligned}\dot{\mathfrak{T}}_{\alpha,\beta}(r,f) &\geqslant \frac{\omega}{\pi}\int\int_{\Omega(\alpha+\delta,\beta-\delta;sr)}\frac{\sin\omega(\theta-\alpha)}{t^{\omega+1}}K(t,\theta)\mathrm{d}t\mathrm{d}\theta\\ &= \frac{\omega}{\pi}\int_{\alpha+\delta}^{\beta-\delta}\int_1^{sr}\frac{\sin\omega(\theta-\alpha)}{t^{\omega+1}}K(t,\theta)\mathrm{d}t\mathrm{d}\theta\\ &\geqslant \frac{\omega}{\pi}\int_{\alpha+\delta}^{\beta-\delta}\int_1^{sr}\frac{\sin\widehat{\omega}(\theta-\alpha-\delta)}{t^{\omega+1}}K(t,\theta)\mathrm{d}t\mathrm{d}\theta\\ &= \frac{\omega}{\pi}\int_1^{sr}\frac{1}{t^{\omega+1}}D_{\alpha+\delta,\beta-\delta}(t)\mathrm{d}t\\ &\geqslant \frac{\omega}{2\pi}\int_1^{sr}\left(\frac{1}{t^{\widehat{\omega}+1}}+\frac{t^{\widehat{\omega}-1}}{r^{2\widehat{\omega}}}\right)D_{\alpha+\delta,\beta-\delta}(t)\mathrm{d}t\\ &= \frac{\omega}{2\widehat{\omega}}\dot{S}_{\alpha+\delta,\beta-\delta}(sr,f),\end{aligned}$$

where we have used, in turn, the inclusion (2.3.2) and the equality

$$D_{\alpha+\delta,\beta-\delta}(t)=\int_{\alpha+\delta}^{\beta-\delta}\sin[\widehat{\omega}(\theta-\alpha-\delta)]K(t,\theta)\mathrm{d}\theta$$

and $\widehat{\omega}>\omega$. □

Finally, we come to compare the integrated counting functions.

Lemma 2.3.3. *Let $f(z)$ be a meromorphic function in $\Omega(\alpha,\beta)$. Then for $\varepsilon>0$, we have*

$$\mathfrak{N}_{\alpha,\beta}(r,f)\leqslant\omega\frac{N(r,\Omega,f)}{r^\omega}+\omega^2\int_1^r\frac{N(t,\Omega,f)}{t^{\omega+1}}\mathrm{d}t,$$

$$C_{\alpha,\beta}(r,f)\geqslant 2\mathfrak{N}_{\alpha,\beta}(r,f)$$

and

$$\mathfrak{N}_{\alpha,\beta}(r,f)\geqslant\omega c^\omega\frac{N(cr,\Omega_\varepsilon,f)}{r^\omega}+\omega^2c^\omega\int_1^{cr}\frac{N(t,\Omega_\varepsilon,f)}{t^{\omega+1}}\mathrm{d}t,$$

$$C_{\alpha+\varepsilon,\beta-\varepsilon}(r,f)\leqslant\frac{4\widehat{\omega}\sigma^\omega}{\omega}\mathfrak{N}_{\alpha,\beta}(\sigma r,f),$$

where $\hat{\omega} = \frac{\pi}{\beta-\alpha-2\varepsilon}$ *and* $\sigma > 1$, *and* $0 < c < 1$ *are two constants depending on* ε, *and* $N(r,\Omega,f)$ *is defined as in Lemma 2.2.2.*

Proof. In view of (2.3.4) and (2.3.2), we have

$$\begin{aligned}\mathfrak{N}_{\alpha,\beta}(r,f) &= \omega\int_1^r \frac{\mathfrak{n}_{\alpha,\beta}(t,f)}{t^{\omega+1}}\mathrm{d}t\\ &\leqslant \omega\int_1^r \frac{n(t,\Omega,f)}{t^{\omega+1}}\mathrm{d}t\\ &= \omega\frac{N(r,\Omega,f)}{r^{\omega}} + \omega^2\int_1^r \frac{N(t,\Omega,f)}{t^{\omega+1}}\mathrm{d}t,\end{aligned}$$

and in virtue of the definitions of $\mathfrak{N}_{\alpha,\beta}(r,f)$ and $C_{\alpha,\beta}(r,f)$, we get

$$\begin{aligned}\frac{1}{2}C_{\alpha,\beta}(r,f) &\geqslant \sum_{b_n\in\Xi(\alpha,\beta;r)}\left(\frac{1}{|b_n|^{\omega}} - \frac{|b_n|^{\omega}}{r^{2\omega}}\right)\sin\omega(\beta_n-\alpha)\\ &= \mathfrak{N}_{\alpha,\beta}(r,f) + \sum_{b_n\in\Xi(\alpha,\beta;r)}\left(\frac{1}{r^{\omega}} - \frac{|b_n|^{\omega}}{r^{2\omega}}\sin\omega(\beta_n-\alpha)\right)\\ &> \mathfrak{N}_{\alpha,\beta}(r,f).\end{aligned}$$

On the other hand, we have

$$\begin{aligned}\mathfrak{N}_{\alpha,\beta}(r,f) &= \omega\int_1^r \frac{\mathfrak{n}_{\alpha,\beta}(t,f)}{t^{\omega+1}}\mathrm{d}t\\ &\geqslant \omega\int_1^r \frac{n(ct,\Omega_\varepsilon,f)}{t^{\omega+1}}\mathrm{d}t\\ &\geqslant \omega c^{\omega}\int_1^{cr} \frac{n(t,\Omega_\varepsilon,f)}{t^{\omega+1}}\mathrm{d}t\\ &= \omega c^{\omega}\frac{N(cr,\Omega_\varepsilon,f)}{r^{\omega}} + \omega^2 c^{\omega}\int_1^{cr}\frac{N(t,\Omega_\varepsilon,f)}{t^{\omega+1}}\mathrm{d}t\end{aligned}$$

for some $0 < c < 1$ depending on ε, and

$$\begin{aligned}C_{\alpha+\varepsilon,\beta-\varepsilon}(r,f) &\leqslant 2\hat{\omega}\int_1^r n(t,\Omega_\varepsilon,f)\left(\frac{1}{t^{\hat{\omega}}} + \frac{t^{\hat{\omega}}}{r^{2\hat{\omega}}}\right)\frac{\mathrm{d}t}{t}\\ &\leqslant 2\hat{\omega}\int_1^r \mathfrak{n}_{\alpha,\beta}(\sigma t,f)\left(\frac{1}{t^{\hat{\omega}}} + \frac{t^{\hat{\omega}}}{r^{2\hat{\omega}}}\right)\frac{\mathrm{d}t}{t}\\ &\leqslant 4\hat{\omega}\int_1^r \frac{\mathfrak{n}_{\alpha,\beta}(\sigma t,f)}{t^{\omega+1}}\mathrm{d}t\\ &\leqslant 4\hat{\omega}\sigma^{\omega}\int_1^{\sigma r}\frac{\mathfrak{n}_{\alpha,\beta}(t,f)}{t^{\omega+1}}\mathrm{d}t\\ &= \frac{4\hat{\omega}\sigma^{\omega}}{\omega}\mathfrak{N}_{\alpha,\beta}(\sigma r,f).\end{aligned}$$

We complete the proof of Lemma 2.3.3. □

2.4 Ahlfors-Shimizu's Characteristic

The Ahlfors-Shimizu characteristic of a meromorphic function can also stems from the second Green formula in the point of view of analysis. We apply the formula to $u(z) = \frac{1}{2}\log(1+|f(z)|^2)$ and $G_D(z,a)$ to obtain for $a \in D$ with $f(a) \neq \infty$

$$\begin{aligned} u(a) &+ \frac{1}{\pi}\int\int_D G_D(z,a)(f^\#(z))^2 \mathrm{d}\sigma(z) \\ &= \frac{1}{2\pi}\int_{\partial D} \frac{1}{2}\log(1+|f(z)|^2)\frac{\partial G_D(z,a)}{\partial n}\mathrm{d}s + \sum_{a_n \in D} G_D(a_n,a) \end{aligned}$$

where a_n is a pole of $f(z)$ in D, appearing often according to its multiplicity. Define

$$\mathscr{T}(D,a,f) = \frac{1}{\pi}\int\int_D G_D(z,a)(f^\#(z))^2 \mathrm{d}\sigma(z).$$

Since

$$\log^+|f(z)| \leqslant \frac{1}{2}\log(1+|f(z)|^2) \leqslant \frac{1}{2}\log 2 + \log^+|f(z)|,$$

we therefore have

$$\begin{aligned} \mathscr{T}(D,a,f) &\leqslant \frac{1}{2}\log 2 + T(D,a,f) - u(a) \\ &\leqslant T(D,a,f) - \log^+|f(a)| + \frac{1}{2}\log 2 \end{aligned}$$

and

$$\begin{aligned} \mathscr{T}(D,a,f) &\geqslant T(D,a,f) - u(a) \\ &\geqslant T(D,a,f) - \log^+|f(a)| - \frac{1}{2}\log 2, \end{aligned}$$

so that

$$\mathscr{T}(D,a,f) = T(D,a,f) - \log^+|f(a)| + C, \tag{2.4.1}$$

where $0 \leqslant |C| \leqslant \frac{1}{2}\log 2$. We are allowed to consider the case $f(a) = \infty$. In this case, we use $v(z) = u(z) - pG_D(z,a)$ in the place of $u(z)$ where p is the multiplicity of pole of $f(z)$ at a. Since $v(a) = \log|c(a)| - p\omega_D(a,a)$, (2.4.1) holds for $\log|c(a)|$ in the place of $\log^+|f(a)|$ where $c(a)$ is the coefficient of the first term of Laurent series of $f(z)$ at a.

We take the disk into account. Define

$$\mathscr{A}(r,f) = \frac{1}{\pi}\int\int_{|z|\leqslant r} (f^\#(z))^2 \mathrm{d}\sigma(z) = \frac{1}{\pi}\int_0^{2\pi}\int_0^r (f^\#(te^{\mathrm{i}\theta}))^2 t\mathrm{d}t\mathrm{d}\theta.$$

Obviously, $\pi\mathscr{A}(r,f)$ is the area of the Riemann surface $F_r = f(\{z : |z| \leqslant r\})$ measured in the spherical metric. When $D = \{z : |z| \leqslant r\}$ and $a = 0$, by the formula for integration by parts we have immediately

$$\begin{aligned}\mathscr{T}(D,0,f) &= \frac{1}{\pi}\int\int_D G_D(z,0)(f^{\#}(z))^2 \mathrm{d}\sigma(z)\\ &= \frac{1}{\pi}\int\int_{|z|\leqslant r} \log\frac{r}{|z|}\,(f^{\#}(z))^2 \mathrm{d}\sigma(z)\\ &= \int_0^r \log\frac{r}{t}\mathrm{d}\mathscr{A}(t,f)\\ &= \int_0^r \frac{\mathscr{A}(t,f)}{t}\mathrm{d}t,\end{aligned}$$

by noting

$$\frac{\mathrm{d}\mathscr{A}(t,f)}{\mathrm{d}t} = \frac{1}{\pi}\int_0^{2\pi}(f^{\#}(t\mathrm{e}^{\mathrm{i}\theta}))^2 t\mathrm{d}\theta.$$

Usually, we write $\mathscr{T}(r,f)$ for $\mathscr{T}(D,0,f)$, that is,

$$\mathscr{T}(r,f) = \int_0^r \frac{\mathscr{A}(t,f)}{t}\mathrm{d}t$$

which is known as the Ahlfors-Shimizu characteristic of $f(z)$ on disk $\{z : |z| \leqslant r\}$. Then in view of (2.4.1) we have

$$|T(r,f) - \mathscr{T}(r,f) - \log^+|f(0)|| \leqslant \frac{1}{2}\log 2 \tag{2.4.2}$$

for $f(0) \neq \infty$, while $\log^+|f(0)|$ will be replaced by $\log|c(0)|$ for $f(0) = \infty$. Since $\frac{\mathrm{d}\mathscr{T}(r,f)}{\mathrm{d}\log r} = \mathscr{A}(r,f)$ is increasing, $\mathscr{T}(r,f)$ is convex with respect to $\log r$.

It is important to notice that we can take the Ahlfors-Shimizu characteristic into account in the point of view of geometry. Let m be the normalized area measure on the Riemann sphere $\mathbb{S}$, which is produced by the sphere metric $|dz|/(1+|z|^2)$. Then consulting Theorem 2.14 of Conway [5], we have

$$\mathscr{A}(t,f) = \int_{\hat{\mathbb{C}}} n(t,f=a)\mathrm{d}m(a),$$

which is therefore the mean covering number of the map $f : \{z : |z| \leqslant r\} \to \mathbb{S}$. Furthermore

$$\mathscr{T}(r,f) = \int_{\hat{\mathbb{C}}} N(r,f=a)\mathrm{d}m(a).$$

The Ahlfors-Shimizu characteristic of $f(z)$ for an angle is important and applicable in the discussion of argument distribution of meromorphic functions and is naturally introduced as above. For $\overline{\Omega} = \{z : \alpha \leqslant \arg z \leqslant \beta\}$, define

$$\mathscr{A}(r,\Omega,f) = \frac{1}{\pi}\int_\alpha^\beta\int_0^r (f^{\#}(t\mathrm{e}^{\mathrm{i}\phi}))^2 t\mathrm{d}t\mathrm{d}\phi$$

and

$$\mathscr{T}(r,\Omega,f) = \int_0^r \frac{\mathscr{A}(t,\Omega,f)}{t}\mathrm{d}t.$$

Then $\mathscr{T}(r,f)$ is $\mathscr{T}(r,\mathbb{C},f)$. Where no confusion seems possible, omitting f we write $\mathscr{T}(r,\Omega)$ for $\mathscr{T}(r,\Omega,f)$. As in the above discussion, we have

$$\mathscr{A}(t,\Omega) = \int_{\hat{\mathbb{C}}} n(t,\Omega,f=a)\mathrm{d}m(a),$$

which is therefore the mean covering number of the map $f:\Omega(r)\to\mathbb{S}$, where $\Omega(r)=\Omega\cap\{z:|z|<r\}$, and

$$\mathscr{T}(r,\Omega) = \int_{\hat{\mathbb{C}}} N(r,\Omega,f=a)\mathrm{d}m(a).$$

$\mathscr{T}(r,\Omega)$ is convex with respect to $\log r$ and hence increases to infinity as r does.

We want to establish the second fundamental inequality for the Ahlfors-Shimizu characteristic in an angular domain corresponding to that of Nevanlinna's, that is, the fundamental inequality for estimation of the Ahlfors-Shimizu characteristic in terms of several quantities $N(r,\Omega,f=a)$. We shall realize this process by employing the Ahlfors theory of covering surfaces, which can be found in Hayman [16], Nevanlinna [26] and Tsuji [31]. The key point of this work is in estimation of error term appeared in the theory. However, the derivative is not considered in the Ahlfors theory of covering surfaces and hence it does not seem to be easy to establish an analogy of the Milloux inequality.

Let $\mathscr{F}$ be a simply connected finitely covering surface of the Riemann sphere $\mathbb{S}$. Given a simply connected domain D on $\mathbb{S}$ bounded by an analytic Jordan curve, we denote the part of $\mathscr{F}$ lying over D by $\mathscr{F}(D)$. $\mathscr{F}(D)$ consists of finitely many connected surfaces which are decomposed into two classes: Island and Tongue (or Peninsula). A connected surface of $\mathscr{F}(D)$ is called an island over D if its boundary lies over the boundary of D; a tongue over D if there exist some parts of its boundary which does not lie over the boundary of D.

Set

$$\mathscr{A} = \frac{|\mathscr{F}|}{\pi},$$

where $|\mathscr{F}|$ is the area of $\mathscr{F}$ counting its sheets on the Riemann sphere. $\mathscr{A}$ is the mean sheet number of $\mathscr{F}$. Then we state the following celebrated result of the Ahlfors theory of covering surfaces, which may be regarded as the Ahlfors unintegrated second fundamental theorem (cf. Theorem VI.3 of Tsuji [31]).

Theorem 2.4.1. *Let $\mathscr{F}$ be a simply connected finitely covering surface of the Riemann sphere $\mathbb{S}$ and D_v $(v=1,2,\cdots,q)$ be q disjoint simply connected domains on $\mathbb{S}$ each of which is bounded by an analytic Jordan curve. Then*

$$(q-2)\mathscr{A} \leqslant \sum_{v=1}^{q} n(D_v) + hL, \tag{2.4.3}$$

where $n(D_\nu)$ stands for the number of simply connected islands over D_ν and h is a constant only depending on D_ν $(\nu = 1,2,\cdots,q)$ and L is the length of the boundary of $\mathscr{F}$.

Theorem 2.4.1 is still true even if D_ν reduces into a single point a_ν.

Let us come to a special case produced by a meromorphic function. Given a simply connected domain $\mathscr{F}$ on $\mathbb{C}$ and a meromorphic function $w = f(z)$ on $\overline{\mathscr{F}}$, we have a finitely covering surface of the Riemann sphere $\mathbb{S}$, denoted still by $\mathscr{F}$, generated by $w = f(z)$. For this $\mathscr{F}$, Theorem 2.4.1 holds.

For any given $\alpha, \beta \in [0, 2\pi)$ such that $\alpha < \beta$, let $f(z)$ be a meromorphic function in $\overline{\Omega} = \overline{\Omega}(\alpha,\beta) = \{z : \alpha \leqslant \arg z \leqslant \beta\}$. Set

$$L(r,\theta) = \frac{1}{2\pi}\int_1^r \frac{|f'(te^{i\theta})|}{1+|f(te^{i\theta})|^2}dt$$

and

$$L(t,\alpha,\beta) = \frac{1}{\pi}\int_\alpha^\beta \frac{|f'(te^{i\theta})|}{1+|f(te^{i\theta})|^2}t d\theta.$$

Therefore, the length of the boundary of the covering surface of $\mathbb{S}$ generated by $w = f(z)$ from $\Omega(\alpha,\beta;r)$ is $L(r,\alpha) + L(r,\beta) + L(r,\alpha,\beta) + L(1,\alpha,\beta)$.

Thus employing Theorem 2.4.1 to $\Omega(r)$ and $w = f(z)$ obtains the following

Theorem 2.4.2. *Let D_ν $(\nu = 1,2,\cdots,q)$ be q disjoint simply connected domains on $\mathbb{S}$ each of which is bounded by an analytic Jordan curve. Then*

$$\begin{aligned}(q-2)(\mathscr{A}(r,\Omega) - \mathscr{A}(1,\Omega)) \leqslant & \sum_{\nu=1}^{q} n(D_\nu) + h(L(r,\alpha) \\ & + L(r,\beta) + L(r,\alpha,\beta) + L(1,\alpha,\beta)). \quad (2.4.4)\end{aligned}$$

When D_ν reduces a single point a_ν, that is, $D_\nu = \{a_\nu\}$, in Theorems 2.4.1 and 2.4.2, we have $n(\{a_\nu\}) = \overline{n}(X, f = a_\nu)$, that is, the number of distinct roots of $f(z) = a_\nu$ in the planar domain $X = \mathscr{F}$ or $\Omega(\alpha,\beta;r)$.

We establish the (integrated) second fundamental inequality for the Ahlfors-Shimizu characteristic in an angular domain. Let us begin with several lemmas.

Lemma 2.4.1. *Let $f(z)$ be a meromorphic function in $\overline{\Omega}(\alpha,\beta)$ $(0 \leqslant \alpha < \beta < 2\pi)$. Then*

$$\frac{1}{\pi}\int_1^r \int_\alpha^\beta \frac{|f'(te^{i\theta})|}{1+|f(te^{i\theta})|^2} d\theta dt \leqslant \sqrt{\frac{\beta-\alpha}{\pi}}\mathscr{A}^{\frac{1}{2}}(r,\Omega)(\log r)^{\frac{1}{2}}. \quad (2.4.5)$$

Proof. From the Schwarz inequality, we have

$$\begin{aligned}\frac{1}{\pi}\int_1^r\int_\alpha^\beta \frac{|f'(te^{i\theta})|}{1+|f(te^{i\theta})|^2}d\theta dt &= \frac{1}{\pi}\int_1^r\int_\alpha^\beta \left(\frac{|f'(te^{i\theta})|}{1+|f(te^{i\theta})|^2}\sqrt{t}\right)\frac{1}{\sqrt{t}}d\theta dt\\ &\leqslant \frac{1}{\pi}\left\{\int_1^r\int_\alpha^\beta\left(\frac{|f'(te^{i\theta})|\sqrt{t}}{1+|f(te^{i\theta})|^2}\right)^2 d\theta dt\int_1^r\int_\alpha^\beta\frac{dtd\theta}{t}\right\}^{\frac{1}{2}}\\ &= \frac{1}{\sqrt{\pi}}\mathscr{A}^{\frac{1}{2}}(r,\Omega)\sqrt{\beta-\alpha}(\log r)^{\frac{1}{2}}.\end{aligned}$$

This is the inequality (2.4.5). □

We compare $\mathscr{A}(r,\Omega)$ to $\mathscr{T}(r,\Omega)$.

Lemma 2.4.2. *Let $f(z)$ be a meromorphic function in $\overline{\Omega}(\alpha,\beta)$ $(0\leqslant\alpha<\beta<2\pi)$. Then for $\varepsilon>0$ we have*

$$\mathscr{A}(r,\Omega)\leqslant e\mathscr{T}(r)(1+(\log\mathscr{T}(r))^{1+\varepsilon}), r\notin F \tag{2.4.6}$$

and

$$\mathscr{A}(r,\Omega)\log r\leqslant\mathscr{T}(r)(\log\mathscr{T}(r))^{1+\varepsilon}, r\notin\hat{F}, \tag{2.4.7}$$

where F has only finite logarithmic measure and $\int_{\hat{F}}\frac{dt}{t\log t}<\infty$ and $\mathscr{T}(r)=\mathscr{T}(r,\Omega)$.

Proof. We establish (2.4.6) by using Corollary 1.1.1. Indeed, we have

$$\begin{aligned}\mathscr{A}(r,\Omega) &\leqslant \left[\log\left(1+\frac{1}{(\log\mathscr{T}(r))^{1+\varepsilon}}\right)\right]^{-1}\int_r^{r+\frac{r}{(\log\mathscr{T}(r))^{1+\varepsilon}}}\frac{\mathscr{A}(t,\Omega)}{t}dt\\ &\leqslant (1+(\log\mathscr{T}(r))^{1+\varepsilon})\mathscr{T}\left(r+\frac{r}{(\log\mathscr{T}(r))^{1+\varepsilon}}\right)\\ &\leqslant e\mathscr{T}(r)(1+(\log\mathscr{T}(r))^{1+\varepsilon}),\ r\notin F.\end{aligned}$$

Now to show (2.4.7). Set

$$\hat{F}=\{r:\mathscr{A}(r,\Omega)\log r>\mathscr{T}(r)(\log\mathscr{T}(r))^{1+\varepsilon}\}.$$

Since $\frac{d\mathscr{T}(r)}{dr}r=\mathscr{A}(r,\Omega)$, for $r\in\hat{F}$ we then have

$$\frac{d\mathscr{T}(r)}{dr}r\log r>\mathscr{T}(r)(\log\mathscr{T}(r))^{1+\varepsilon}$$

and hence

$$\int_{\hat{F}}\frac{dt}{t\log t}<\int_{\hat{F}}\frac{d\mathscr{T}(t)}{\mathscr{T}(t)(\log\mathscr{T}(t))^{1+\varepsilon}}\leqslant\int_1^\infty\frac{dt}{t(\log t)^{1+\varepsilon}}<\infty.$$

Thus we complete the proof of Lemma 2.4.2. □

Throughout this book, we mean by $F(f,\Omega)$ the exceptional set F outside which the inequality in Corollary 1.1.1 holds $\omega_1+hc=1$ for $\mathscr{T}(r,\Omega)$. Then the F in

Lemma 2.4.2 is $F(f,\Omega)$. Where no confusion seems to be possible, we write $F(\Omega)$ for $F(f,\Omega)$ and $F(f)$ for $F(f,\mathbb{C})$. Now we can establish the main result of this section.

Theorem 2.4.3. *Let $f(z)$ be a meromorphic function in $\overline{\Omega}=\overline{\Omega}(\alpha,\beta)$ and $a_1,a_2,\cdots,a_q$ be q distinct points on the extended complex plane $\widehat{\mathbb{C}}$. Then for any $\varepsilon>0$ with $0<\varepsilon<\frac{\beta-\alpha}{2}$, we have*

$$(q-2)\mathscr{T}(r,\Omega_\varepsilon)\leqslant\sum_{v=1}^{q}\overline{N}(r,\Omega,f=a_v)+\frac{2\pi h^2}{(q-2)\varepsilon}(\log r)^2 +H(r,\Omega), \tag{2.4.8}$$

where
(1) $H(r,\Omega)=\mathscr{T}^{1/2}(r,\Omega)\log\mathscr{T}(r,\Omega)+O(\log r),\ r\notin\hat{F}$;
(2) $H(r,\Omega)=\mathscr{T}^{3/4}(r,\Omega)\log\mathscr{T}(r,\Omega),\ r\notin F$.
Here $\Omega_\varepsilon=\Omega(\alpha+\varepsilon,\beta-\varepsilon)$ and $F=F(\Omega)$ is a set of finite logarithmic measure and $\hat{F}$ is a set with $\int_{\hat{F}}\frac{dt}{t\log t}<\infty$.

We shall call Theorem 2.4.3 the Ahlfors' second main theorem in an angular domain. The inequality (2.4.8) with (1) was attained by Zhang X. L. [40] and (2.4.8) with (2) by the author [41] in a different way from that of Zhang [40] under the assumption that $\mathscr{T}(r,\Omega)>(\log r)^p,\ p>2$.

Proof. By reducing the islands $D_1,D_2,\cdots,D_q$ to $a_1,a_2,\cdots,a_q$ in Theorem 2.4.2, we can immediately deduce that for $1<t\leqslant r$,

$$\begin{aligned}(q-2)(\mathscr{A}(t,\Omega_\varepsilon)-\mathscr{A}(1,\Omega_\varepsilon))&\leqslant\sum_{v=1}^{q}\overline{n}(t,\Omega_\varepsilon,f=a_v)+h[L(t,\alpha+\varepsilon,\beta-\varepsilon)\\&\quad+L(t,\alpha+\varepsilon)+L(t,\beta-\varepsilon)+L(1,\alpha+\varepsilon,\beta-\varepsilon)]\\&\leqslant\sum_{v=1}^{q}\overline{n}(t,\Omega,f=a_v)+h[L(t,\alpha,\beta)\\&\quad+L(t,\alpha+\varepsilon)+L(t,\beta-\varepsilon)+L(1,\alpha,\beta)],\end{aligned} \tag{2.4.9}$$

where h is a constant depending only on $\{a_1,a_2,\cdots,a_q\}$.

Now to proceed following Zhang [40] (also see [16]). Set

$$\begin{aligned}\psi(\tau)&=(q-2)(\mathscr{A}(t,\Omega_\tau)-\mathscr{A}(1,\Omega))\\&\quad-\sum_{v=1}^{q}\overline{n}(t,\Omega,f=a_v)-h(L(t,\alpha,\beta)+L(1,\alpha,\beta))\end{aligned}$$

for fixed t. $\psi(\tau)$ is a decreasing function of τ and

$$\frac{d\psi(\tau)}{d\tau}=(q-2)\frac{d\mathscr{A}(r,\Omega_\tau)}{d\tau}$$

and in view of (2.4.9)

$$\psi(\tau) \leqslant h[L(t,\alpha+\tau)+L(t,\beta-\tau)].$$

From the Schwarz inequality it follows that

$$\begin{aligned}(L(t,\alpha+\tau) &+ L(t,\beta-\tau))^2 \\ &\leqslant 2(L^2(t,\alpha+\tau)+L^2(t,\beta-\tau)) \\ &\leqslant 2\int_1^t \frac{\mathrm{d}\rho}{\rho}\left[\int_1^t (f^\#(\rho e^{\mathrm{i}(\alpha+\tau)}))^2\rho\mathrm{d}\rho + \int_1^t (f^\#(\rho e^{\mathrm{i}(\beta-\tau)}))^2\rho\mathrm{d}\rho\right] \\ &= -2\pi\log t\frac{\mathrm{d}\mathscr{A}(t,\Omega_\tau)}{\mathrm{d}\tau} \\ &= -\frac{2\pi}{q-2}\log t\frac{\mathrm{d}\psi(\tau)}{\mathrm{d}\tau}.\end{aligned}$$

Assume that for a ε_0 with $0<\varepsilon_0<\frac{\beta-\alpha}{2}$, we have $\psi(\varepsilon_0)>0$ and hence for $0\leqslant\tau\leqslant\varepsilon_0$, $\psi(\tau)>0$. Thus

$$\psi(\tau)^2 \leqslant -\frac{2h^2\pi}{q-2}\log t\frac{\mathrm{d}\psi(\tau)}{\mathrm{d}\tau}, \text{ namely, } 1 \leqslant -\frac{2h^2\pi}{q-2}\log t\frac{\psi'(\tau)}{\psi(\tau)^2}.$$

For $0<\varepsilon\leqslant\varepsilon_0$, we achieve

$$\varepsilon = \int_0^\varepsilon \mathrm{d}\tau \leqslant -\frac{2h^2\pi}{q-2}\log t\int_0^\varepsilon \frac{\psi'(\tau)}{\psi(\tau)^2}\mathrm{d}\tau = \frac{2h^2\pi}{q-2}\log t\left(\frac{1}{\psi(\varepsilon)}-\frac{1}{\psi(0)}\right)$$

so that

$$\psi(\varepsilon) < \frac{2h^2\pi}{(q-2)\varepsilon}\log t.$$

If $\psi(\varepsilon)\leqslant 0$, then the above inequality is still true and consequently for any ε with $0<\varepsilon<\frac{\beta-\alpha}{2}$, we have

$$\begin{aligned}(q-2)\mathscr{A}(t,\Omega_\varepsilon) \leqslant &\sum_{v=1}^q \overline{n}(t,\Omega,f=a_v)+hL(t,\alpha,\beta) \\ &+\frac{2h^2\pi}{(q-2)\varepsilon}\log t+O(1).\end{aligned} \tag{2.4.10}$$

By noticing that $\int_1^r \frac{L(t,\alpha,\beta)}{t}dt$ is exactly the form in the left side of (2.4.5), then it follows by dividing t both sides of the inequality (2.4.10) and integrating them from 1 to r and from (2.4.5) that

$$\begin{aligned}(q-2)\mathscr{T}(r,\Omega_\varepsilon) \leqslant &\sum_{v=1}^q \overline{N}(r,\Omega,f=a_v)+h\sqrt{\frac{\beta-\alpha}{\pi}}\mathscr{A}^{1/2}(r,\Omega)(\log r)^{1/2} \\ &+\frac{2h^2\pi}{(q-2)\varepsilon}(\log r)^2+O(\log r).\end{aligned} \tag{2.4.11}$$

Then in view of (2.4.7) we achieve (2.4.8) with (1).

Now we come to prove (2.4.8) with (2). Obviously, we can assume that $\mathscr{T}(r,\Omega_\varepsilon) > \frac{2\pi h^2}{(q-2)^2\varepsilon}(\log r)^2, r \notin F(\Omega)$, and then $\mathscr{T}(r) > \frac{2\pi h^2}{(q-2)^2\varepsilon}(\log r)^2$. In light of (2.4.6) we have

$$\begin{aligned}\mathscr{A}^{1/2}(r,\Omega)(\log r)^{1/2} &\leqslant \sqrt{2e}\mathscr{T}^{1/2}(r)(\log \mathscr{T}(r))^{(1+\varepsilon)/2}(\log r)^{1/2}\\ &\leqslant \sqrt{2e}K\mathscr{T}^{3/4}(r)(\log \mathscr{T}(r))^{(1+\varepsilon)/2},\end{aligned} \tag{2.4.12}$$

for $r \notin F(\Omega)$, where $K = \sqrt{\frac{q-2}{h}}\sqrt[4]{\frac{\varepsilon}{2\pi}}$, and for all sufficient large r, we have

$$h\sqrt{\frac{\beta-\alpha}{\pi}}\sqrt{2e}K\mathscr{T}^{3/4}(r)(\log \mathscr{T}(r))^{(1+\varepsilon)/2} + O(\log r) \leqslant \mathscr{T}^{3/4}(r)\log \mathscr{T}(r).$$

Consequently, we attain our propose for $F = [1,r_0]\cup F(\Omega)$ and for some sufficient large r_0 by noticing that $[1,r_0]\cup F(\Omega)$ has also finite logarithmic measure. □

We emphasize that (2.4.11) holds for all r. The following is a direct consequence of Theorem 2.4.3.

Corollary 2.4.1. *The same assumption as in Theorem 2.4.3 is given. Then for an unbounded sequence $\{r_n\}$ of positive real numbers outside $F(f,\Omega)$ such that $T(r_n,\Omega)/(\log r_n)^2 \to \infty$ as $n\to\infty$, we have*

$$(q-2)\mathscr{T}(r,\Omega_\varepsilon) \leqslant \sum_{v=1}^{q}\overline{N}(r,\Omega,f=a_v) + o(\mathscr{T}(r,\Omega)),\ r=r_n,\ \text{as } n\to\infty. \tag{2.4.13}$$

The proof of Corollary 2.4.1 is easily completed by employing the fact that for all sufficient large r_n, r_n will be outside the set F in Theorem 2.4.3 and $H(r,\Omega) = o(\mathscr{T}(r,\Omega)), r=r_n\to\infty$.

In some sense, Theorem 2.4.3 is a generalization of Lemma 3 in [14] and Theorem VII.3 in [31], which is stated as follows and which was used to deduce Theorem 3.1 of [14] for the case when $T(r,f)$ is of the slow growth.

Theorem 2.4.4. *Let $f(z)$ be meromorphic on the whole complex plane. Then for any three distinct points a_1,a_2 and a_3 on $\widehat{\mathbb{C}}$ and any small $\varepsilon>0$, we have*

$$\mathscr{T}(r,\Omega_\varepsilon) \leqslant 3\sum_{v=1}^{3}\overline{N}(2r,\Omega,f=a_v) + O((\log r)^2),$$

where $\Omega = \{z:\alpha<\arg z<\beta\}$ and $\Omega_\varepsilon = \{z:\alpha+\varepsilon<\arg z<\beta-\varepsilon\}$.

We think the inequality in Theorem 2.4.4 seems to be rude. Naturally, we wish we would be able to drop "3" before the sum symbol and could consider $q(\geqslant 3)$ distinct values a_v.

The following is applicable in discussion of angular distribution of a meromorphic function dealing with small functions, which is Theorem VIII in [31].

Theorem 2.4.5. *Let $f(z)$ and $a_j(z)(j=1,2,3,4)$ be meromorphic functions in the complex plane and*

$$g(z)=\frac{a_1(z)f(z)+a_2(z)}{a_3(z)f(z)+a_4(z)}.$$

Consider an angle $\Omega(\alpha,\beta)$ with $0<\beta-\alpha\leqslant 2\pi$, then for any $\varepsilon>0$, we have

$$\mathscr{A}(r,\Omega_\varepsilon,f)\leqslant 27\mathscr{A}(64r,\Omega,g)+O\left(\int_1^{128r}\frac{T(t,a)}{t}\mathrm{d}t\right), \tag{2.4.14}$$

where $T(t,a)=\sum_{j=1}^4 T(t,a_j)$.

Next we establish an analogue for the Valiron-Mohon'ko theorem, which is proved in [22].

Theorem 2.4.6. *Let $R(z)$ and $Q(z)$ be two rational functions and $f(z)$ be a meromorphic function on an angle $\overline{\Omega}$. Then*

$$\mathscr{T}(r,\Omega,R(f))\leqslant K_R\mathscr{T}(r,\Omega,f)$$

and if $(R/Q)(\infty)\neq 1$, and $R(z)+Q(z)$ and $R(z)$ have the same poles with the same multiplicities, we have

$$\mathscr{T}(r,\Omega,R(f)+Q(f))\leqslant L_{R,Q}(\mathscr{T}(r,\Omega,R(f))+\mathscr{T}(r,\Omega,Q(f))) \tag{2.4.15}$$

for two positive constants K_R depending only on $R(z)$ and $L_{R,Q}$ on $R(z)$ and $Q(z)$.

Proof. We first of all prove the second inequality (2.4.15). Set $\tau=(R/Q)(\infty)$, $g=R(f)$ and $h=Q(f)$, and assume $|\tau|>1$. A simple calculation yields

$$(g+h)^\#\leqslant\frac{1+|g|^2}{1+|g+h|^2}g^\#+\frac{1+|h|^2}{1+|g+h|^2}h^\#. \tag{2.4.16}$$

When $|g|\geqslant d|h|$, $1<d<|\tau|$, we have $|g+h|\geqslant|g|-|h|\geqslant\frac{d-1}{d}|g|\geqslant(d-1)|h|$ and so

$$\frac{1+|g|^2}{1+|g+h|^2}\leqslant\frac{1+|g|^2}{1+\left(\frac{d-1}{d}\right)^2|g|^2}\leqslant\left(\frac{d}{d-1}\right)^2\frac{1+|g|^2}{\left(\frac{d}{d-1}\right)^2+|g|^2}\leqslant\left(\frac{d}{d-1}\right)^2$$

and

$$\frac{1+|h|^2}{1+|g+h|^2}\leqslant\left(\frac{1}{d-1}\right)^2\frac{1+|h|^2}{(d-1)^{-2}+|h|^2}\leqslant 1+\left(\frac{1}{d-1}\right)^2.$$

When $|g|\leqslant d|h|$, then $f(z)$ is bounded, that is, for some $K>0$, $|f(z)|\leqslant K$. If $R(z)\sim\frac{b}{(z-a)^m}$ and $Q(z)\sim\frac{c}{(z-a)^n}$ as $z\to a$, then

$$\frac{1+|R(z)|^2}{1+|R(z)+Q(z)|^2}\sim\left(\frac{b}{c}\right)^2|z-a|^{2(n-m)}\ \text{ for } n>m;$$

~ 1 for $n < m$ and $\sim \left(\frac{b}{b+c}\right)^2$ for $n = m$ where $b+c \neq 0$ for $R(z)$ and $R(z)+Q(z)$ have same poles with the same multiplicities. Thus for $|f(z)| \leqslant K$, we have

$$\frac{1+|g|^2}{1+|g+h|^2} \leqslant K_1$$

for a positive constant K_1. It is easy to see that the poles of $Q(z)$ must be ones of $R(z)+Q(z)$. Therefore,

$$\frac{1+|h|^2}{1+|g+h|^2} \leqslant K_2$$

for a positive constant K_2.

We use the above estimations to (2.4.16) to obtain

$$((g+h)^{\#})^2 \leqslant K\{(g^{\#})^2+(h^{\#})^2\}$$

for a positive constant K. Thus (2.4.15) follows.

In order to prove the first inequality in Theorem 2.4.6, we begin with one simple case. For a non-zero complex number a and an integer k, we have

$$\begin{aligned}(af^k)^{\#} &= |a||k|\frac{|f^{k-1}|(1+|f|^2)}{1+|af^k|^2}f^{\#}\\ &= |a|^{-1}|k|\frac{|f|^{k-1}+|f|^{k+1}}{1+|f|^{2k}}\frac{1+|f|^{2k}}{|a|^{-2}+|f|^{2k}}f^{\#}\\ &\leqslant \max\{|a|,|a|^{-1}\}|k|f^{\#},\end{aligned}$$

where we used the inequalities $(|f|^{k-1}-1)(|f|^{k+1}-1) = |f|^{2k}-(|f|^{k-1}+|f|^{k+1})+1 \geqslant 0$, and $|a|^{-2}+x^2 \geqslant 1+x^2$ for $|a|<1$; $|a|^{-2}+x^2 \geqslant |a|^{-2}(1+x^2)$ for $|a| \geqslant 1$. Thus

$$\mathscr{T}(r,\Omega,af^k) \leqslant (\max\{|a|,|a|^{-1}\}|k|)^2\mathscr{T}(r,\Omega,f).$$

We write $R(z) = \frac{P(z)}{H(z)}$ with two relatively prime polynomials $P(z)$ and $H(z)$, and $P(z) = a_p z^p + \cdots + a_1 z + a_0$. Then in view of (2.4.15), we have

$$\begin{aligned}\mathscr{T}(r,\Omega,R(f)) &\leqslant L_1\sum_{k=0}^{p}\mathscr{T}\left(r,\Omega,\frac{a_k f^k}{H(f)}\right) = L_1\sum_{k=0}^{p}\mathscr{T}\left(r,\Omega,\frac{H(f)}{a_k f^k}\right)\\ &\leqslant L_2\sum_{k=0}^{\max\{p,q\}}\mathscr{T}(r,\Omega,f^k) \leqslant L\mathscr{T}(r,\Omega,f),\end{aligned}$$

where L_1, L_2 and L are constants and $q = \deg H$. We have attained the desired inequality. □

If $\Omega = \mathbb{C}$, then we have known that the constant K_R can be replaced by $\deg R + o(1)$ from Theorem 2.1.3, and (2.4.15) holds for $R(z)$ and $Q(z)$ on which no conditions are imposed. Then it is an important question of whether we could replace K_R with $\deg R + o(1)$ in Theorem 2.4.6.

Finally, we conclude this section with discussion about relations between $\mathscr{T}(r,\Omega)$ and other characteristics for an angle. Obviously it suffices to reveal the relation between $\mathscr{T}(r,\Omega)$ and $\dot{S}_{\alpha,\beta}(r,f)$, from Lemma 2.2.1, Lemma 2.3.2 and Theorem 2.3.3.

Theorem 2.4.7. *Let $f(z)$ be a function meromorphic on $\overline{\Omega}(\alpha,\beta)$. Then*

$$\dot{S}_{\alpha,\beta}(r,f) \leqslant 2\omega\frac{\mathscr{T}(r,\Omega)}{r^{\omega}} + \omega^2\int_1^r \frac{\mathscr{T}(t,\Omega)}{t^{\omega+1}}\mathrm{d}t$$

and for $\varepsilon > 0$

$$\dot{S}_{\alpha,\beta}(r,f) \geqslant \omega\sin(\omega\varepsilon)\frac{\mathscr{T}(r,\Omega_\varepsilon)}{r^{\omega}} + \omega^2\sin(\omega\varepsilon)\int_1^r \frac{\mathscr{T}(t,\Omega_\varepsilon)}{t^{\omega+1}}\mathrm{d}t - \omega\mathscr{T}(1,\Omega_\varepsilon).$$

Proof. It is obvious that $D_{\alpha,\beta}(t) \leqslant \pi\mathscr{A}(t,\Omega)$ and $D_{\alpha,\beta}(t) \geqslant \sin(\omega\varepsilon)\pi\mathscr{A}(t,\Omega_\varepsilon)$. Thus in view of (2.2.12) and the formula for integration by parts, we have

$$\begin{aligned}
\dot{S}_{\alpha,\beta}(r,f) &\leqslant \omega\int_1^r\left(\frac{1}{t^{\omega+1}}+\frac{t^{\omega-1}}{r^{2\omega}}\right)\mathscr{A}(t,\Omega)\mathrm{d}t\\
&= \omega\int_1^r\left(\frac{1}{t^{\omega}}+\frac{t^{\omega}}{r^{2\omega}}\right)\mathrm{d}\mathscr{T}(t,\Omega)\\
&< 2\omega\frac{\mathscr{T}(r,\Omega)}{r^{\omega}} + \omega^2\int_1^r \frac{\mathscr{T}(t,\Omega)}{t^{\omega+1}}\mathrm{d}t.
\end{aligned}$$

This implies the first desired inequality. The second desired inequality follows from the following inequality

$$\dot{S}_{\alpha,\beta}(r,f) \geqslant \omega\sin(\omega\varepsilon)\int_1^r\left(\frac{1}{t^{\omega+1}}+\frac{t^{\omega-1}}{r^{2\omega}}\right)\mathscr{A}(t,\Omega_\varepsilon)\mathrm{d}t$$

and the following estimation

$$\begin{aligned}
&\int_1^r\left(\frac{1}{t^{\omega+1}}+\frac{t^{\omega-1}}{r^{2\omega}}\right)\mathscr{A}(t,\Omega_\varepsilon)\mathrm{d}t\\
&= \int_1^r\left(\frac{1}{t^{\omega}}+\frac{t^{\omega}}{r^{2\omega}}\right)\mathrm{d}\mathscr{T}(t,\Omega_\varepsilon)\\
&= 2\frac{\mathscr{T}(r,\Omega_\varepsilon)}{r^{\omega}} - \left(1+\frac{1}{r^{2\omega}}\right)\mathscr{T}(1,\Omega_\varepsilon) + \omega\int_1^r\left(\frac{1}{t^{\omega+1}}-\frac{t^{\omega-1}}{r^{2\omega}}\right)\mathscr{T}(t,\Omega_\varepsilon)\mathrm{d}t\\
&= \frac{\mathscr{T}(r,\Omega_\varepsilon)}{r^{\omega}} + \omega\int_1^r \frac{\mathscr{T}(t,\Omega_\varepsilon)}{t^{\omega+1}}\mathrm{d}t - \mathscr{T}(1,\Omega_\varepsilon)\\
&\quad + \frac{\mathscr{T}(r,\Omega_\varepsilon)}{r^{\omega}} - \frac{\mathscr{T}(1,\Omega_\varepsilon)}{r^{2\omega}} - \omega\int_1^r \frac{t^{\omega-1}}{r^{2\omega}}\mathscr{T}(t,\Omega_\varepsilon)\mathrm{d}t\\
&\geqslant \frac{\mathscr{T}(r,\Omega_\varepsilon)}{r^{\omega}} + \omega\int_1^r \frac{\mathscr{T}(t,\Omega_\varepsilon)}{t^{\omega+1}}\mathrm{d}t - \mathscr{T}(1,\Omega_\varepsilon).
\end{aligned}$$

□

2.5 Estimates of the Error Terms

In this section, we take into account the various error terms appearing in (2.1.14), (2.2.6) and (2.3.5) and in other places, but the proofs of the coming results which can be easily found in other books will be omitted.

It is crucial in the theory of value distribution of a meromorphic function to estimate the error terms. Obviously, Theorems 2.1.4 and 2.1.5 make sense only if $S(D,a,f)$ is less than $T(D,a,f)$. Indeed we wish that $S(D,a,f) = o(T(D,a,f))$ as D becomes larger and larger in the sense of inclusion. We knew that $m\left(D,a,\frac{f^{(p)}}{f}\right)$ is main ingredient of $S(D,a,f)$ and hence it suffices to compare $m\left(D,a,\frac{f^{(p)}}{f}\right)$ to $T(D,a,f)$.

Let us begin with the case when $D = \{z : |z| < R\}$. The following is the lemma for the logarithmic derivative for a disk (see Lemma 1.3 and Lemma 4.3 of [36]).

Lemma 2.5.1. *Let $f(z)$ be meromorphic in $\{z : |z| < R\}$ $(0 < R \leqslant +\infty)$. If $f(0) \neq 0$, ∞, then for $0 < r < \rho < R$,*

$$\log^+\left|\frac{f'(z)}{f(z)}\right| \leqslant 10\log 2 + 2\log^+\log^+\frac{1}{|f(0)|} + \log^+\frac{1}{r} + 3\log^+\frac{1}{\rho - r} + 2\log^+ T(\rho,f) + \log^+\frac{r}{\delta(z)},$$

where $\delta(z)$ is the distance of z from the zeros and poles of f, and for a positive constant C_p,

$$m\left(r,\frac{f^{(p)}}{f}\right) < C_p\left\{1 + \log^+\log^+\frac{1}{|f(0)|} + \log^+\frac{1}{r} + \log^+\frac{1}{\rho - r} + \log^+\rho + \log^+ T(\rho,f)\right\}.$$

Lemma 2.5.1 is an improving version, due to Valiron [33], of the lemma for the logarithmic derivative for a disk which was established by Nevanlinna and the final inequality there was obtained for $p > 1$ by Hiong K. L. [20].

Applying the Poisson-Jensen formula (2.1.6) to $\frac{f^{(k)}(z)}{f(z)}$ yields the following result, which was proved by Yang and Zhang [37] for $k = 1$.

Lemma 2.5.2. *Let $f(z)$ be meromorphic in $\{z : |z| < R\}$ $(0 < R \leqslant +\infty)$. Then for any positive integer k and $0 < r \leqslant t < R$, we have*

$$\log\left|\frac{f^{(k)}(z)}{f(z)}\right| \leqslant \frac{R+r}{R-r}m\left(R,\frac{f^{(k)}}{f}\right) - \frac{R-r}{R+r}m\left(R,\frac{f}{f^{(k)}}\right) - \frac{(R-t)^2}{R^2+t^2}n(t,f^{(k)} = 0) + [n(R,f=0) + k\overline{n}(R,f=\infty)]\log\frac{R+r}{H}$$

for z in $\{z : |z| \leqslant r\}$ outside (γ), and (γ) is the set of the Boutroux-Cartan exceptional disks for the zeros (counted with multiplicities) and poles (counted exactly k times) of f in $|z| < R$ and H.

Proof. Set $D = \{z : |z| < R\}$. For $z \in D$ with $|z| \leqslant r$, we have

$$m\left(D, z, \frac{f^{(k)}}{f}\right) \leqslant \frac{R+r}{R-r} m\left(R, \frac{f^{(k)}}{f}\right)$$

and

$$m\left(D, z, \frac{f}{f^{(k)}}\right) \geqslant \frac{R-r}{R+r} m\left(R, \frac{f}{f^{(k)}}\right).$$

In view of (2.1.6), it is easy to get

$$\begin{aligned} & N\left(D, z, \frac{f^{(k)}}{f}\right) - N\left(D, z, \frac{f}{f^{(k)}}\right) \\ &= \frac{1}{2\pi} \int_{\partial D} \log \left| \frac{f(\zeta)}{f^{(k)}(\zeta)} \right| \frac{\partial G_D(\zeta, z)}{\partial n} ds + \log \left| \frac{f^{(k)}(z)}{f(z)} \right| \\ &= N(D, z, f^{(k)}) + N\left(D, z, \frac{1}{f}\right) - N(D, z, f) - N\left(D, z, \frac{1}{f^{(k)}}\right) \\ &= k\overline{N}(D, z, f) + N\left(D, z, \frac{1}{f}\right) - N\left(D, z, \frac{1}{f^{(k)}}\right). \end{aligned}$$

Let $\{a_n\}$ be the sequence of zeros and poles of $f(z)$ where zeros appear often according to their multiplicities and poles are counted k times. Then we have

$$\begin{aligned} N\left(D, z, \frac{1}{f}\right) + k\overline{N}(D, z, f) &= \sum_{a_n \in D} G_D(z, a_n) \\ &\leqslant \sum_{a_n \in D} \log \frac{R+r}{|z - a_n|} \\ &< [n(R, f = 0) + k\overline{n}(R, f = \infty)] \log \frac{R+r}{H} \end{aligned}$$

for z outside (γ). Let $\{b_n\}$ be the sequence of zeros of $f^{(k)}(z)$ appearing often according to their multiplicities. It follows that

$$
\begin{aligned}
N\left(D, z, \frac{1}{f^{(k)}}\right) &= \sum_{b_n \in D} G_D(z, b_n) = \sum_{b_n \in D} \log \left| \frac{R^2 - \overline{b_n} z}{R(z - b_n)} \right| \\
&= \sum_{b_n \in D} \frac{1}{2} \log \left(\frac{(R^2 - |b_n|^2)(R^2 - |z|^2)}{R^2 |z - b_n|^2} + 1 \right) \\
&\geqslant \sum_{b_n \in D} \frac{1}{2} \log \left(\frac{(R^2 - |b_n|^2)(R^2 - |z|^2)}{R^2 (|z| + |b_n|)^2} + 1 \right) \\
&\geqslant \sum_{b_n \in D} \log \frac{R^2 + |b_n||z|}{R(|z| + |b_n|)} \\
&= \sum_{b_n \in D} \log \left(1 + \frac{(R - |z|)(R - |b_n|)}{R(|z| + |b_n|)} \right) \\
&\geqslant \sum_{b_n \in D} \frac{(R - |z|)(R - |b_n|)}{R^2 + |b_n||z|} \\
&> n(t, f^{(k)} = 0) \frac{(R - t)^2}{R^2 + t^2},
\end{aligned}
$$

where the inequality $\log(1+x) \geqslant \frac{x}{1+x}$ has been used. In view of (2.1.6), we get

$$
\begin{aligned}
\log \left| \frac{f^{(k)}(z)}{f(z)} \right| &= T\left(D, z, \frac{f^{(k)}}{f}\right) - T\left(D, z, \frac{f}{f^{(k)}}\right) \\
&= m\left(D, z, \frac{f^{(k)}}{f}\right) - m\left(D, z, \frac{f}{f^{(k)}}\right) \\
&\quad + N\left(D, z, \frac{f^{(k)}}{f}\right) - N\left(D, z, \frac{f}{f^{(k)}}\right).
\end{aligned}
$$

This together with the above inequalities imply the desired inequality. □

Thus employing the Borel Lemma 1.1.5, we straightly obtain the following consequence of Lemma 2.5.1.

Corollary 2.5.1. *Let $f(z)$ be a meromorphic function in the complex plane. If $f(z)$ is of finite order, then the error terms appearing in Section 2.1*

$$
S(r, f) = O(\log r);
$$

If $f(z)$ is of infinite order, then

$$
S(r, f) = O(\log r + \log T(r, f)),
$$

outside certain possible exceptional set E of r with finite measure, as $r \to \infty$. Here $S(r, f) = S(\{|z| < r\}, 0, f)$.

Proof. When $f(z)$ is of finite order, it is easily seen that $\log T(2r,f) = O(\log r)$. Taking $\rho = 2r$ in Lemma 2.5.1 yields the desired result for the case. Assume $f(z)$ is of infinite order. Taking $\rho = r + 1/T(r,f)$ we have $\log T(\rho,f) \leqslant \log T(r,f) + \log 2$ for $r \notin E(f)$ and in view of Lemma 1.1.5 we deduce our desired result. □

Throughout this book, we mean by $E(f)$ the set appeared in Corollary 2.5.1. Actually, $E(f)$ is determined by Borel Lemma 1.1.5 for $T(r,f)$ and hence we mean by $E_\Omega(f)$ the set which is determined for $\mathscr{T}(r,\Omega,f)$ in an angle Ω.

We remark that in order to prove Corollary 2.5.1 it is sufficient to show Corollary 2.5.1 for $m\left(r,\frac{f'}{f}\right)$ in the place of $S(r,f)$. The reason is that in terms of the result for $m\left(r,\frac{f'}{f}\right)$, we can prove that Corollary 2.5.1 holds for $m\left(r,\frac{f^{(p)}}{f}\right)$ and so does for $S(r,f)$, whose proof will be provided below. Therefore, we emphasize that it is crucial to estimate the first order logarithmic derivative, and this is also available about the error terms associated with other characteristics.

Assume that Corollary 2.5.1 holds for $m\left(r,\frac{f'}{f}\right)$ in the place of $S(r,f)$. By induction, we assume that for $p \geqslant 1$

$$m\left(r,\frac{f^{(p)}}{f}\right) = O(\log rT(r,f)), r \notin E.$$

Now we consider the case for $p+1$. Since

$$\begin{aligned} T(r,f^{(p)}) &\leqslant N(r,f^{(p)}) + m(r,f) + m\left(r,\frac{f^{(p)}}{f}\right) \\ &\leqslant p\overline{N}(r,f) + T(r,f) + O(\log rT(r,f)) \\ &\leqslant (p+1)T(r,f) + O(\log rT(r,f)), r \notin E, \end{aligned}$$

we have

$$\begin{aligned} m\left(r,\frac{f^{(p+1)}}{f}\right) &\leqslant m\left(r,\frac{f^{(p+1)}}{f^{(p)}}\right) + m\left(r,\frac{f^{(p)}}{f}\right) \\ &\leqslant O(\log rT(r,f^{(p)})) + O(\log rT(r,f)) \\ &= O(\log T(r,f) + \log r) \end{aligned} \tag{2.5.1}$$

for all but a set of r with finite measure.

It is important and interesting to seek a precise estimate of $m\left(r,\frac{f^{(p)}}{f}\right)$. For the detail discussion, the reader is refereed to Cherry and Ye's book [3].

Another consequence of Lemma 2.5.1 is to be able to estimate the error term $R_{\alpha,\beta}(r,f)$ for the case of an angular domain in terms of $T(r,f)$ when the function considered is meromorphic in the whole complex plane.

Lemma 2.5.3. *Let $f(z)$ be a meromorphic function in the complex plane. For any $r < R$,*

$$B_{\alpha,\beta}\left(r,\frac{f'}{f}\right) \leqslant \frac{4\omega}{r^{\omega}} m_{\alpha,\beta}\left(r,\frac{f'}{f}\right)$$

and

$$A_{\alpha,\beta}\left(r,\frac{f'}{f}\right) \leqslant K\left[\left(\frac{R}{r}\right)^{\omega}\int_1^R \frac{\log T(t,f)}{t^{1+\omega}}\mathrm{d}t + \log\frac{r}{R-r} + \log\frac{R}{r}\right], \tag{2.5.2}$$

where $\omega = \frac{\pi}{\beta-\alpha}$ and K is a constant independent of r and R.

Furthermore, $R_{\alpha,\beta}(r,f) = O(\log T(r,f) + \log r)$, as $r \to \infty$, possibly outside a set of r with finite linear measure. If, in addition, $\int_1^{\infty} t^{-\omega-1}\log^+ T(t,f)\mathrm{d}t < \infty$, then $R_{\alpha,\beta}(r,f) = O(1)$.

Lemma 2.5.3 was established in [11]. As we did before Lemma 2.5.3, in view of (2.5.2), we can estimate $A_{\alpha,\beta}\left(r,\frac{f^{(p)}}{f}\right)$ by the quantity in the right-side of (2.5.2) with suitable K depending on p so that we can obtain the estimation of $R_{\alpha,\beta}(r,f)$ stated in Lemma 2.5.3. However, whether or not can we estimate the error term $R_{\alpha,\beta}(r,f)$ in terms of $S_{\alpha,\beta}(r,f)$? The difficulty we encounter is that generally on the boundary of a domain in question, we cannot obtain an estimation of the derivative $(\log f(z))'$ in terms of $\log f(z)$. However, it is possible to establish such a estimation of its derivative inside the domain in terms of the values of $\log f(z)$ on the boundary, this is what the formula (2.1.2) presents and therefore it is possible to control $R_{\alpha,\beta}(r,f)$ in terms of $S_{\alpha-\delta,\beta+\delta}(r,f)$, which will be realized as follows.

Consider the upper half disk, that is, $\Omega(0,\pi;R,0)$. Set $\Gamma = \partial\Omega(0,\pi;R,0)$. It is easy from (2.1.2) and (2.1.19) to see that

$$\begin{aligned}\log f(z) = &\frac{1}{2\pi\mathrm{i}}\int_{\Gamma}\log|f(\zeta)|\left(\frac{1}{\zeta-z}-\frac{z}{R^2-\zeta z}-\frac{1}{\zeta-\bar{z}}+\frac{\bar{z}}{R^2-\zeta\bar{z}}\right)\mathrm{d}\zeta\\ &-\sum_{\substack{|a_m|<R\\ \mathrm{Im}a_m>0}}\log\left[\frac{R^2-\bar{a}_m z}{R(z-a_m)}\frac{R(z-\bar{a}_m)}{R^2-a_m z}\right]\\ &+\sum_{\substack{|b_m|<R\\ \mathrm{Im}b_m>0}}\log\left[\frac{R^2-\bar{b}_m z}{R(z-b_m)}\frac{R(z-\bar{b}_m)}{R^2-b_m z}\right]+C(\bar{z}),\end{aligned} \tag{2.5.3}$$

where $C(\bar{z})$ is a function only in $\bar{z}$. Find partial derivative of both sides of (2.5.3) in z to obtain, by noting that $f_z'(z) = f'(z)$ as $f(z)$ is analytic,

$$\begin{aligned}\frac{f'(z)}{f(z)} &= \frac{1}{2\pi i}\int_{\Gamma}\log|f(\zeta)|\left[\frac{1}{(\zeta-z)^2}-\frac{R^2}{(R^2-\zeta z)^2}\right]d\zeta\\ &\quad-\sum_{\substack{|a_m|<R\\ \operatorname{Im}a_m>0}}\left[\frac{(a_m-\bar{a}_m)R^2}{(R^2-a_m z)(R^2-\bar{a}_m z)}+\frac{(\bar{a}_m-a_m)}{(z-a_m)(z-\bar{a}_m)}\right]\\ &\quad+\sum_{\substack{|b_m|<R\\ \operatorname{Im}b_m>0}}\left[\frac{(b_m-\bar{b}_m)R^2}{(R^2-b_m z)(R^2-\bar{b}_m z)}+\frac{(\bar{b}_m-b_m)}{(z-b_m)(z-\bar{b}_m)}\right].\end{aligned} \tag{2.5.4}$$

Below we often use the following equality

$$(R^2-\zeta z)^2-R^2(\zeta-z)^2=(R^2-\zeta^2)(R^2-z^2).$$

For $-R\leqslant t\leqslant R$ and $z=re^{i\phi}\in\Omega$, we have

$$\begin{aligned}\left|\frac{1}{(t-z)^2}-\frac{R^2}{(R^2-tz)^2}\right| &= \left|\frac{(R^2-t^2)(R^2-z^2)}{(t-z)^2(R^2-tz)^2}\right|\\ &\leqslant \frac{1}{\sin^2\phi}\left(\frac{1}{t^2}-\frac{1}{R^2}\right)\left|\frac{R^2(R^2-z^2)}{(R^2-tz)^2}\right|\\ &\leqslant \frac{1}{\sin^2\phi}\frac{R^2+r^2}{(R-r)^2}\left(\frac{1}{t^2}-\frac{1}{R^2}\right).\end{aligned} \tag{2.5.5}$$

For $z=re^{i\phi}$ and $\zeta=Re^{i\theta}$, we then get

$$\begin{aligned}\left|\frac{1}{(Re^{i\theta}-z)^2}-\frac{R^2}{(R^2-Re^{i\theta}z)^2}\right| &= \left|\frac{(R^2-\zeta^2)(R^2-z^2)}{(\zeta-z)^2(R^2-\zeta z)^2}\right|\\ &\leqslant 2\frac{R^2+r^2}{(R-r)^4}\sin\theta.\end{aligned} \tag{2.5.6}$$

For $R>1$ and $|z|=r$, employing (2.5.5) and (2.5.6), we estimate the integral in (2.5.4) and have

$$\begin{aligned}&\left|\frac{1}{2\pi i}\int_{\Gamma}\log|f(\zeta)|\left[\frac{1}{(\zeta-z)^2}-\frac{R^2}{(R^2-\zeta z)^2}\right]d\zeta\right|\\ &\leqslant \frac{1}{2}\frac{1}{\sin^2\phi}\frac{R^2+r^2}{(R-r)^2}\left[A(R,f)+A\left(R,\frac{1}{f}\right)+O(1)\right]\\ &\quad+\frac{1}{2}\frac{R^2(R^2+r^2)}{(R-r)^4}\left[B(R,f)+B\left(R,\frac{1}{f}\right)\right],\end{aligned} \tag{2.5.7}$$

for $|\log|f(\zeta)||\leqslant\log^+|f(\zeta)|+\log^+|1/f(\zeta)|$.

Now we estimate the terms in the brackets in (2.5.4). It is obvious that

$$\left|\frac{(a_m-\bar{a}_m)R^2}{(R^2-a_m z)(R^2-\bar{a}_m z)}\right|\leqslant\frac{2R}{(R-r)^2}.$$

Thus using the above inequalities to (2.5.4) we have

$$\begin{aligned}\left|\frac{f'(z)}{f(z)}\right| \leqslant & \frac{1}{2\sin^2\phi}\left(\frac{R+r}{R-r}\right)^4\left\{(A+B)\left[(R,f)+\left(R,\frac{1}{f}\right)\right]+O(1)\right\} \\ & +\left[n(R,\Omega,f)+n\left(R,\Omega,\frac{1}{f}\right)\right]\frac{2R}{(R-r)^2} \\ & +\sum_{\substack{|a_m|<R \\ \mathrm{Im}a_m>0}}\left[\frac{1}{|z-a_m|}+\frac{1}{|z-\bar{a}_m|}\right] \\ & +\sum_{\substack{|b_m|<R \\ \mathrm{Im}b_m>0}}\left[\frac{1}{|z-b_m|}+\frac{1}{|z-\bar{b}_m|}\right]. \end{aligned} \tag{2.5.8}$$

Now we can establish the following result essentially about estimate of the error term associated to the Nevanlinna characteristic for an angle.

Theorem 2.5.1. *Let $f(z)$ be a meromorphic function on $\Omega(\alpha-\varepsilon,\beta+\varepsilon)$ for $\varepsilon>0$ and $0<\alpha<\beta<2\pi$. Then for $R>r>1$*

$$A_{\alpha,\beta}\left(r,\frac{f'}{f}\right) \leqslant K(\log^+ S_{\alpha-\varepsilon,\beta+\varepsilon}(R,f)+\log R+\log\frac{R}{R-r}+1) \tag{2.5.9}$$

and

$$B_{\alpha,\beta}\left(r,\frac{f'}{f}\right) \leqslant \frac{K}{r^{\omega}}(\log^+ S_{\alpha-\varepsilon,\beta+\varepsilon}(R,f)+\log R+\log\frac{R}{R-r}+1). \tag{2.5.10}$$

And furthermore

$$(A+B)_{\alpha,\beta}\left(r,\frac{f'}{f}\right) \leqslant K(\log^+ S_{\alpha-\varepsilon,\beta+\varepsilon}(r,f)+\log r+1) \tag{2.5.11}$$

for $r>1$ possibly except a set with finite linear measure.

Proof. For the simplicity, we assume that $\alpha-\varepsilon/2=0$ and $\beta+\varepsilon/2=\pi$, that is, we consider the upper half plane, denoted by Ω_0. To establish our desired result, we need the following basic inequalities. For $0<s<1$ and $1<a<r$, it is easy to see that

$$\int_1^r \frac{\mathrm{d}t}{|t-a|^s} = \frac{1}{1-s}(a-1)^{-s+1}+\frac{1}{1-s}(r-a)^{-s+1} \leqslant \frac{2}{1-s}r^{-s+1}$$

and in view of (1.2.6), we therefore have

$$\begin{aligned}
&\int_1^r \log^+ \sum_{a_m\in\Omega_0} \frac{2}{|te^{\mathrm{i}\alpha}-a_m|^s}\frac{\mathrm{d}t}{t^{\omega+1}}\\
&\leqslant A\log^+\left(\frac{2}{A}\sum_{a_m\in\Omega_0}\int_1^r \frac{1}{|t-|a_m||^s}\frac{\mathrm{d}t}{t^{\omega+1}}\right)+\log 2,\ A=\omega(1-r^{-\omega})\\
&\leqslant A\log^+\left(\frac{2}{A}\sum_{a_m\in\Omega_0}\frac{2}{1-s}r^{1-s}\right)+\log 2\\
&\leqslant K(\log N+\log^+ r),
\end{aligned}$$

where N is the number of a_m in the above sum. In view of Lemma 2.2.2, setting $\hat{R}=\frac{1}{2}(R+r)$, we have

$$\begin{aligned}
n(\hat{R},\Omega_0,f=0,\infty) &\leqslant \frac{R}{R-\hat{R}}\int_{\hat{R}}^R \frac{n(t,\Omega_0,f=0,\infty)}{t}\mathrm{d}t\\
&\leqslant \frac{2R}{R-r}N(R,\Omega_0,f=0,\infty)\\
&\leqslant \frac{2R}{R-r}R^{\omega'}(2\omega'\sin(\omega'\varepsilon/2))^{-1}C_{\alpha-\varepsilon,\beta+\varepsilon}(R,f=0,\infty),
\end{aligned}$$

where $\omega'=\frac{\pi}{\beta-\alpha+2\varepsilon}$.

From (2.5.8) and in view of Lemma 2.2.1 it follows that

$$\begin{aligned}
\int_1^r \frac{1}{t^\omega}\log^+\left|\frac{f'(te^{\mathrm{i}\alpha})}{f(te^{\mathrm{i}\alpha})}\right|\frac{\mathrm{d}t}{t} &= \frac{1}{s}\int_1^r \log^+\left|\frac{f'(te^{\mathrm{i}\alpha})}{f(te^{\mathrm{i}\alpha})}\right|^s\frac{\mathrm{d}t}{t^{\omega+1}}\\
&\leqslant \int_1^r \left[4\log\frac{\hat{R}+r}{\hat{R}-r}+\log^+ S_{0,\pi}(\hat{R},f)+\log^+\frac{2\hat{R}}{(\hat{R}-r)^2}\right.\\
&\quad \left.+\log^+ n(\hat{R},\Omega_0,f=0,\infty)\right]\frac{\mathrm{d}t}{t^{\omega+1}}+O(1)\\
&\quad +\frac{1}{s}\int_1^r \log^+\sum_{a_m\in\Omega_0}\frac{2}{|te^{\mathrm{i}\alpha}-a_m|^s}\frac{\mathrm{d}t}{t^{\omega+1}}\\
&\quad +\frac{1}{s}\int_1^r \log^+\sum_{b_m\in\Omega_0}\frac{2}{|te^{\mathrm{i}\alpha}-b_m|^s}\frac{\mathrm{d}t}{t^{\omega+1}}\\
&\leqslant C_1\left[\log^+ S_{0,\pi}(\hat{R},f)+\log^+ R+\log\frac{R}{R-r}\right.\\
&\quad \left.+\log^+ n(\hat{R},\Omega_0,f=0,\infty)+1\right]\\
&\leqslant C_2\left[\log^+ S_{\alpha-\varepsilon,\beta+\varepsilon}(R,f)+\log^+ R+\log\frac{R}{R-r}\right.\\
&\quad \left.+\log^+ C_{\alpha-\varepsilon,\beta+\varepsilon}(R,f=0,\infty)+1\right]\\
&\leqslant C_3\left[\log^+ S_{\alpha-\varepsilon,\beta+\varepsilon}(R,f)+\log^+ R+\log\frac{R}{R-r}+1\right],
\end{aligned}$$

where the inequality $(\sum a_j)^s \leqslant \sum a_j^s$ for positive numbers a_j has been used.

Therefore in view of the definition of $A_{\alpha,\beta}(r,f)$, we have

$$A_{\alpha,\beta}\left(r,\frac{f'}{f}\right) \leqslant \frac{\omega}{s\pi}\int_1^r \frac{1}{t^\omega}\left(\log^+\left|\frac{f'(te^{i\alpha})}{f(te^{i\alpha})}\right|^s + \log^+\left|\frac{f'(te^{i\beta})}{f(te^{i\beta})}\right|^s\right)\frac{dt}{t}$$
$$\leqslant C_4\left[\log S_{\alpha-\varepsilon,\beta+\varepsilon}(R,f) + \log R + \log\frac{R}{R-r} + 1\right].$$

This has shown the inequality (2.5.9).

Now to prove (2.5.10). For $0 < d < 1$, we have

$$\int_0^{2\pi} \frac{d\theta}{|re^{i\theta}-a|^d} \leqslant \int_0^{2\pi}\frac{d\theta}{|r\sin\theta|^d} = \frac{4}{r^d}\int_0^{\pi/2}\frac{d\theta}{\sin^d\theta}$$
$$\leqslant \frac{4}{r^d}\left(\frac{\pi}{2}\right)^d\int_0^{\pi/2}\frac{d\theta}{\theta^d} = \frac{2\pi}{r^d(1-d)}.$$

In view of (2.5.8) and the above inequality, it is easy to see that

$$B_{\alpha,\beta}\left(r,\frac{f'}{f}\right) \leqslant \frac{2\omega}{d\pi r^\omega}\int_\alpha^\beta \log^+\left|\frac{f'(re^{i\theta})}{f(re^{i\theta})}\right|^d d\theta$$
$$\leqslant K_1\frac{2\omega}{d\pi r^\omega}\left[\log^+ S_{\alpha-\varepsilon,\beta+\varepsilon}(R,f) + \log^+ R + \log\frac{R}{R-r} + 1\right]$$
$$+\frac{2\omega}{d\pi r^\omega}\int_\alpha^\beta \log^+ \sum_{a_m\in\Omega_0}\frac{2}{|re^{i\theta}-a_m|^d}d\theta$$
$$+\frac{2\omega}{d\pi r^\omega}\int_\alpha^\beta \log^+ \sum_{b_m\in\Omega_0}\frac{2}{|re^{i\theta}-b_m|^d}d\theta$$
$$\leqslant K_2\frac{2\omega}{d\pi r^\omega}\left[\log^+ S_{\alpha-\varepsilon,\beta+\varepsilon}(R,f) + \log^+ R + \log\frac{R}{R-r} + 1\right].$$

This is (2.5.10).

Thus (2.5.11) follows from Lemma 2.2.1 which asserts that $S_{\alpha,\beta}(r,f)$ is increasing up to a bounded quantity, and the Borel Lemma 1.1.5. □

About the error terms for Tsuji characteristic, we have the following result, which can be proved by the method similar to the proof of Theorem 2.5.1 to some extent (see [11]). In this case it is the disk $\{z : |z-\frac{1}{2}iR| < \frac{1}{2}R\}$ that is considered in our implication and crucially, the disks for different R have the common frontier only at the origin.

Lemma 2.5.4. *Assume that $f(z)$ is a meromorphic function in $\Omega(\alpha,\beta)$. Then for $0 < r < R$, we have*

$$\mathfrak{m}_{\alpha,\beta}\left(r,\frac{f^{(p)}}{f}\right) \leqslant K\left[\log^+ \mathfrak{T}_{\alpha,\beta}(R,f) + \log\frac{R}{R-r} + 1\right].$$

Furthermore, $Q_{\alpha,\beta}(r,f) = O(\log r + \log^+ \mathfrak{T}_{\alpha,\beta}(r,f))$ *as* $r \to \infty$ *possibly except a set of* r *with finite linear measure.*

However we should mention that the error term appeared in Theorem 2.3.1 satisfies instead $Q_{\alpha,\beta}(r,f) = O(\log r) + o(\mathfrak{T}_{\alpha,\beta}(r,f))$ as $r \to \infty$ possibly except a set of r with finite linear measure, because we consider there small functions as targets.

2.6 Characteristic of Derivative of a Meromorphic Function

When $f(z)$ is a meromorphic function in a domain, it is easy to see that $f^{(p)}(z)$, the pth order derivative of $f(z)$, is also a meromorphic function in the same domain. Then we can consider various characteristics of $f^{(p)}(z)$. In this section, we mainly compare characteristics of a meromorphic function with those of its derivative. In view of the basic inequality of the proximate function (consult the paragraph before Theorem 2.1.2) and Lemma 2.5.1, for $\tau > 1$ we immediately get

$$\begin{aligned} T(r,f^{(p)}) &\leqslant N(r,f^{(p)}) + m(r,f) + m\left(r, \frac{f^{(p)}}{f}\right) \\ &= p\overline{N}(r,f) + T(r,f) + m\left(r, \frac{f^{(p)}}{f}\right) \\ &\leqslant (p+1)T(r,f) + K_{\tau,p}[\log^+ T(\tau r, f) + \log^+ r + 1], \end{aligned} \tag{2.6.1}$$

where $K_{\tau,p}$ is a positive constant.

On the other hand, we have the following Chuang's inequality when $p = 1$.

Theorem 2.6.1. *Let* $f(z)$ *be a meromorphic function in* $\{z : |z| < R\}$ *with* $f(0) \neq \infty$. *Then for* $\tau > 1$ *and* $0 < r < \tau^{-1}R$, *we have*

$$T(r,f) < C_\tau T(\tau r, f^{(p)}) + \frac{p(p+1)}{2}\log^+(2\tau r) + \sum_{j=0}^{p-1} \log^+ |f^{(j)}(0)| + \log 2p, \tag{2.6.2}$$

where C_τ *is a positive constant.*

Proof. We shall use Lemma 2.1.3 to complete our proof. Set $\tilde{R} = \frac{R+r}{2}$ and $h = \frac{R-r}{12e}$. Application of Lemma 2.1.3 to $f^{(p)}$ yields

$$\begin{aligned} \log^+ |f^{(p)}(z)| &\leqslant \left(\frac{\tilde{R}+\rho}{\tilde{R}-\rho} + \left(\log \frac{R}{\tilde{R}}\right)^{-1} \log \frac{\tilde{R}}{h}\right) T(R, f^{(p)}) \\ &\leqslant \left[\frac{5R+7r}{R-r} + \frac{2R}{R-r}\left(\log 6e + \log \frac{R+r}{R-r}\right)\right] T(R, f^{(p)}) \end{aligned} \tag{2.6.3}$$

for $z \notin (\gamma)$ with $|z| = \rho \leqslant \frac{R+2r}{3}$ by noticing that $\log\left(1 + \frac{R-r}{R+r}\right) > \frac{R-r}{2R}$. We denote the coefficient of $T(R, f^{(p)})$ in (2.6.3) by K below for simplification of statement.

It is obvious that we can find a ρ between r and $\frac{R+2r}{3}$ such that $\{z:|z|=\rho\}\cap(\gamma)=\varnothing$. Take a fixed point $z_0\notin(\gamma)$ with $|z_0|<\frac{R-r}{2}$ and set $z_1=\rho e^{\mathrm{i}\arg z_0}\notin(\gamma)$. From the segment $\overline{z_0z_1}$ we construct a curve Γ by replacing the parts of $\overline{z_0z_1}$ lying in (γ) with the minor arcs of two parts of the circles which are cut into by $\overline{z_0z_1}$. Then the length of Γ does not exceed $\rho+\frac{\pi}{2}eh<2R$. We have (2.6.3) for all $z\in\Gamma$.

In view of the formula for the integration by parts, we have for $|z|=\rho$

$$f(z)=\frac{1}{(p-1)!}\int_{z_0}^{z}(z-\zeta)^{p-1}f^{(p)}(\zeta)\mathrm{d}\zeta+\sum_{j=0}^{p-1}\frac{f^{(j)}(z_0)}{j!}(z-z_0)^j, \tag{2.6.4}$$

where the path, denoted by $\widehat{z_0z}$, of the integral is from z_0 to z_1 along Γ and then from z_1 to z along $|z|=\rho$. For simplification of statement, we use H to denote the sum of the second term in the right side of the above equation. Then we have

$$\log^+|H|\leqslant\sum_{j=0}^{p-1}\log^+|f^{(j)}(z_0)|+\frac{p(p-1)}{2}\log^+(2R)+K_1$$

for a positive constant K_1.

We estimate the module of the integral

$$\begin{aligned}\left|\int_{z_0}^{z}(z-\zeta)^{p-1}f^{(p)}(\zeta)\mathrm{d}\zeta\right|&\leqslant\int_{z_0}^{z}|z-\zeta|^{p-1}|f^{(p)}(\zeta)||\mathrm{d}\zeta|\\&\leqslant\max_{\zeta\in\widehat{z_0z}}|f^{(p)}(\zeta)|\int_{z_0}^{z}|z-\zeta|^{p-1}|\mathrm{d}\zeta|\\&\leqslant\max_{\zeta\in\widehat{z_0z}}|f^{(p)}(\zeta)|(2R)^{p-1}\int_{z_0}^{z}|\mathrm{d}\zeta|\\&\leqslant\max_{\zeta\in\widehat{z_0z}}|f^{(p)}(\zeta)|(2R)^{p}.\end{aligned}$$

Therefore on $|z|=\rho$ in virtue of (2.6.4) and (2.6.3) we have

$$\log^+|f(z)|\leqslant KT(R,f^{(p)})+p\log^+(2R)+\log^+|H|+\log 2,$$

and so

$$\begin{aligned}T(r,f)&\leqslant m(\rho,f)+N(\rho,f)\\&\leqslant N(R,f^{(p)})+KT(R,f^{(p)})+p\log^+(2R)+\log 2+\log^+|H|\\&\leqslant(1+K)T(R,f^{(p)})+\frac{p(p+1)}{2}\log^+(2R)\\&\quad+\sum_{j=0}^{p-1}\log^+|f^{(j)}(z_0)|+K_1+\log 2.\end{aligned}\tag{2.6.5}$$

Since $f(0)\neq\infty$, we can choose $z_0=0$. Letting $R=\tau r$ in (2.6.5) immediately deduces (2.6.2) from the above inequality. □

Furthermore in view of Corollary 1.1.1 and from (2.6.5) we can establish the following, which is essentially due to Edrei and Fuchs [8].

Theorem 2.6.2. *Let $f(z)$ be a transcendental meromorphic function. Then for a $\varepsilon > 0$*

$$T(r,f) < (\log T(r,f^{(p)}))^{1+\varepsilon} T(r,f^{(p)}) \tag{2.6.6}$$

possibly outside a set of r with finite logarithmic measure.

Proof. Set

$$R = r\mathrm{e}^{\alpha(r)}, \quad \alpha(r) = (\log T(r,f^{(p)}))^{-1-\bar{\varepsilon}}, \quad 0 < \bar{\varepsilon} < \varepsilon.$$

Assume that $T(r,f^{(p)}) > \mathrm{e}$ for $r \geqslant r_0$ and so $0 < \alpha(r) < 1$. Then

$$\frac{5R+7r}{R-r} = 5 + \frac{12r}{R-r} < 5 + \frac{12}{\alpha(r)} = 5 + 12(\log T(r,f^{(p)}))^{1+\bar{\varepsilon}}$$

and

$$\begin{aligned} \log\frac{R+r}{R-r} &= \log\frac{\mathrm{e}^{\alpha(r)}+1}{\mathrm{e}^{\alpha(r)}-1} \leqslant \log(\mathrm{e}+1) - \log\alpha(r) \\ &= \log(\mathrm{e}+1) + (1+\bar{\varepsilon})\log\log T(r,f^{(p)}) \end{aligned}$$

so that for K appeared in the proof of Theorem 2.6.1 we have

$$K+1 \leqslant C\log\log T(r,f^{(p)})(\log T(r,f^{(p)}))^{1+\bar{\varepsilon}} \leqslant \frac{1}{2\mathrm{e}}(\log T(r,f^{(p)}))^{1+\varepsilon}$$

for $0 < \bar{\varepsilon} < \varepsilon$ and all sufficiently large r and for some constant $C > 0$.

From Corollary 1.1.1 with $c = 1$ it follows that

$$T(R,f^{(p)}) \leqslant \mathrm{e}T(r,f^{(p)})$$

for all r possibly outside a set E with finite logarithmic measure. Now in view of (2.6.5) we have

$$\begin{aligned} T(r,f) &\leqslant (K+1)\mathrm{e}T(r,f^{(p)}) + \frac{p(p+1)}{2}(\log^+ r + \alpha(r)) + O(1) \\ &\leqslant \frac{1}{2}(\log T(r,f^{(p)}))^{1+\varepsilon} T(r,f^{(p)}) + \frac{p(p+1)}{2}(\log^+ r + 1) + O(1) \\ &\leqslant (\log T(r,f^{(p)}))^{1+\varepsilon} T(r,f^{(p)}), \end{aligned}$$

for all sufficiently large $r \notin E$, where we have used the fact that $\log r = o(T(r,f^{(p)}))$ as $r \to +\infty$. This is (2.6.6). □

We mean by a Pólya peak sequence of its Nevanlinna characteristic the Pólya peak sequence of a meromorphic function. Yang Lo posed a problem on existence of common Pólya peak sequences of a meromorphic function with finite lower order and its derivatives. This problem is still open, however we have the following in virtue of the Chuang's inequality (2.6.2).

Theorem 2.6.3. *Assume that $f(z)$ is a transcendental meromorphic function with finite lower order. Then there exist a sequence of positive numbers which is common relaxed Pólya peaks of order β of f and $f^{(j)}$ for positive integer j and $\mu(f) \leqslant \beta \leqslant \lambda(f)$.*

Proof. Let $\{r_n\}$ be a sequence of Pólya peak for $T(r,f)$. It is easy to see that $\{2r_n\}$ is a sequence of relaxed Pólya peak for $T(r,f)$. Now we prove that $\{2r_n\}$ is a sequence of relaxed Pólya peak for $T(r,f^{(j)})$. Actually, in view of Definition 1.1.1 we have for $r_n' \leqslant t \leqslant r_n''$

$$\begin{aligned} T(t,f^{(j)}) &\leqslant (j+1)T(t,f) + O(\log tT(2t,f)) \\ &\leqslant K_1\left(\frac{t}{r_n}\right)^{\beta} T(r_n,f) \\ &\leqslant K_1\left(\frac{t}{r_n}\right)^{\beta} K_2 T(2r_n,f^{(j)}) \end{aligned}$$

and the same argument yields 4) in Definition 1.1.1 for $\{2r_n\}$.

Thus we complete the proof of Theorem 2.6.3. □

Here the order and lower order of a transcendental meromorphic function mean those of its Nevanlinna characteristic. The type function of a meromorphic function is defined as that of its Nevanlinna characteristic. It is not difficult to see that a meromorphic function and its derivatives can have the same type functions up to a positive constant.

The following is due to Hayman and Miles [17].

Theorem 2.6.4. *Let $f(z)$ be a meromorphic function in the complex plane. Then for a given $K > 1$, there exists a set $M(K)$ with $\overline{\log\mathrm{dens}}M(K) \leqslant \delta(K)$, $\delta(K) = \min\{(2\mathrm{e}^{K-1}-1)^{-1}, (1+\mathrm{e}(K-1))\exp(\mathrm{e}(1-K))\}$, such that*

$$\limsup_{\substack{r\to+\infty \\ r\notin M(K)}} \frac{T(r,f)}{T(r,f^{(p)})} \leqslant 3\mathrm{e}K. \tag{2.6.7}$$

If $f(z)$ is entire, the bound $3\mathrm{e}K$ in (2.6.7) can be replaced by $2\mathrm{e}K$.

Here let us outline the proof of Theorem 2.6.4, and the reader is referred to Hayman and Miles [17] for the detail. It was first proved that for a positive function $T(r)$ with the positive continuous first and second order derivatives and for $K > 1$, there exists a set $M(K)$ with $\overline{\log\mathrm{dens}}M(K) \leqslant \delta(K)$ appearing in Theorem 2.6.4 such that for $r \notin M(K)$, we have a $\rho \in (1,r)$ with $\rho T'(\rho) \geqslant \frac{T(r)}{\mathrm{e}K\log r/\rho}$. Using this to the Ahlfors-Shimizu characteristic $\mathscr{T}(r,f)$ implies that

$$\mathscr{A}(\rho,f) \geqslant \frac{\mathscr{T}(r,f)}{\mathrm{e}K\log r/\rho}. \tag{2.6.8}$$

To establish relation between $f(z)$ and $f^{(p)}(z)$, we need a lemma of Hall and Ruscheweyh [15] which says that for a closed analytic curve $\Gamma(t)$, $a \leqslant t \leqslant b$, with $\Gamma'(t) \neq 0$, we have

$$\int_\Gamma |\mathrm{d}\phi| \leqslant \int_\Gamma |\mathrm{d}\psi|, \tag{2.6.9}$$

where for a fixed point $P_0 \notin \Gamma(t)$ and a fixed ray L starting from P_0, ψ and ϕ are the angles made respectively by the tangent vector to Γ at a point $P \in \Gamma$ and the radius vector $\overrightarrow{P_0P}$ with L.

Consider the curve $\Gamma(\theta) = f(\rho e^{i\theta})$, $0 \leqslant \theta \leqslant 2\pi$, with $f(\rho e^{i\theta}) \neq \infty$ and $f'(\rho e^{i\theta}) \neq 0$. For an arbitrary complex number a with $f(\rho e^{i\theta}) \neq a$, i.e., $a \notin \Gamma$, and the ray L starting from a paralleling to the positive real axis, we have $\psi(\theta) = \arg \Gamma'(\theta)$ and

$$\int_\Gamma |\mathrm{d}\psi| = \int_0^{2\pi} \left| \frac{\mathrm{d}}{\mathrm{d}\theta} \arg \Gamma'(\theta) \right| \mathrm{d}\theta.$$

Let us calculate $\frac{\mathrm{d}}{\mathrm{d}\theta} \arg \Gamma'(\theta)$. First we have $\Gamma'(\theta) = i\rho e^{i\theta} f'(\rho e^{i\theta})$ and then

$$\frac{\mathrm{d}}{\mathrm{d}\theta} \arg \Gamma'(\theta) = \mathrm{Im} \frac{\mathrm{d}}{\mathrm{d}\theta} \log \Gamma'(\theta) = \mathrm{Im} \frac{\Gamma''(\theta)}{\Gamma'(\theta)} = 1 + \mathrm{Re} \frac{\rho e^{i\theta} f''(\rho e^{i\theta})}{f'(\rho e^{i\theta})},$$

whence

$$\int_\Gamma |\mathrm{d}\psi| = \int_0^{2\pi} \left| 1 + \mathrm{Re} \frac{\rho e^{i\theta} f''(\rho e^{i\theta})}{f'(\rho e^{i\theta})} \right| \mathrm{d}\theta. \tag{2.6.10}$$

It is obvious that $\phi(\theta) = \arg(\Gamma(\theta) - a)$ and

$$\frac{\mathrm{d}}{\mathrm{d}\theta} \arg(\Gamma(\theta) - a) = \mathrm{Im} \frac{\Gamma'(\theta)}{\Gamma(\theta) - a},$$

and therefore we have

$$\int_\Gamma |\mathrm{d}\phi| = \int_0^{2\pi} \left| \mathrm{Re} \frac{\rho e^{i\theta} f'(\rho e^{i\theta})}{f(\rho e^{i\theta}) - a} \right| \mathrm{d}\theta. \tag{2.6.11}$$

Then combining (2.6.9), (2.6.10) and (2.6.11) deduces

$$\frac{1}{2\pi} \int_0^{2\pi} \left| \mathrm{Re} \frac{\rho e^{i\theta} f''(\rho e^{i\theta})}{f'(\rho e^{i\theta})} \right| \mathrm{d}\theta \geqslant \frac{1}{2\pi} \int_0^{2\pi} \left| \mathrm{Re} \frac{\rho e^{i\theta} f'(\rho e^{i\theta})}{f(\rho e^{i\theta}) - a} \right| \mathrm{d}\theta - 1.$$

Generally it follows from the above step that

$$\frac{1}{2\pi} \int_0^{2\pi} \left| \mathrm{Re} \frac{\rho e^{i\theta} f^{(p+1)}(\rho e^{i\theta})}{f^{(p)}(\rho e^{i\theta})} \right| \mathrm{d}\theta \geqslant \frac{1}{2\pi} \int_0^{2\pi} \left| \mathrm{Re} \frac{\rho e^{i\theta} f'(\rho e^{i\theta})}{f(\rho e^{i\theta}) - a} \right| \mathrm{d}\theta - p. \tag{2.6.12}$$

To estimate the left-side integral in (2.6.12) we need the following result, which is Lemma 4 of [17], that for a meromorphic function $h(z)$ in $|z| < r$ with $h(z) \sim cz^q$ as $z \to 0$ for some $c \neq 0$, we have for $1 < \rho < r$

$$\frac{1}{2\pi}\int_0^{2\pi}\left|\mathrm{Re}\frac{\rho e^{i\theta}h'(\rho e^{i\theta})}{h(\rho e^{i\theta})}\right| d\theta \leqslant \frac{2T(r,h)-\log|c|}{\log(r/\rho)}. \tag{2.6.13}$$

Since $\mathscr{A}(\rho,f)$ is an average of $n(\rho,f=a)$ in a, with the help of (2.6.12) and (2.6.13) we can get an a such that

$$\begin{aligned}
\mathscr{A}(\rho,f) &\leqslant n(\rho,f=a) = n(\rho,f=\infty)+n(\rho,f=a)-n(\rho,f=\infty)\\
&= n(\rho,f=\infty)+\frac{1}{2\pi i}\int_{|z|=\rho}\frac{f'(z)}{f(z)-a}dz\\
&= n(\rho,f=\infty)+\frac{1}{2\pi}\int_0^{2\pi}\frac{\rho e^{i\theta}f'(\rho e^{i\theta})}{f(\rho e^{i\theta})-a}d\theta\\
&\leqslant n(\rho,f=\infty)+\frac{1}{2\pi}\int_0^{2\pi}\left|\mathrm{Re}\frac{\rho e^{i\theta}f'(\rho e^{i\theta})}{f(\rho e^{i\theta})-a}\right|d\theta\\
&\leqslant n(\rho,f=\infty)+\frac{1}{2\pi}\int_0^{2\pi}\left|\mathrm{Re}\frac{\rho e^{i\theta}f^{(p+1)}(\rho e^{i\theta})}{f^{(p)}(\rho e^{i\theta})}\right|d\theta+p\\
&\leqslant n(\rho,f=\infty)+\frac{2T(r,f^{(p)})-\log|c|}{\log(r/\rho)}+p.
\end{aligned}$$

Thus by noting that $n(\rho,f=\infty)\log\frac{r}{\rho}\leqslant N(r,f)$ and from (2.6.8), we have

$$\begin{aligned}
\mathscr{T}(r,f) &< eK\left(n(\rho,f=\infty)\log\frac{r}{\rho}+2T(r,f^{(p)})-\log|c|+p\log\frac{r}{\rho}\right)\\
&< eK\{[2+o(1)]T(r,f^{(p)})+N(r,f)\}\\
&\leqslant eK[3+o(1)]T(r,f^{(p)}).
\end{aligned}$$

This produces (2.6.7).

Now we compare the characteristics of a meromorphic function and its derivative in an angular domain. By means of Theorem 2.2.1, we establish the following

Theorem 2.6.5. *Let $f(z)$ be a meromorphic function in $\overline{\Omega}(\alpha,\beta)$. Then for $\tau>1$ and a natural number p, we have*

$$S_{\alpha+\delta,\beta-\delta}(r,f)\leqslant K(S_{\alpha,\beta}(\tau r,f^{(p)})+\log^+ r+1), \tag{2.6.14}$$

where δ is such that $0<2\delta<\beta-\alpha$ and K is a constant only depending on τ,δ,α and β.

Proof. Set $\sigma=\sqrt{\tau}$. For $1\leqslant t\leqslant r$, in view of Theorem 2.2.1 for $R=\tau r$ and $R_0=\sigma t$, we have for $z=te^{i\theta}\in\Omega_{\delta/3}\setminus(\gamma)$

$$\log^+|f^{(p)}(z)|\leqslant Kt^{\omega_1}(S_{\alpha,\beta}(\tau r,f^{(p)})+1), \tag{2.6.15}$$

where $\omega_1=\frac{\pi}{\beta-\alpha-2\delta/3}$ and here and below K stands for a positive constant independent of t and it may not be same at each occurrence. From (2.6.4) it follows that for

$z = te^{i\theta} \in \Omega_{\delta/3} \setminus (\gamma)$,

$$\log^+ |f(z)| \leqslant Kt^{\omega_1}(S_{\alpha,\beta}(\tau r, f^{(p)}) + \log r + 1). \tag{2.6.16}$$

We can assume that (2.6.16) holds on the boundary of $\Omega(\alpha+\delta, \beta-\delta; r)$ for $S_{\alpha+\delta,\beta-\delta}(r, f)$ is decreasing in δ and increasing in r up to a bounded quantity. Employing (2.6.16), noting $\omega_1 < \tilde{\omega} = \frac{\pi}{\beta-\alpha-2\delta}$ and in view of the definition of $A_{\alpha,\beta}(r, f)$, we have

$$\begin{aligned} A_{\alpha+\delta,\beta-\delta}(r, f) &\leqslant K(S_{\alpha,\beta}(\tau r, f^{(p)}) + \log r + 1)\frac{\tilde{\omega}}{\pi}\int_1^r \left(\frac{1}{t^{\tilde{\omega}}} - \frac{t^{\tilde{\omega}}}{r^{2\tilde{\omega}}}\right) t^{\omega_1}\frac{dt}{t} \\ &< \frac{\tilde{\omega}}{\pi(\tilde{\omega}-\omega_1)} K(S_{\alpha,\beta}(\tau r, f^{(p)}) + \log r + 1) \end{aligned}$$

and in view of the definition of $B_{\alpha,\beta}(r, f)$

$$\begin{aligned} B_{\alpha+\delta,\beta-\delta}(r, f) &\leqslant K(S_{\alpha,\beta}(\tau r, f^{(p)}) + \log r + 1)\frac{2\tilde{\omega}}{\pi r^{\tilde{\omega}}} r^{\omega_1} \int_{\alpha+\delta}^{\beta-\delta} \sin\tilde{\omega}(\theta - \alpha - \delta)d\theta \\ &\leqslant K(S_{\alpha,\beta}(\tau r, f^{(p)}) + \log r + 1). \end{aligned}$$

In view of Lemma 2.2.1 we estimate

$$\begin{aligned} C_{\alpha+\delta,\beta-\delta}(r, f) &\leqslant C_{\alpha+\delta,\beta-\delta}(r, f^{(p)}) \\ &\leqslant S_{\alpha+\delta,\beta-\delta}(r, f^{(p)}) \\ &\leqslant K(S_{\alpha,\beta}(\tau r, f^{(p)}) + 1). \end{aligned}$$

Thus we have completed the proof of Theorem 2.6.5. □

Finally, we come to the case of the Tsuji characteristic and to establish an analogy of Theorem 2.6.5 for the Tsuji characteristic.

Theorem 2.6.6. *Let $f(z)$ be a meromorphic function in $\Omega(\alpha, \beta)$. Then for $\tau > 1$ and a natural number p, we have*

$$\mathfrak{T}_{\alpha+\delta,\beta-\delta}(r, f) \leqslant K(\mathfrak{T}_{\alpha,\beta}(\sigma\tau r, f^{(p)}) + \log r + 1), \tag{2.6.17}$$

where $\sigma > 1$ depends on δ, δ is such that $0 < 2\delta < \beta - \alpha$ and K is a constant only depending on τ, δ, α and β.

Proof. The inequality (2.6.17) we intend to prove follows directly from the implication

$$
\begin{aligned}
\mathfrak{T}_{\alpha+\delta,\beta-\delta}(r,f) &= \dot{\mathfrak{T}}_{\alpha+\delta,\beta-\delta}(r,f)+O(1) \quad \text{(by Lemma 2.3.2)}\\
&< \dot{S}_{\alpha+\delta,\beta-\delta}(r,f)+O(1) \quad \text{(by Theorem 2.3.3)}\\
&= S_{\alpha+\delta,\beta-\delta}(r,f)+O(1) \quad \text{(by Lemma 2.2.1)}\\
&\leqslant K_0(S_{\alpha+\delta/2,\beta-\delta/2}(\tau r,f^{(p)})+\log r+1) \quad \text{(by Theorem 2.6.5)}\\
&\leqslant K_1(\dot{\mathfrak{T}}_{\alpha,\beta}(\sigma\tau r,f^{(p)})+\log r+1) \quad \text{(by Theorem 2.3.3)}\\
&= K(\mathfrak{T}_{\alpha,\beta}(\sigma\tau r,f^{(p)})+\log r+1).
\end{aligned}
$$

□

There is a problem which is worth to discuss. Could we have the inequality (2.6.17) with $\mathfrak{T}_{\alpha,\beta}(r,f)$ in the place of $\mathfrak{T}_{\alpha+\delta,\beta-\delta}(r,f)$?

2.7 Meromorphic Functions in an Angular Domain

Let $f(z)$ be a transcendental meromorphic function. The lower order μ and the order λ of $f(z)$ are defined to be respectively those of the monotone increasing real function $T(r,f)$. In view of Lemma 2.1.3, if $f(z)$ is entire, then the order and lower order of $T(r,f)$ and $\log M(r,f)$ coincide.

If $f(z)$ is a meromorphic function in an angular domain $\Omega=\Omega(\alpha,\beta)$, then define the lower order and order of $f(z)$ in Ω respectively by

$$\mu_\Omega=\mu_\Omega(f)=\liminf_{r\to\infty}\frac{\log\mathscr{T}(r,\Omega,f)}{\log r}$$

and

$$\lambda_\Omega=\lambda_\Omega(f)=\limsup_{r\to\infty}\frac{\log\mathscr{T}(r,\Omega,f)}{\log r}.$$

Sometimes, we also write $\mu_{\alpha,\beta}$ and $\lambda_{\alpha,\beta}$ for μ_Ω and λ_Ω in the context. In view of (2.4.2), the definition of order in an angular domain is reasonable in the point of view of the case of the complex plane.

We say $f(z)$ to be transcendental (in the Ahlfors-Shimizu's sense) in Ω if $\mathscr{T}(r,\Omega,f)/\log r\to\infty$ $(r\to\infty)$. We make a remark on the transcendental definition in an angular domain. It is well-known that a meromorphic function on $\mathbb{C}$ is transcendental if and only if $T(r,f)/\log r\to\infty$ as $r\to\infty$ and so the transcendental definition in an angular domain is compatible with that on the complex plane. However, a transcendental meromorphic function assumes infinitely often all but at most two values on $\widehat{\mathbb{C}}$, while we cannot confirm the result for a transcendental meromorphic function in an angular domain. In terms of Theorem 2.4.4, the result holds if $\limsup_{r\to\infty}\mathscr{T}(r,\Omega,f)/(\log r)^2=\infty$.

Define

$$\rho_\Omega(a)=\rho_\Omega(f;a)=\limsup_{r\to\infty}\frac{\log N(r,\Omega,f=a)}{\log r},$$

which is called convergent exponent of a-value points of $f(z)$ in Ω, and then an $a \in \hat{\mathbb{C}}$ is called a Borel exceptional value of $f(z)$ in Ω provided that $\rho_\Omega(f;a) < \lambda_\Omega(f)$.

We can obtain the following result which is a version of the Borel Theorem for an angular domain.

Theorem 2.7.1. *Let $f(z)$ be a transcendental and meromorphic function in $\Omega(\alpha,\beta)$. Set*

$$\lambda(\varepsilon) = \limsup_{r\to+\infty} \frac{\log \mathcal{T}(r,\Omega_\varepsilon,f)}{\log r} \leqslant \lambda_\Omega(f).$$

Then there exist at most two $a \in \hat{\mathbb{C}}$ such that $\rho_\Omega(f;a) < \lim_{\varepsilon\to 0} \lambda(\varepsilon)$.

Proof. Suppose on contrary that there exist three $a \in \hat{\mathbb{C}}$ such that $\rho_\Omega(f;a) < \lim_{\varepsilon\to 0} \lambda(\varepsilon)$. Then we have a $\varepsilon_0 > 0$ with $\lambda(\varepsilon_0) > \rho_\Omega(f;a)$ for these three a and $\lambda(\varepsilon)$ is continuous at $\varepsilon_0 > 0$. It follows from (2.4.8) with (2) for Ω_{ε_0} and $\varepsilon > \varepsilon_0$ that

$$\lambda(\varepsilon) \leqslant \max\{\max_a\{\rho_\Omega(f;a)\}, \frac{3}{4}\lambda(\varepsilon_0)\}.$$

This derives a contradiction by letting $\varepsilon \to \varepsilon_0$, from which the proof of Theorem 2.7.1 is completed. □

Here we do not know if $\lim_{\varepsilon\to 0} \lambda(\varepsilon) = \lambda_\Omega(f)$.

We consider the inverse of Theorem 2.7.1. This leads us to ask a question

Question 2.7.1. *Should we have $\lambda_\Omega(f) \geqslant \lambda$ if $\rho_{\Omega_\varepsilon}(f;a) \geqslant \lambda$ for three distinct points a in $\hat{\mathbb{C}}$?*

From the definition of the Nevanlinna characteristic for a disk and the first fundamental theorem it is natural to control $N(r,f=a)$ in terms of $T(r,f)$, indeed we have $N(r,f=a) \leqslant T(r,f)+O(1) = \mathcal{T}(r,\mathbb{C},f)+O(1)$. Thus Question 2.7.1 is true for $\Omega = \mathbb{C}$. However, we have no such inequality for the case of an angular domain and hence we propose a question:

Question 2.7.2. *Does*

$$\limsup_{r\to\infty} \frac{N(r,\Omega,f=a)}{\mathcal{T}(r,\Omega,f)} < \infty \tag{2.7.1}$$

or

$$\liminf_{r\to\infty} \frac{N(r,\Omega,f=a)}{\mathcal{T}(r,\Omega,f)} < \infty \tag{2.7.2}$$

hold for $a \in \mathbb{C}$ possibly outside a set of a with measure zero?

The following is available to these two questions to a certain extent.

Theorem 2.7.2. *Let $f(z)$ be a transcendental and meromorphic function in $\Omega(\alpha,\beta)$. Then*

$$\frac{N(r,\Omega_\varepsilon,f=a)}{r^\omega}+\int_1^r\frac{N(t,\Omega_\varepsilon,f=a)}{t^{\omega+1}}\mathrm{d}t \leqslant K\left(\frac{\mathscr{T}(r,\Omega,f)}{r^\omega}+\int_1^r\frac{\mathscr{T}(t,\Omega,f)}{t^{\omega+1}}\mathrm{d}t\right)+O(1)$$

and

$$\frac{N(r,\Omega_\varepsilon,f=a)}{r^\omega}+\int_1^r\frac{N(t,\Omega_\varepsilon,f=a)}{t^{\omega+1}}\mathrm{d}t\leqslant K\int_1^{kr}\frac{\mathscr{T}(t,\Omega,f)}{t^{\omega+1}}\mathrm{d}t+O(1)$$

for a $\varepsilon>0$ and for a $k>1$ and a constant $K>1$ only depending on ε, ω and k.

Proof. According to the definition of $\dot{S}_{\alpha,\beta}(r,f)$, by the formula for integration by parts, we have

$$\begin{aligned}\dot{S}_{\alpha,\beta}(r,f) &\leqslant \frac{1}{\pi}\int_1^r\int_\alpha^\beta\left(\frac{1}{t^\omega}-\frac{t^\omega}{r^{2\omega}}\right)(f^\sharp(te^{i\theta}))^2t\mathrm{d}t\mathrm{d}\theta\\ &=\int_1^r\left(\frac{1}{t^\omega}-\frac{t^\omega}{r^{2\omega}}\right)\mathrm{d}\mathscr{A}(t,\Omega)\\ &<\omega\int_1^r\left(\frac{1}{t^{\omega+1}}+\frac{t^{\omega-1}}{r^{2\omega}}\right)\mathscr{A}(t,\Omega)\mathrm{d}t\\ &<2\omega\int_1^r\frac{\mathscr{A}(t,\Omega)}{t^{\omega+1}}\mathrm{d}t\\ &<2\omega\frac{\mathscr{T}(r,\Omega)}{r^\omega}+2\omega^2\int_1^r\frac{\mathscr{T}(t,\Omega)}{t^{\omega+1}}\mathrm{d}t.\end{aligned}$$

On the other hand, in view of Lemma 2.2.1 and Lemma 2.2.2 in turn it follows that

$$\begin{aligned}\dot{S}_{\alpha,\beta}(r,f) &= S_{\alpha,\beta}(r,f)+O(1)\\ &\geqslant C_{\alpha,\beta}(r,f=a)+O(1)\\ &\geqslant 2\omega\sin(\omega\varepsilon)\frac{N(r,\Omega_\varepsilon,f=a)}{r^\omega}\\ &\quad+2\omega^2\sin(\omega\varepsilon)\int_1^r\frac{N(t,\Omega_\varepsilon,f=a)}{t^{\omega+1}}\mathrm{d}t+O(1).\end{aligned}$$

Combination of above two inequalities yields the first desired inequality, and the second one follows from the first inequality and the following inequality

$$\int_1^{kr}\frac{\mathscr{T}(t,\Omega,f)}{t^{\omega+1}}\mathrm{d}t\geqslant\int_r^{kr}\frac{\mathscr{T}(t,\Omega,f)}{t^{\omega+1}}\mathrm{d}t\geqslant\frac{1}{\omega}\left(1-\frac{1}{k^\omega}\right)\frac{\mathscr{T}(r,\Omega,f)}{r^\omega}.$$

□

The second inequality in Theorem 2.7.2 was proved in Tsuji's book [31] in a different method. In view of Lemma 1.1.2 we have the following consequence of Theorem 2.7.2.

Corollary 2.7.1. *Let $f(z)$ be as in Theorem 2.7.2. If*

$$\liminf_{r\to\infty} \frac{\mathscr{T}(dr,\Omega)}{\mathscr{T}(r,\Omega)} > d^{\omega} \tag{2.7.3}$$

for some $d > 1$, then

$$N(r,\Omega_\varepsilon, f = a) \leqslant K\mathscr{T}(r,\Omega)$$

for a constant $K > 1$ only depending on ε and ω.

The proof of this corollary can be completed by using the first inequality in Theorem 2.7.2 and then by noticing that $\mathscr{T}(r,\Omega)$ satisfies the condition of Lemma 1.1.2 which is used to estimate the integral in the first inequality.

We remark that the inequality (2.7.3) implies that the lower order $\mu_\Omega(f) > \omega$. In fact, from (2.7.3) we can assume that for all natural number n and some $\delta > 0$

$$T(d^n,\Omega) \geqslant d^{n(\omega+\delta)}T(1,\Omega).$$

Then for $r > d$ we have $d^n \leqslant r < d^{n+1}$ for some n, and

$$\mathscr{T}(r,\Omega) \geqslant \mathscr{T}(d^n,\Omega) \geqslant d^{n(\omega+\delta)}\mathscr{T}(1,\Omega) \geqslant d^{-\omega-\delta}\mathscr{T}(1,\Omega)r^{\omega+\delta}.$$

This implies that $\mu_\Omega(f) \geqslant \omega+\delta$.

Theorem 2.7.3. *If for some $a \in \hat{\mathbb{C}}$, $\rho_{\Omega_\varepsilon}(a) > \omega$, then for all but at most two values of $b \in \hat{\mathbb{C}}$, we have*

$$\rho_\Omega(b) \geqslant \rho_{\Omega_\varepsilon}(a).$$

Proof. Under the assumption that $\rho_{\Omega_\varepsilon}(a) > \omega$, we choose a $\omega < \rho < \rho_{\Omega_\varepsilon}(a)$. From Theorem 2.7.2 it follows that

$$\begin{aligned}\frac{N(r,\Omega_\varepsilon, f = a)}{r^\rho} &\leqslant Kr^{\omega-\rho}\int_1^{kr}\frac{T(t,\Omega_{\varepsilon/2},f)}{t^{\omega+1}}\mathrm{d}t + O(1)\\ &\leqslant Kk^{\rho-\omega}\int_1^{kr}\frac{T(t,\Omega_{\varepsilon/2},f)}{t^{\rho+1}}\mathrm{d}t + O(1),\end{aligned}$$

where we have used the inequality

$$\frac{r^{\omega-\rho}}{t^\omega} = \left(\frac{t}{r}\right)^{\rho-\omega}\frac{1}{t^\rho} \leqslant k^{\rho-\omega}\frac{1}{t^\rho}.$$

Since the quantity in the left side is unbounded, we have

$$\int_1^\infty \frac{T(t,\Omega_{\varepsilon/2},f)}{t^{\rho+1}}\mathrm{d}t = \infty$$

and hence in view of Lemma 1.1.1, $\lambda_{\Omega_{\varepsilon/2}}(f) \geqslant \rho > \omega$. Letting $\rho \to \rho_{\Omega_\varepsilon}(a)^-$ immediately implies that $\lambda_{\Omega_{\varepsilon/2}}(f) \geqslant \rho_{\Omega_\varepsilon}(a)$. Thus we complete our proof by employing Theorem 2.7.1. □

It is natural to ask if the condition "$\rho_{\Omega_\varepsilon}(a) > \omega$" could be removed in Theorem 2.7.3. The question is true for $\Omega = \mathbb{C}$, while the example constructed in Hayman and Yang [19] asserts that this question is not always true (for the detail see Theorem 2.7.10 in the sequel). Then what condition imposed on suffices to confirm this question? We shall confirm this question if $f(z)$ is a transcendental meromorphic solution of a linear differential equation with polynomial coefficients in Chapter 3. Here we establish the following

Theorem 2.7.4. *Assume that for two distinct $a_i \in \hat{\mathbb{C}}$ $(i = 1,2)$ with $0 < \rho_{\Omega_\varepsilon}(a_i) \leqslant \omega$, $n(r, \Omega_\varepsilon, f = a_i)$ have a sequence of common (relaxed) Pólya peaks with order $\rho_{\Omega_\varepsilon}(a_i) > 0$. Then for all but at most two values of $b \in \hat{\mathbb{C}}$, we have*

$$\rho_\Omega(b) \geqslant \rho = \min_i\{\rho_{\Omega_\varepsilon}(a_i)\}.$$

Proof. Let $\{r_n\}$ be a sequence of common (relaxed) Pólya peaks of $n(r, \Omega_\varepsilon, f = a_i)$ $(i = 1,2)$ with order $\rho_{\Omega_\varepsilon}(a_i) > 0$ (When the relaxed Pólya peak is considered, we need to replace 2 with a large positive number in the below statement). Then

$$n(r_n/2, \Omega_\varepsilon, f = a_i) \leqslant (1 + o(1))2^{-\rho} n(r_n, \Omega_\varepsilon, f = a_i)$$

and so for large n and some $d > 1$, we have

$$n(r_n, \Omega_\varepsilon, f = a_i) > dn(r_n/2, \Omega_\varepsilon, f = a_i), \quad i = 1,2.$$

Now suppose on the contrary that there exist three distinct values b_i, $i = 1,2,3$, such that for some $\rho_0 < \rho$ and for all large r,

$$N(r) = \sum_{i=1}^{3} n(r, \Omega, f = b_i) < r^{\rho_0}.$$

Consider the closed domain $\hat{\Omega}_n = \Omega \cap \{z : r_n/20 \leqslant |z| \leqslant 20r_n\}$. It is easy to see that we can use a finite number of disks to cover the domain $\Omega_n = \Omega_\varepsilon \cap \{z : r_n/2 \leqslant |z| \leqslant 2r_n\}$ and the number of these disks is independent of n and the disks enlarged by five time still lie in $\hat{\Omega}_n$.

Take a ρ_1 with $\rho_0 < \rho_1 < \rho$. The following inequalities are taken into account provided that n is sufficiently large. Assume that for some $z_0 \in \Omega_n \setminus (\gamma)_{a_1}$, we have

$$\log \frac{1}{|f(z_0), a_1|} \geqslant r_n^{\rho_1}.$$

Employing Theorem 2.1.7 yields that for each $z \in \Omega_n \setminus (\gamma)$ and some positive constant K_1, $\log \frac{1}{|f(z), a_1|} \geqslant K_1 r_n^{\rho_1}$. Thus for each $z \in \Omega_n \setminus (\gamma)$

$$\log \frac{1}{|f(z), a_2|} \leqslant \log \frac{2}{|a_1, a_2|}.$$

Now we apply Lemma 2.1.7 to obtain

$$n(\Omega_n, f = a_2) \leqslant K_2 r_n^{\rho_1}.$$

An absurd inequality is derived as follows:

$$\begin{aligned} n(\Omega_n, f = a_2) &\geqslant n(r_n, \Omega_\varepsilon, f = a_2) - n(r_n/2, \Omega_\varepsilon, f = a_2) \\ &\geqslant \frac{d-1}{d} n(r_n, \Omega_\varepsilon, f = a_2) \\ &\geqslant r_n^{\rho + o(1)} \\ &\geqslant K_2^{-1} r_n^{\rho - \rho_1 + o(1)} n(\Omega_n, f = a_2), \end{aligned}$$

for $K_2^{-1} r_n^{\rho-\rho_1+o(1)} \to \infty$ as $n \to \infty$.

Therefore we obtain that for $z \in \Omega_n \setminus (\gamma)_{a_1}$, $\log \frac{1}{|f(z), a_1|} \leqslant r_n^{\rho_1}$. In this case, employing Lemma 2.1.7 yields

$$n(\Omega_n, f = a_1) \leqslant K_3 r_n^{\rho_1}.$$

The same argument as in above can derive an absurd inequality. Thus we complete the proof of Theorem 2.7.4. □

If there exists a $K > 1$ such that

$$K^{-1} n(r, \Omega_\varepsilon, f = a_1) \leqslant n(r, \Omega_\varepsilon, f = a_2) \leqslant K n(r, \Omega_\varepsilon, f = a_1),$$

then $n(r, \Omega_\varepsilon, f = a_i)$ $(i = 1, 2)$ have a sequence of common relaxed Pólya peaks with order $\rho = \rho_{\Omega_\varepsilon}(a_1) = \rho_{\Omega_\varepsilon}(a_2)$.

To establish a modified version of an important result Valiron [33] obtained in 1938, we formulate the following result, which is of independent significance and will be often used in the sequel, by consulting the proof of Theorem 3.9 of Yang [36].

Lemma 2.7.1. *Let $f(z)$ be a meromorphic function in an angular domain Ω and a_j $(j = 1, 2, 3)$ be three distinct complex numbers or ∞ in $\widehat{\mathbb{C}}$ and $\varepsilon > 0$. Then for a fixed $a \in \mathbb{C}$ and positive integer m, we have*

$$\begin{aligned} N(r, \Omega_\varepsilon, f = az^m + b) \leqslant K \sum_{j=1}^{3} N(2r, \Omega, f = a_j) + O((\log r)^2 \log^+ |ar|) \\ + O((\log r)^2 \log\log r) \end{aligned} \tag{2.7.4}$$

for all $b \in \widehat{\mathbb{C}}$ possibly outside a set E of b with measure zero, where K is a constant and does not depend on b.

Proof. Set

$$r_n = \left(1 + \frac{\varepsilon}{4}\right)^n.$$

We use $|z| = r_n$ $(n = 0, 1, 2, \cdots)$ to divide Ω_ε into a sequence of curvilinear quadrangle $A_n = \{z : r_n \leqslant |z| \leqslant r_{n+1}, z \in \Omega_\varepsilon\}$. We can use finitely many disks A_{jn} to cover

A_n such that the number s of these disks is independent of n and the resulting disks B_{jn} produced by enlarging A_{jn} five times are in $\Omega \cap \{z : r_n/2 \leqslant |z| \leqslant 2r_n\}$ (Notice that $2r_n > r_{n+1}$).

Applying Lemma 2.1.7 to B_{jn} we obtain

$$n(A_{jn}, f = az^m + b) \leqslant K_1 \left(\sum_{i=1}^{3} n(B_{jn}, f = a_i) + \log^+ |ar_n| + \log \frac{1}{\mathrm{e}^{-2\log n}} \right) \quad (2.7.5)$$

for $b \in \widehat{\mathbb{C}}$ possibly outside a disk E_{jn} with sphere radius $\mathrm{e}^{-2\log n}$.

Set

$$E_n = \bigcup_j E_{jn} \text{ and } E = \bigcap_{k=1}^{\infty} \left(\bigcup_{n=k}^{\infty} E_n \right).$$

Then E has zero measure, for

$$\mathrm{mes}E = \lim_{k\to\infty} \mathrm{mes} \left(\bigcup_{n=k}^{\infty} E_n \right) \leqslant s \lim_{k\to\infty} \sum_{n=k}^{\infty} \frac{\pi}{n^2} = 0.$$

Below let us check that E satisfies the requirement of Lemma 2.7.1. Given $b \notin E$, then $b \notin E_n$ for each $n \geqslant n_0 > 0$, and hence (2.7.5) is available for b and $n \geqslant n_0 > 0$. For $r \geqslant r_{n_0}$, we have $r_N \leqslant r < r_{N+1}$ for some N, and thus

$$\begin{aligned}
n(r, \Omega_\varepsilon, f = az^m + b) &\leqslant \sum_{n=n_0}^{N} \sum_j n(A_{jn}, f = az^m + b) + O(1) \\
&\leqslant K_1 \sum_{n=n_0}^{N} \sum_j \left(\sum_{i=1}^{3} n(B_{jn}, f = a_i) + \log^+ |ar_n| + 2\log n \right) + O(1) \\
&\leqslant K_2 \left(\sum_{i=1}^{3} n(2r, \Omega, f = a_i) + N\log^+ |ar_N| + 2N\log N \right) + O(1) \\
&\leqslant K \left(\sum_{i=1}^{3} n(2r, \Omega, f = a_i) + (\log r)(\log^+ |ar|) + (\log r)\log\log r \right) + O(1).
\end{aligned}$$

This immediately yields the desired inequality (2.7.4). □

Lemma 2.7.1 is actually a consequence of Valiron Lemma 2.1.7, while the important thing is to formulate this result. The following is due to Valiron [33](cf. Theorem 3.9 of Yang [36]), which follows from Lemma 2.7.1.

Theorem 2.7.5. *Let $f(z)$ be a meromorphic function in an angular domain $\Omega(\alpha, \beta)$ and such that $\rho_\Omega(a) \leqslant \rho$ for three distinct values of a. Then $\rho_{\Omega'}(a) \leqslant \rho$ for every $\Omega'(\alpha', \beta')$, $\alpha < \alpha' < \beta' < \beta$ and all complex number a outside a set of measure zero.*

If, in addition, for some c, $\rho_{\Omega_\varepsilon}(c) > \omega$ in Theorem 2.7.5, then we can obtain more. In fact, in view of Theorem 2.7.3, $\rho_\Omega(a) \geqslant \rho_{\Omega_\varepsilon}(c) > \omega$ for all but at most two values of a. Therefore under the assumption of Theorem 2.7.5 we have $\rho > \omega$. Suppose that for some b and some Ω', $\rho_{\Omega'}(b) > \rho > \omega$ and then by means of Theorem 2.7.3 $\rho_\Omega(a) \geqslant \rho_{\Omega'}(b) > \rho$ for at least one of three values of a in question. This implies $\rho > \rho$, impossible, whence $\rho_{\Omega'}(b) \leqslant \rho$ for every Ω' and all complex number b.

Theorem 2.7.1, Theorem 2.7.4 and Theorem 2.7.5 are important in the discussion of argument distribution of meromorphic functions, for the results do not deal with the opening magnitude of the angular domain considered.

In what follows, we consider the function $f(z)$ analytic in an angular domain Ω. Define

$$M_\Omega(f) = \limsup_{r\to\infty} \frac{\log^+\log^+ M(r,\Omega,f)}{\log r}, \tag{2.7.6}$$

where $M(r,f,\Omega) = \sup\limits_{\alpha\leqslant\theta\leqslant\beta} |f(re^{i\theta})|$. $M_\Omega(f)$ is called the order of $f(z)$ on Ω in the sense of maximum modulus.

If $f(z)$ is only assumed to be analytic in the angular domain, there are few explicit relations between $\mathcal{T}(r,\Omega)$ and $\log M(r,\Omega)$. Observe the exponential function $f(z) = e^z$. In the angular domain $\Omega = \{z : -\frac{\pi}{4} \leqslant \arg z \leqslant \frac{\pi}{4}\}$, it is easy to see that $\log M(r,\Omega,f) = r$ and $\frac{1}{8}e^{-2}\log r < \mathcal{T}(r,\Omega) < \frac{\sqrt{2}+1}{4}e^{-\sqrt{2}}\log r$, and therefore it is impossible that we estimate the order of $\log M(r,\Omega,f)$ in terms of that of $\mathcal{T}(r,\Omega)$. However it is well-known that $\log M(r,\Omega,f)$ and $\mathcal{T}(r,\Omega)$ have the same growth for $\Omega = \mathbb{C}$. Then we ask

Question 2.7.3. *Under what conditions may* $\log M(r,\Omega,f)$ *and* $\mathcal{T}(r,\Omega)$ *have the same growth for* $\Omega \neq \mathbb{C}$?

Since for any $a \in \hat{\mathbb{C}}$, $e^z = a$ has only finitely many roots in $\Omega(-\frac{\pi}{4},\frac{\pi}{4})$, it is thus impossible to use the order of $N(r,\Omega,f=a)$ to estimate that of $\log M(r,\Omega,f)$. However from Corollary 2.2.2, Theorem 2.4.7 and Lemma 2.2.2, we can show the following

Theorem 2.7.6. *Let $f(z)$ be an analytic function in $\overline{\Omega}(\alpha,\beta)$. Then for any $\varepsilon > 0$ we have*

$$\log M(r,\Omega,f) \leqslant K\frac{2\omega}{2^\omega}\mathcal{T}(2r,\Omega) + K\omega^2 r^\omega \int_1^{2r} \frac{\mathcal{T}(t,\Omega)}{t^{\omega+1}}dt + O(r^\omega),$$

$$\frac{\mathcal{T}(r,\Omega_\varepsilon)}{r^\omega} \leqslant K\left(\int_1^r \frac{\log^+ M(t,\Omega,f)}{t^{\omega+1}}dt + \frac{\log^+ M(r,\Omega,f)}{r^\omega}\right) + O(1)$$

and for each $a \in \widehat{\mathbb{C}}$

$$\frac{N(r,\Omega_\varepsilon,f=a)}{r^\omega} \leqslant K\left(\int_1^r \frac{\log^+ M(t,\Omega,f)}{t^{\omega+1}}dt + \frac{\log^+ M(r,\Omega,f)}{r^\omega}\right) + O(1).$$

In 1924, R. Nevanlinna [25](see Lemma 2.11 of [39]) investigated the convergent exponent of a-value points of a function which is restricted to be analytic only in an angular domain $\overline{\Omega}(\alpha,\beta)$ and established the following result, which can be directly obtained from Theorem 2.7.6.

Theorem 2.7.7. *Let $f(z)$ be an analytic function in $\overline{\Omega}(\alpha,\beta)$ with the order $M_\Omega(f)$. If for some $\varepsilon > 0$, $M_{\Omega_\varepsilon}(f) > \omega = \frac{\pi}{\beta-\alpha}$, then*

$$M_\Omega(f) \geqslant \lim_{\delta\to 0^+} \rho_{\Omega_\delta}(a) = \lim_{\delta\to 0^+} M_{\Omega_\delta}(f) \tag{2.7.7}$$

for each $a \in \mathbb{C}$ possibly except at most one value of a.

Proof. As in the proof of Theorem 2.7.3, in view of Theorem 2.7.6 for $\delta < \varepsilon$ we can deduce $\lambda_{\Omega_\delta}(f) \geqslant M_{\Omega_\delta}(f) \geqslant M_{\Omega_\varepsilon}(f) > \omega$. From Theorem 2.7.1 it follows that $\rho_{\Omega_{\delta/2}}(a) \geqslant \lambda_{\Omega_\delta}(f)$ for all but at most one value of a. Finally, applying the last inequality in Theorem 2.7.6 yields $M_{\Omega_{\delta/3}}(f) \geqslant \rho_{\Omega_{\delta/2}}(a)$ for each a. Thus we easily get (2.7.7). □

Theorem 2.7.7 is still true if the condition "$M_{\Omega_\varepsilon}(f) > \omega$" is replaced by "$\rho_{\Omega_\varepsilon}(a) > \omega$" for some a. Basically, under the new condition, in view of the last inequality in Theorem 2.7.6 we can deduce that $M_{\Omega_{\varepsilon/2}}(f) \geqslant \rho_{\Omega_\varepsilon}(a)$ and then we immediately have the result of Theorem 2.7.7.

Observing the exponential function $w = e^z$ implies that the condition "$M_{\Omega_\varepsilon(f)} > \frac{\pi}{\beta-\alpha}$" is necessary. In fact, the significance of many theorems dealing with an angle, such as Theorem 2.7.2 and Theorem 2.7.6, relies on a similar condition. However, Littlewood posed the following, which kicks out this condition by adding another assumption.

Conjecture. *Assume that for some positive number λ, we have*

$$M_{\Omega_\varepsilon}(f) \geqslant \lambda \quad \text{and} \quad \rho_{\Omega_\varepsilon}(0) \geqslant \lambda. \tag{2.7.8}$$

Then for every $a \in \mathbb{C}$ with at most one exception or at least for most values of a, we have $\rho_\Omega(a) \geqslant \lambda$.

This stimulates us to pose a question in view of Theorem 2.7.1.

Question 2.7.4. *Under the condition (2.7.8) of Littlewood conjecture, do we have $\lambda_\Omega(f) \geqslant \lambda$?*

If the question 2.7.4 is confirmed, then the Littlewood conjecture is true. This is because actually we can have $\lambda_{\Omega_{\varepsilon/2}}(f) \geqslant \lambda$. Let us observe the function $f(z) = K(e^z - e)$ for a positive number K. In the right half plane $\Omega = \{z : \mathrm{Re}z \geqslant 0\}$, $\rho_\Omega(0) = 1$ and $M_{\Omega_\varepsilon}(f) = 1$ for arbitrary $0 < \varepsilon < \frac{\pi}{2}$, but $f(z) \neq -Ke + a$ for $|a| < K$ and $z \in \Omega$. Hence Littlewood conjecture is not true provided that $\rho_{\Omega_\varepsilon}(0) \geqslant \lambda$ in (2.7.8) is replaced with $\rho_\Omega(0) \geqslant \lambda$.

Hayman and Yang [19] carefully investigated this conjecture. Following some of their ideas, we can prove the following

Theorem 2.7.8. *Assume that there exist a sequence of (relaxed) Pólya peaks $\{r_n\}$ of $n(r,\Omega_\varepsilon, f=0)$ with order $\rho \geqslant \lambda$ and a sequence of positive numbers $\{R_n\}$ such that*

$$\limsup_{n\to\infty}\frac{\log\log M(R_n,\Omega_\varepsilon,f)}{\log R_n}\geqslant\lambda. \tag{2.7.9}$$

If $\lim_{n\to\infty}\frac{\log r_n}{\log R_n}=1$, *then for every* $a\in\mathbb{C}$ *with at most one exception, we have* $\rho_\Omega(a)\geqslant\lambda$.

Proof. The proof we offer here is similar to that of Theorem 2.7.4. We assume without any loss of generalities that the limit in (2.7.9) exists, otherwise we consider a subsequence of $\{R_n\}$. Suppose on the contrary that there exist two distinct values a and b such that for some $\lambda_0<\lambda$ and for all large r,

$$N(r)=n(r,\Omega,f=a)+n(r,\Omega,f=b)<r^{\lambda_0}.$$

Set $K_n=\max\{r_n/R_n,R_n/r_n\}=R_n^{o(1)}$. Then

$$\Omega_\varepsilon\cap\{z:r_n/2\leqslant|z|\leqslant 2r_n\}\subset\Omega_\varepsilon\cap\{z:R_n/(2K_n)\leqslant|z|\leqslant 2R_nK_n\}.$$

We can use a finite number of disks to cover the domain $\Omega_n=\Omega_\varepsilon\cap\{z:R_n/(2K_n)\leqslant|z|\leqslant 2R_nK_n\}$ and the number of these disks is at most $O(\log K_n)$ and the disks enlarged by five times are still in the domain $\hat{\Omega}_n=\Omega\cap\{z:R_n/(10K_n)<|z|<10R_nK_n\}$.

For sufficiently large n, in view of (2.7.9) we have a point z_1 with $|z_1|=R_n$ and $\alpha+\varepsilon\leqslant\arg z_1\leqslant\beta-\varepsilon$ such that

$$\log|f(z_1)|>R_n^{\lambda+o(1)}.$$

Noticing that for a fixed number d

$$d^{O(\log K_n)}=K_n^{O(1)}=R_n^{o(1)},$$

therefore, in view of Theorem 2.1.7, we can find a Jordan curve Γ such that $\Omega_n\subset\mathrm{int}\Gamma\subset\hat{\Omega}_n$ and on Γ, $\log|f(z)|>R_n^{\lambda+o(1)}$. According to the Rouché Theorem, the number of zeros of $f(z)$ in Ω_n is at most $(10K_nR_n)^{\lambda_0}=10^{\lambda_0}r_n^{\lambda_0+o(1)}$. On the other hand, we have

$$n(\Omega_n,f=0)\geqslant n(r_n,\Omega_\varepsilon,f=0)-n(r_n/2,\Omega_\varepsilon,f=0)\geqslant r_n^{\rho+o(1)}.$$

However, $r_n^{\rho+o(1)}/r_n^{\lambda_0}\to\infty$ as $n\to\infty$ because $\rho>\lambda_0$. Thus a contradiction has been derived, from which Theorem 2.7.8 follows. □

The following is Theorem 1 of Hayman and Yang [19](it was obtained from their Theorem 3) which can be proved in view of Theorem 2.7.8.

Theorem 2.7.9. *Suppose that* $\rho_{\Omega_\varepsilon}(0)\geqslant\lambda$ *and*

$$\liminf_{\nu\to\infty}\frac{\log\log M(r_\nu,\Omega_\varepsilon,f)}{\log r_\nu}\geqslant\lambda \tag{2.7.10}$$

for an increasing unbounded sequence of positive numbers $\{r_\nu\}$ *with* $\frac{\log r_{\nu+1}}{\log r_\nu} \to 1$ *as* $\nu \to \infty$. *Then for every* $a \in \mathbb{C}$ *with at most one exception, we have* $\rho_\Omega(a) \geqslant \lambda$.

Proof. It suffices to prove that the conditions of Theorem 2.7.8 holds for the case $\rho_{\Omega_\varepsilon}(0) < \infty$ in view of Theorem 1.1.3. Actually, if $\rho_{\Omega_\varepsilon}(0) = \infty$, we need to do nothing in terms of Theorem 2.7.3. With the help of Theorem 1.1.3 there exist a sequence of Pólya peaks $\{r'_n\}$ of $n(r, \Omega_\varepsilon, f = 0)$ with order $\rho_{\Omega_\varepsilon}(0)$. For each large n, we have $r_{\nu_n} \leqslant r'_n < r_{\nu_n+1}$ and then $r'_n = r_{\nu_n}^{1+o(1)}$. Employing Theorem 2.7.8 for $R_n = r_{\nu_n}$ and $r_n = r'_n$ yields the desired result of Theorem 2.7.9. □

Actually, (2.7.10) is equivalent to

$$\liminf_{r\to\infty} \frac{\log\log M(r, \Omega_\varepsilon, f)}{\log r} \geqslant \lambda$$

to a great extent. If $M(r, \Omega_\varepsilon, f)$ is non-decreasing, then they are definitely equivalent. Indeed, assume without loss of generalities that the liminf of (2.7.10) for $\{r_\nu\}$ is finite, denoted by μ. For all large r, we have $r_\nu \leqslant r < r_{\nu+1}$ and hence

$$\log M(r, \Omega_\varepsilon, f) \geqslant \log M(r_\nu, \Omega_\varepsilon, f) \geqslant r_\nu^{\mu+o(1)} = r_{\nu+1}^{\mu+o(1)} > r^{\mu+o(1)}.$$

This attains our purpose.

The hard work in Hayman and Yang [19] is the proof of their Theorem 2 which is stated as follows.

Theorem 2.7.10. *Suppose that we are given* δ, η *and* λ *such that* $0 < \delta < \frac{\pi}{2}, 1 < \eta < 2, \frac{2}{3} < \lambda < 1$, *and a sequence* r_n *tending to* ∞ *as n does. There exists an entire function* $f(z)$ *with order* 1, *mean type, such that if* $\Omega = \Omega(-\frac{\pi}{2} + \frac{3}{4}\delta, \frac{\pi}{2} - \frac{3}{4}\delta)$, *then* $M_\Omega(f) = \lambda < 1$ *and*

$$\liminf \frac{\log|f(z)|}{|z|^\lambda} > 0$$

as $z \to \infty$ *in* Ω *outside the sequence of annuli*

$$r_n < |z| < r_n^\eta, \text{ for } n = 1, 2, \cdots.$$

Further, $\rho_{\Omega_{\frac{1}{4}\delta}}(0) = \lambda$, *but for all* $a \neq 0$,

$$\rho_\Omega(a) \leqslant \lambda - \frac{(1-\lambda)(\eta-1)}{5} < \lambda.$$

2.8 Deficiency and Deficient Values

Let $f(z)$ be a meromorphic function in the complex plane. In this section, we discuss the Nevanlinna deficiency $\delta(a, f)$ of $f(z)$ with respect to a value $a \in \widehat{\mathbb{C}}$ or a small function $a(z)$ of $f(z)$, which is defined by

$$\delta(a,f) = \liminf_{r\to\infty} \frac{m\left(r,\frac{1}{f-a}\right)}{T(r,f)} = 1 - \limsup_{r\to\infty} \frac{N\left(r,\frac{1}{f-a}\right)}{T(r,f)}$$

for $a \not\equiv \infty$ and $\delta(\infty, f)$ is defined by the above formula with $m(r,f)$ and $N(r,f)$ in place of $m\left(r,\frac{1}{f-a}\right)$ and $N\left(r,\frac{1}{f-a}\right)$. The a with $\delta(a,f) > 0$ is called Nevanlinna deficient value or function of $f(z)$. The Valiron deficiency of $f(z)$ for $a \in \widehat{\mathbb{C}}$ is defined by the formula

$$\Delta(a,f) = \limsup_{r\to\infty} \frac{m\left(r,\frac{1}{f-a}\right)}{T(r,f)}$$

and if $\Delta(a,f) > 0$, then a is called the Valiron deficient value of $f(z)$. Obviously, $0 \leqslant \delta(a,f) \leqslant \Delta(a,f) \leqslant 1$.

Define

$$\Theta(a,f) = 1 - \limsup_{r\to\infty} \frac{\overline{N}\left(r,\frac{1}{f-a}\right)}{T(r,f)}$$

and

$$\theta(a,f) = \liminf_{r\to\infty} \frac{N\left(r,\frac{1}{f-a}\right) - \overline{N}\left(r,\frac{1}{f-a}\right)}{T(r,f)},$$

and $\Theta(\infty, f)$ and $\theta(\infty, f)$ are defined by the above formula with $\overline{N}(r,f)$ and $N(r,f)$ in place of $\overline{N}\left(r,\frac{1}{f-a}\right)$ and $N\left(r,\frac{1}{f-a}\right)$. It is easy to see that $\delta(a,f) + \theta(a,f) \leqslant \Theta(a,f)$ in view of the first Nevanlinna fundamental theorem.

Using the second fundamental theorem 2.1.8 for small functions as targets, we easily verify that

$$\sum_a \{\delta(a,f) + \theta(a,f)\} \leqslant \sum_a \Theta(a,f) \leqslant 2, \tag{2.8.1}$$

where $\sum$ is taken over all small functions a of $f(z)$, and hence there are at most countable number of Nevanlinna deficient values and deficient functions.

We introduce Tsuji deficiency of $f(z)$ meromorphic in an angle $\Omega(\alpha,\beta)$ and transcendental with respect to the Tsuji characteristic, that is, $\limsup_{r\to\infty} \frac{\mathfrak{T}_{\alpha,\beta}(r,f)}{\log r} = \infty$ Set

$$\delta_T(a,f;\alpha,\beta) = \liminf_{r\to\infty} \frac{\mathfrak{m}_{\alpha,\beta}\left(r,\frac{1}{f-a}\right)}{\mathfrak{T}_{\alpha,\beta}(r,f)} = 1 - \limsup_{r\to\infty} \frac{\mathfrak{N}_{\alpha,\beta}\left(r,\frac{1}{f-a}\right)}{\mathfrak{T}_{\alpha,\beta}(r,f)}$$

for $a \not\equiv \infty$ and $\delta_T(\infty, f;\alpha,\beta)$ is defined by the above formula with $\mathfrak{m}_{\alpha,\beta}(r,f)$ and $\mathfrak{N}_{\alpha,\beta}(r,f)$ in place of $\mathfrak{m}_{\alpha,\beta}\left(r,\frac{1}{f-a}\right)$ and $\mathfrak{N}_{\alpha,\beta}\left(r,\frac{1}{f-a}\right)$. If no confusion occur in the context, then we simply write $\delta_T(a,f)$ for $\delta_T(a,f;\alpha,\beta)$. $\delta_T(a,f)$ is called the Tsuji deficiency of $f(z)$ at a and if $\delta_T(a,f) > 0$, then a is said to be a Tsuji deficient value of $f(z)$. It is also obvious that the total sum of all Tsuji deficiencies does not exceed 2, and there are at most countable number of Tsuji deficient values.

The Nevanlinna deficient value has attracted interests of many mathematicians. Actually it is an important quantity in the value distribution of meromorphic functions, and a great number of interesting and deep results about it have been established, some of which will be collected in the below chapters. Here it would be good to exhibit the results related to it, which will be often used in the sequel. Let us begin with three results concerning approximation of a meromorphic function to its Nevanlinna deficient values.

Lemma 2.8.1. *Let $f(z)$ be a meromorphic function in $\widehat{\mathbb{C}}$ with a as its Nevanlinna deficient value and $\{r_n\}$ be a sequence of positive numbers monotonically tending to infinity such that $T(dr_n, f) \leqslant KT(r_n, f)$ for a $d > 1$ and a positive constant K. Then there exist a sequence of $\{R_n\}$ such that $r_n \leqslant R_n < dr_n$ and for $n \geqslant n_0 > 0$*

$$\mathrm{mes}E_n(a) \geqslant t(a) > 0,$$

where $t(a)$ is independent of n, and

$$E_n(a) = \left\{\theta \in [0, 2\pi) : \log^+ \frac{1}{|f(R_n e^{i\theta}) - a|} > \frac{\delta}{4} T(R_n, f)\right\}, \ a \neq \infty,$$

and

$$E_n(a) = \left\{\theta \in [0, 2\pi) : \log^+ |f(R_n e^{i\theta})| > \frac{\delta}{4} T(R_n, f)\right\}, \ a = \infty,$$

and $\delta = \delta(a, f)$.

Proof. Assume here $a \neq \infty$. The below method will also show Lemma 2.8.1 for $a = \infty$. From Lemma 2.1.3, there exists an $R_n \in (r_n, \sqrt{d} r_n)$ such that

$$\log^+ \frac{1}{|f(z) - a|} \leqslant K_0 T\left(dr_n, \frac{1}{f(z) - a}\right) \leqslant 2K_0 K T(R_n, f)$$

on $|z| = R_n$. Since a is a Nevanlinna deficient value of $f(z)$, for sufficiently large r_n we have

$$\begin{aligned}
\frac{\delta}{2} T(R_n, f) &\leqslant m\left(R_n, \frac{1}{f(z) - a}\right) \\
&\leqslant \frac{1}{2\pi}\left(\int_{E_n(a)} + \int_{[0,2\pi)\setminus E_n(a)}\right) \log^+ \frac{1}{|f(R_n e^{i\theta}) - a|} d\theta \\
&\leqslant \frac{K_0 K}{\pi} T(R_n, f) \mathrm{mes}E_n(a) + \frac{\delta}{4} T(R_n, f),
\end{aligned}$$

so that $\mathrm{mes}E_n(a) \geqslant \frac{\delta\pi}{4K_0K} > 0$. Lemma 2.8.1 follows. □

If $\{r_n\}$ is a Pólya peak sequence with order β, then could we get a precise estimate from below of $\text{mes}E_n(a)$? We shall discuss this question in terms of Baernstein's method in the end of this section.

The following is an improvement of Lemma 2 of Edrei [6].

Lemma 2.8.2. *Let $f(z)$ be a meromorphic function with a as its Nevanlinna deficient value. Let $E(r;a)$ be a subset of $\theta \in [0,2\pi)$ defined by using the expression of $E_n(a)$ with r in the place of R_n. Then*

$$\text{mes}E(r;a) \geqslant (\log T(r,f))^{-1-\varepsilon}$$

for $\varepsilon > 0$ and all r possibly outside a set with finite logarithmic measure.

Proof. Here we only offer a proof of Lemma 2.8.2 for $a = \infty$. As in the proof of Theorem 2.6.2, in view of Lemma 2.1.3 we easily demonstrate

$$\log^+ |f(z)| \leqslant (\log T(r,f))^{1+\varepsilon} T(r,f), \ r \notin E$$

for a set E with finite logarithmic measure and $z = re^{i\theta} \notin (\gamma)$ with

$$h = \frac{r}{2\pi e}\left(\frac{\pi\delta}{70\sqrt{e}}\frac{\alpha(r)}{\alpha(r)+1}\right)^2, \ \alpha(r) = (\log T(r,f))^{-1-\varepsilon},$$

and $\delta = \delta(a,f)$. Set

$$I(r) = \{\theta \in [0,2\pi) : z = re^{i\theta} \in (\gamma)\}.$$

Then $\text{mes}I(r) \leqslant \frac{2\pi eh}{r}$. By noting that $x(1-\log x) < \frac{2}{\sqrt{e}}\sqrt{x}$ for $x > 0$ and applying Lemma 2.1.5 with $R = re^{\alpha(r)}$, we have

$$\begin{aligned}
\frac{1}{2\pi}\int_{I(r)} \log^+ |f(re^{i\theta})| d\theta &\leqslant \frac{7R}{\pi(R-r)} T(R,f) \text{mes}I(r)\,[1-\log \text{mes}I(r)] \\
&\leqslant \frac{7(\alpha(r)+1)}{\pi\alpha(r)} eT(r,f)\frac{2}{\sqrt{e}}\sqrt{\text{mes}I(r)} \\
&\leqslant \frac{14\sqrt{e}}{\pi}\frac{\alpha(r)+1}{\alpha(r)}\sqrt{\frac{2\pi eh}{r}}T(r,f) \\
&= \frac{\delta}{5}T(r,f).
\end{aligned}$$

For all sufficiently large r outside E, combining the above inequalities yields

$$\begin{aligned}
\frac{\delta}{2}T(r,f) \leqslant m(r,f) &= \frac{1}{2\pi}\left[\int_{E(r;a)\setminus I(r)} + \int_{I(r)} + \int_{[0,2\pi)\setminus E(r;a)}\right] \log^+ |f(re^{i\theta})| d\theta \\
&\leqslant \frac{1}{2\pi}(\log T(r,f))^{1+\varepsilon} T(r,f) \text{mes}E(r;a) + \frac{\delta}{5}T(r,f) + \frac{\delta}{4}T(r,f) \\
&= \frac{1}{2\pi}(\log T(r,f))^{1+\varepsilon} T(r,f) \text{mes}E(r;a) + \frac{9\delta}{20}T(r,f), \ r \notin E.
\end{aligned}$$

This immediately deduces the desired result by changing ε a little bit. □

In order to solve some of problems about the deficiency, certain new methods, tools and concepts have been introduced, created and developed. Here, we introduced the Baernstein's celebrated result which solves the Edrei spread conjecture.

Given a positive function $\Lambda(r)$ with $\Lambda(r) \to 0$ as $r \to +\infty$, we define for $r > 0$ and $a \in \mathbb{C}$

$$D_\Lambda(r,a) = \left\{ \theta \in [-\pi,\pi) : \log^+ \frac{1}{|f(re^{i\theta}) - a|} > \Lambda(r)T(r,f) \right\}$$

and

$$D_\Lambda(r,\infty) = \left\{ \theta \in [-\pi,\pi) : \log^+ |f(re^{i\theta})| > \Lambda(r)T(r,f) \right\}.$$

Baernstein [2] proved the following, which was conjectured by Edrei [7].

Theorem 2.8.1. *Let $f(z)$ be a transcendental and meromorphic function in $\mathbb{C}$ with the finite lower order μ and the order $0 < \lambda \leqslant \infty$ and for some $a \in \widehat{\mathbb{C}}$, $\delta = \delta(a,f) > 0$. Then for arbitrary sequence of Pólya peaks $\{r_n\}$ of order $\beta > 0$, $\mu \leqslant \beta \leqslant \lambda$ and arbitrary positive function $\Lambda(r)$ with $\Lambda(r) \to 0$ as $r \to +\infty$, we have*

$$\liminf_{n\to\infty} \operatorname{mes} D_\Lambda(r_n,a) \geqslant \min\left\{ 2\pi, \frac{4}{\beta}\arcsin\sqrt{\frac{\delta}{2}} \right\}.$$

This inequality is called spread relation. We do not know if Theorem 2.8.1 is true for a sequence of the relaxed Pólya peaks. Let us sketch out the proof of Baernstein Theorem 2.8.1. Here it is sufficient to consider the case $a = \infty$. Baernstein first introduced the powerful T^* function. Set for $z = re^{i\theta}$ with $0 < \theta < \pi$

$$m^*(z) = \sup_E \frac{1}{2\pi} \int_E \log|f(re^{i\phi})| d\phi$$

where the supremum is taken over all measurable set E in $(-\pi,\pi)$ with measure 2θ and

$$T^*(z) = m^*(z) + N(|z|,f).$$

It was verified that $T^*(z)$ is continuous and subharmonic on the upper half plane. Set $\sigma_n = \operatorname{mes} D_\Lambda(r_n,\infty)$. It is easy to prove that

$$\begin{aligned} m(r_n,f) &= \frac{1}{2\pi}\left(\int_{D_\Lambda(r_n,\infty)} + \int_{[0,2\pi)\setminus D_\Lambda(r_n,\infty)} \right) \log^+ |f(r_n e^{i\phi})| d\phi \\ &\leqslant m^*(r_n \exp(\frac{1}{2} i\sigma_n)) + \Lambda(r_n)T(r_n,f), \end{aligned}$$

so that

$$T(r_n,f) \leqslant T^*(r_n \exp(\frac{1}{2} i\sigma_n)) + \Lambda(r_n)T(r_n,f).$$

The hard work is to estimate $T^*(r_n\exp(\frac{1}{2}\mathrm{i}\sigma_n))$ in term of $T(r_n,f)$. Since $T^*(z)$ is continuous and subharmonic on the upper half plane, in term of the Poisson formula on the upper half disk centered at the origin and the basic inequality of Pólya peak sequence, we have either

$$T^*(r_n\exp(\frac{1}{2}\mathrm{i}\sigma_n)) \leqslant T(r_n,f)\left[\cos\left(\pi-\frac{\sigma_n}{2\gamma}\right)\gamma\beta+o(1)\right]$$

or $2\gamma\pi \leqslant \sigma_n$ where

$$\gamma=\frac{2}{\pi\beta}\arcsin\sqrt{\frac{\delta(\infty,f)}{2}}.$$

If $2\gamma\pi > \sigma_n$, then we have

$$1 \leqslant \cos\left(\pi-\frac{\sigma_n}{2\gamma}\right)\gamma\beta+o(1)$$

and further $\pi-\frac{\sigma_n}{2\gamma}=o(1)$ as $n\to\infty$. Therefore, we always have $\sigma_n \geqslant 2\gamma\pi+o(1)$ as $n\to\infty$.

This immediately deduces Theorem 2.8.1.

If we take $\Lambda(r)\equiv d<\delta=\delta(\infty,f)$, then from the Baernstein's proof of Theorem 2.8.1, it follows that either $2\gamma\pi \leqslant \sigma_d(\infty)$ or

$$1-d \leqslant \cos\left(\pi-\frac{\sigma_d(\infty)}{2\gamma}\right)\gamma\beta, \tag{2.8.2}$$

where $\sigma_d(\infty)=\liminf\limits_{n\to\infty}D_d(r_n,\infty)$. Assume that $2\gamma\pi \geqslant \sigma_d(\infty)$ and then in view of (2.8.2),

$$0 \leqslant \left(\pi-\frac{\sigma_d(\infty)}{2\gamma}\right)\gamma\beta \leqslant \frac{\pi}{2}-\arcsin(1-d).$$

This implies that

$$\begin{aligned}\sigma_d(\infty) &\geqslant 2\gamma\pi-\frac{\pi}{\beta}+\frac{2}{\beta}\arcsin(1-d)\\ &=\frac{2}{\beta}\left(2\arcsin\sqrt{\frac{\delta}{2}}-\frac{\pi}{2}+\arcsin(1-d)\right)\\ &=\frac{2}{\beta}(-\arcsin(1-\delta)+\arcsin(1-d))\\ &\geqslant \frac{2}{\beta}(\delta-d).\end{aligned}$$

Since $\arcsin\sqrt{\frac{\delta}{2}} \geqslant \sqrt{\frac{\delta}{2}} > \frac{\delta}{2}$, if $\sigma_d(\infty) > 2\gamma\pi$ we have $\sigma_d(\infty) \geqslant \frac{2}{\beta}\delta$. Therefore we always have $\sigma_d(\infty) \geqslant \frac{2}{\beta}(\delta-d)$ and now taking $d=\delta/2$ implies

$$\sigma_{\delta/2}(\infty) \geqslant \frac{\delta}{\beta}.$$

Thus we have proved the following result.

Theorem 2.8.2. *Under the same assumption as in Theorem 2.8.1, we have*

$$\liminf_{n\to\infty} D_{\delta/2}(r_n,\infty) \geqslant \frac{\delta}{\beta}.$$

This improves Lemma 2.8.1 for the sequence $\{r_n\}$ which is chosen to be a sequence of Pólya peaks. A precise result was established by Anderson and Baernstein in [1], which is stated as follows.

Theorem 2.8.3. *Under the same assumption as in Theorem 2.8.1, for $d \in (0,\delta)$, there exists a $R > 0$ such that*

$$\liminf_{n\to\infty} D_d(Rr_n,\infty) \geqslant \min\{2a, 2\pi\}, \tag{2.8.3}$$

where a is the smallest positive solution of

$$d = \frac{\pi\beta(\cos\beta a - (1-\delta))}{\sin\beta a + \beta(\pi - a)\cos\beta a}$$

and $a \in (0,\beta)$.

For $0 < \mu < \frac{1}{2}$ and $1 - \delta(\infty, f) < \cos\pi\mu$, Ostrovskii in [29] got

$$\limsup_{r\to\infty} \frac{L(r,f)}{T(r,f)} \geqslant \kappa(\mu,\delta),$$

where $L(r,f) = \min_{|z|=r} \log|f(z)|$ and $\kappa(\mu,\delta) = \frac{\pi\mu(\cos\pi\mu-(1-\delta))}{\sin\pi\mu}$. However, the quantity on the left of (2.8.3) equals 2π if and only if $0 < \mu < \frac{1}{2}$ and $d \leqslant \kappa(\mu,\delta)$.

Actually, the above results were established for δ-subharmonic functions (see Chapter 7 for basic knowledge of δ-subharmonic functions), while $\log|f(z)|$ for a meromorphic function f is a δ-subharmonic function.

It is natural to consider the spread relation for the Tsuji deficiency. For a meromorphic function $f(z)$ on $\Omega(\alpha,\beta)$, set

$$W_\Lambda(r,a) = \{\theta \in (\arcsin r^{-\omega}, \pi - \arcsin r^{-\omega}) :$$

$$\log|f(re^{i(\alpha+\omega^{-1}\theta)}\sin^{\omega^{-1}}\theta)| \geqslant \Lambda(r) r^{\omega}\sin^2\theta\,\mathfrak{T}_{\alpha,\beta}(r,f)\}.$$

We believe that it is worth to investigate the estimate of $\mathrm{mes}W_\Lambda(r,a)$ from below. The reason is this will be useful for us to understand the deficiency of a function restricted to be meromorphic in an angle.

2.9 Uniqueness of Meromorphic Functions Related to Some Angular Domains

Let X be a subset of $\hat{\mathbb{C}}$. Such a problem is interesting:

Question 2.9.1. *Under what conditions, must two meromorphic functions on X be identical?*

We shall call such result Unique Theorem about meromorphic functions. We first introduce crucial concepts. An $a \in \hat{\mathbb{C}}$ is called an IM (ignoring multiplicities) shared value in X of two functions $f(z)$ and $g(z)$ meromorphic on X provided that in X, $f(z) = a$ if and only if $g(z) = a$ and a CM (counting multiplicities) shared value in X if $f(z)$ and $g(z)$ assume a at the same points in X with the same multiplicities. It is R. Nevanlinna [25] who proved the first unique theorem, called the Five Value Theorem, which says that two meromorphic functions $f(z)$ and $g(z)$ in $\mathbb{C}$ are identical if they have five distinct IM shared values in $X = \mathbb{C}$. After his works on this subject, a great number of unique theorems were established for $X = \mathbb{C}$ which are characterized in terms of some value-points of meromorphic functions and so which relies heavily on the Nevanlinna theory of value distribution of meromorphic functions. The reader is referred to the book [38] of Yi and Yang.

The present author in [42] suggested first time to investigate uniqueness of a meromorphic function in a precise subset of $\hat{\mathbb{C}}$, that is, Question 2.9.1 for $X \neq \mathbb{C}$, and believes this would be an interesting topic. A number of unique theorems in angular domains were established in Zheng [42] and [43] in terms of Nevanlinna characteristic for angular domains. In this section, we use the Tsuji's characteristic to our discussion of this subject instead. To the end, we introduce a concept. A meromorphic function f in an angular domain $\Omega = \{z : \alpha < \arg z < \beta\}$ with $0 \leqslant \alpha < \beta \leqslant 2\pi$ is called transcendental in the Tsuji's sense if

$$\limsup_{r \to \infty} \frac{\mathfrak{T}_{\alpha,\beta}(r,f)}{\log r} = \infty. \tag{2.9.1}$$

Hence we say that f is a transcendental meromorphic function in an angular domain (in the Tsuji's sense) if f is meromorphic in the angular domain and transcendental there in the Tsuji's sense.

In view of the Tsuji second fundamental theorem, that is, the inequality (2.3.5), we extend the five value theorem of Nevanlinna's to an angular domain.

Theorem 2.9.1. *Let $f(z)$ and $g(z)$ be both meromorphic functions in an angular domain $\Omega = \{z : \alpha < \arg z < \beta\}$ with $0 \leqslant \alpha < \beta \leqslant 2\pi$ and $f(z)$ be transcendental in the Tsuji's sense. Assume that $f(z)$ and $g(z)$ have five distinct IM shared values a_j $(j = 1, 2, \cdots, 5)$ in Ω. Then $f(z) \equiv g(z)$.*

Proof. Suppose $f(z) \not\equiv g(z)$. By applying the Tsuji second fundamental theorem to f, we have

$$\begin{aligned}
3\mathfrak{T}_{\alpha,\beta}(r,f) &\leqslant \sum_{j=1}^{5} \overline{\mathfrak{N}}_{\alpha,\beta}\left(r,\frac{1}{f-a_j}\right) + Q_{\alpha,\beta}(r,f) \\
&\leqslant \mathfrak{N}_{\alpha,\beta}\left(r,\frac{1}{f-g}\right) + Q_{\alpha,\beta}(r,f) \\
&\leqslant \mathfrak{T}_{\alpha,\beta}(r,f-g) + Q_{\alpha,\beta}(r,f) \\
&\leqslant \mathfrak{T}_{\alpha,\beta}(r,f) + \mathfrak{T}_{\alpha,\beta}(r,g) + Q_{\alpha,\beta}(r,f),
\end{aligned}$$

so that

$$2\mathfrak{T}_{\alpha,\beta}(r,f) - Q_{\alpha,\beta}(r,f) \leqslant \mathfrak{T}_{\alpha,\beta}(r,g) \tag{2.9.2}$$

and further, from Lemma 2.5.4 we have

$$(2+o(1))\mathfrak{T}_{\alpha,\beta}(r,f) - O(\log r) \leqslant \mathfrak{T}_{\alpha,\beta}(r,g)$$

for all $r > 0$ possibly except a subset F of positive real axis with finite measure. In view of the condition (2.9.1), there exists a sequence of unbounded increasing positive numbers $\{r_n\}$ outside F such that

$$\lim_{n\to\infty} \frac{\mathfrak{T}_{\alpha,\beta}(r_n,f)}{\log r_n} = \lim_{n\to\infty} \frac{\mathfrak{T}_{\alpha,\beta}(r_n,g)}{\log r_n} = \infty.$$

We therefore have

$$(2+o(1))\mathfrak{T}_{\alpha,\beta}(r_n,f) \leqslant \mathfrak{T}_{\alpha,\beta}(r_n,g).$$

The same argument as above implies that

$$(2+o(1))\mathfrak{T}_{\alpha,\beta}(r_n,g) \leqslant \mathfrak{T}_{\alpha,\beta}(r_n,f).$$

Thus $2+o(1) \leqslant (2+o(1))^{-1}$, as $n \to \infty$. This is impossible and so we complete the proof of Theorem 2.9.1. □

We shall call the uniqueness of meromorphic functions for an angle determined in terms of their five-value points the Five-value Theorem for the angle. It is natural to have the following consequence of Theorem 2.9.1.

Corollary 2.9.1. *Let the same assumptions of Theorem 2.9.1 be given with the exception of that $f(z)$ is transcendental. Assume that for some $a \in \widehat{\mathbb{C}}$ and $\varepsilon > 0$,*

$$\limsup_{r\to\infty} \frac{N(r,\Omega_\varepsilon, f=a)}{r^\omega \log r} = \infty, \tag{2.9.3}$$

$\omega = \frac{\pi}{\beta-\alpha}$. *Then $f(z) \equiv g(z)$.*

In view of Lemma 2.3.3, (2.9.3) implies (2.9.1) and hence Corollary 2.9.1 follows from Theorem 2.9.1. Actually, the condition (2.9.3) is a criterion of that f is transcendental in the Tsuji's sense. Therefore, all of later theorems still hold if the transcendental assumption is replaced with (2.9.3) for an $a \in \widehat{\mathbb{C}}$ and a $\varepsilon > 0$.

It is easy to see that if

$$\limsup_{r\to\infty}\frac{\log n(r,\Omega_\varepsilon,f=a)}{\log r}>\omega=\frac{\pi}{\beta-\alpha},\tag{2.9.4}$$

then (2.9.3) holds and thus this condition guarantees that meromorphic functions in the angular domain in question can be uniquely determined by their five-value points in the angular domain, which has been stated in Zheng [43]. According to the existence of the Borel directions of transcendental meromorphic functions (for its definition, please see the next chapter), a meromorphic function on $\mathbb{C}$ with order $\lambda(f)>\frac{1}{2}$ has an angle $\Omega(\alpha,\beta)$ such that $\beta-\alpha>\frac{\pi}{\lambda(f)}$ in which (2.9.4) holds for all but at most two values of a and furthermore, the function is uniquely determined by its five-value points in the angle. It is well known that a meromorphic function $f(z)$ with the infinite order has at least one direction $\arg z=\theta$, $0\leqslant\theta<2\pi$ from the origin such that for arbitrary small $\varepsilon>0$, we have

$$\limsup_{r\to\infty}\frac{\log n(r,Z_\varepsilon(\theta),f=a)}{\log r}=\infty,\tag{2.9.5}$$

for all but at most two $a\in\widehat{\mathbb{C}}$, where $Z_\varepsilon(\theta)=\{z:\theta-\varepsilon<\arg z<\theta+\varepsilon\}$, which is usually known as the Borel direction of infinite order. In view of Theorem 2.9.1, any meromorphic function $g(z)$ having five distinct IM shared values with such $f(z)$ in one fixed $Z_\varepsilon(\theta)$ coincides with $f(z)$ and therefore any five-value points of a meromorphic function in any angle containing one of its Borel directions of infinite order completely determine this function. And we have also known that there exists a meromorphic function $f(z)$ with the infinite order which has any direction from the origin as its Borel direction of infinite order (see Theorem 3.6.2 below). From this clear is the significance of Theorem 2.9.1 and Corollary 2.9.1. We emphasize that the above angles produced from the Borel directions are also suitable to all of later theorems.

When the case of the CM is considered , we have the following four-value theorem.

Theorem 2.9.2. *Let $f(z)$ and $g(z)$ be both meromorphic functions in an angular domain $\Omega=\{z:\alpha<\arg z<\beta\}$ with $0\leqslant\alpha<\beta\leqslant 2\pi$. Assume that $f(z)$ and $g(z)$ share four functions $a_j(z)$ $(j=1,2,3,4)$ CM in Ω which are small with respect to $f(z)$ and $g(z)$ in the Tsuji sense and $f(z)$ is transcendental in the Tsuji's sense. Then $f(z)$ is a quasi-Möbius transformation of $g(z)$. Furthermore, if, in addition, $a_j(j=1,2,3,4)$ are constant and $f(z)\not\equiv g(z)$, then two of these four values are the Picard exceptional values of $f(z)$ as well as $g(z)$ in Ω and their cross ratio is -1.*

The quasi-Möbius transformation means a fraction with coefficients of meromorphic functions, that is to say, $M(z,w)=(b_1(z)w+b_2(z))(b_3(z)w+b_4(z))^{-1}$ for meromorphic functions $b_j(z)(j=1,2,3,4)$ in the angle. Then we say that $f(z)$ is a quasi-Möbius transformation of $g(z)$, provided that $f(z)=M(z,g(z))$ for some $M(z,w)$ whose coefficients are small with respect to $f(z)$ and $g(z)$ in the Tsuji sense. Theorem 2.9.2 was proved by Li and Yang [23] for the case of the complex plane. Here we provide a proof of Theorem 2.9.2 in order to make the reader understand a possible extension of a great number of unique theorems for the complex plane to an angular domain.

To prove Theorem 2.9.2, we need the following result which can be proved in view of the proof of Theorem 5.13 of Yi and Yang [38]:

If $f(z)$ and $g(z)$ share $0, 1, \infty$ CM in Ω and $f(z)$ is not a quasi-Möbius transformation of $g(z)$, then

$$\mathfrak{m}_{\alpha,\beta}\left(r, \frac{1}{f-a}\right) = Q_{\alpha,\beta}(r)$$

for each small function $a(z)(\neq 0, 1, \infty)$ with respect to $f(z)$ and $g(z)$ in Ω, where $Q_{\alpha,\beta}(r) = Q_{\alpha,\beta}(r, f)$ is the error term associated with the Tsuji characteristic and $Q_{\alpha,\beta}(r) = o(\mathfrak{T}_{\alpha,\beta}(r, f))$ when $a(z)$ is not a complex number.

Now we are in the position of the proof of Theorem 2.9.2.

Proof of Theorem 2.9.2 Under a quasi-Möbius transformation, we can assume without any loss of generalities that $f(z)$ and $g(z)$ share $0, 1, \infty$ CM and a function $\alpha(z)$ CM and α is not $0, 1, \infty$ but is small with respect to $f(z)$ and $g(z)$. Certainly we can assume that $\mathfrak{N}_{\alpha,\beta}(r, f) \neq Q_{\alpha,\beta}(r, f)$.

In view of the Tsuji second fundamental theorem, it is easy to get

$$\begin{aligned}
\mathfrak{T}_{\alpha,\beta}(r, f) &\leqslant \mathfrak{N}_{\alpha,\beta}(r, f) + \mathfrak{N}_{\alpha,\beta}\left(r, \frac{1}{f}\right) + \mathfrak{N}_{\alpha,\beta}\left(r, \frac{1}{f-1}\right) + Q_{\alpha,\beta}(r, f) \\
&= \mathfrak{N}_{\alpha,\beta}(r, g) + \mathfrak{N}_{\alpha,\beta}\left(r, \frac{1}{g}\right) + \mathfrak{N}_{\alpha,\beta}\left(r, \frac{1}{g-1}\right) + Q_{\alpha,\beta}(r, f) \\
&\leqslant 3\mathfrak{T}_{\alpha,\beta}(r, g) + Q_{\alpha,\beta}(r, f)
\end{aligned}$$

and then $Q_{\alpha,\beta}(r, f) = Q_{\alpha,\beta}(r, g)$. Alternately, we have $Q_{\alpha,\beta}(r, g) = Q_{\alpha,\beta}(r, f)$. Below we shall briefly write $Q_{\alpha,\beta}(r)$ for $Q_{\alpha,\beta}(r, f)$ or $Q_{\alpha,\beta}(r, g)$.

Set

$$H(z) = \frac{g(z)(f(z)-1)}{f(z)(g(z)-1)} - 1.$$

Suppose $H(z) \not\equiv 0$, otherwise there is nothing to do. A simple computation implies that $H(z)$ has no poles in Ω and each pole of $f(z)$ with multiplicity p is a zero of $H(z)$ with multiplicity at least p and therefore,

$$\mathfrak{N}_{\alpha,\beta}\left(r, \frac{1}{H}\right) \geqslant \mathfrak{N}_{\alpha,\beta}(r, f). \tag{2.9.6}$$

By noting another form of $H(z) = \left(1 + \frac{1}{g-1}\right)\left(1 - \frac{1}{f}\right) - 1$, we have

$$\mathfrak{T}_{\alpha,\beta}(r, H) = \mathfrak{m}_{\alpha,\beta}(r, H) \leqslant \mathfrak{m}_{\alpha,\beta}\left(r, \frac{1}{g-1}\right) + \mathfrak{m}_{\alpha,\beta}\left(r, \frac{1}{f}\right) + \log 8. \tag{2.9.7}$$

Suppose that $f(z)$ is not a quasi-Möbius transformation of $g(z)$. Since $f(z)$ and $g(z)$ share $1, \alpha(z), \infty$ and $0, \alpha(z), \infty$, from the result stated after Theorem 2.9.2 in turn, we therefore have

$$\mathfrak{m}_{\alpha,\beta}\left(r,\frac{1}{f}\right)=Q_{\alpha,\beta}(r) \text{ and } \mathfrak{m}_{\alpha,\beta}\left(r,\frac{1}{g-1}\right)=Q_{\alpha,\beta}(r)$$

and hence

$$\mathfrak{m}_{\alpha,\beta}\left(r,\frac{1}{f}\right)+\mathfrak{m}_{\alpha,\beta}\left(r,\frac{1}{g-1}\right)=Q_{\alpha,\beta}(r).$$

Substituting the above equality into (2.9.7) and then combining the resulting inequality with (2.9.6) yields that

$$\mathfrak{N}_{\alpha,\beta}(r,f)=Q_{\alpha,\beta}(r).$$

A contradiction is derived and it follows that $f(z)$ is a quasi-Möbius transformation of $g(z)$.

We leave to the reader the remainder of the proof of Theorem 2.9.2. □

It is well known that four shared values are not sufficient to uniquely determine a meromorphic function. Even if two meromorphic functions $f(z)$ and $g(z)$ have four distinct IM shared values in the whole complex plane and one of the four shared values is the Picard exceptional value of $f(z)$ and $g(z)$, we cannot assert $f(z)\equiv g(z)$. Hence one considers some additional condition.

Theorem 2.9.3. *Let $f(z)$ and $g(z)$ be meromorphic functions in an angular $\Omega=\{z:\alpha<\arg z<\beta\}$ with $0\leqslant\alpha<\beta\leqslant 2\pi$ and $f(z)$ is transcendental in the Tsuji's sense. Assume that $f(z)$ and $g(z)$ have four distinct IM shared values a_j $(j=1,2,3,4)$ in Ω. If for some $a\in\hat{\mathbb{C}}\setminus\{a_j : j=1,2,3,4\}$, a is a Tsuji deficient value of $f(z)$ in Ω or $\rho_\Omega(a)<\lim_{\varepsilon\to 0}\lambda_{\Omega_\varepsilon}(f)$. Then $f(z)\equiv g(z)$.*

Proof. Suppose $f(z)\not\equiv g(z)$. By applying the Tsuji second fundamental theorem, we have

$$\begin{aligned}
2\mathfrak{T}_{\alpha,\beta}(r,f) &\leqslant \sum_{j=1}^{4}\overline{\mathfrak{N}}_{\alpha,\beta}\left(r,\frac{1}{f-a_j}\right)+Q_{\alpha,\beta}(r,f)\\
&\leqslant \mathfrak{N}_{\alpha,\beta}\left(r,\frac{1}{f-g}\right)+Q_{\alpha,\beta}(r,f)\\
&\leqslant \mathfrak{T}_{\alpha,\beta}(r,f-g)+Q_{\alpha,\beta}(r,f)\\
&\leqslant \mathfrak{T}_{\alpha,\beta}(r,f)+\mathfrak{T}_{\alpha,\beta}(r,g)+Q_{\alpha,\beta}(r,f),
\end{aligned}\tag{2.9.8}$$

so that

$$\mathfrak{T}_{\alpha,\beta}(r,f)-Q_{\alpha,\beta}(r,f)\leqslant\mathfrak{T}_{\alpha,\beta}(r,g).\tag{2.9.9}$$

The same argument shows that

$$\mathfrak{T}_{\alpha,\beta}(r,g)-Q_{\alpha,\beta}(r,g)\leqslant\mathfrak{T}_{\alpha,\beta}(r,f).\tag{2.9.10}$$

This implies that $Q_{\alpha,\beta}(r,g)=Q_{\alpha,\beta}(r,f)$. Assume without any loss of generality that $a\in\mathbb{C}$ and indeed the same argument is available to complete the proof for the case when $a=\infty$. Using the Tsuji second fundamental theorem again and then combining

(2.9.8) together with (2.9.9) and (2.9.10), we deduce

$$\begin{aligned} 3\mathfrak{T}_{\alpha,\beta}(r,f) &\leqslant \sum_{j=1}^{4} \overline{\mathfrak{N}}_{\alpha,\beta}\left(r,\frac{1}{f-a_j}\right) + \overline{\mathfrak{N}}_{\alpha,\beta}\left(r,\frac{1}{f-a}\right) + Q_{\alpha,\beta}(r,f) \\ &\leqslant 2\mathfrak{T}_{\alpha,\beta}(r,f) + \overline{\mathfrak{N}}_{\alpha,\beta}\left(r,\frac{1}{f-a}\right) + Q_{\alpha,\beta}(r,f). \end{aligned}$$

Thus

$$\mathfrak{m}_{\alpha,\beta}\left(r,\frac{1}{f-a}\right) = Q_{\alpha,\beta}(r,f) \tag{2.9.11}$$

and further a cannot be a Tsuji deficient value of $f(z)$.

Now suppose $\rho_\Omega(a) < \lim_{\varepsilon\to 0} \lambda_{\Omega_\varepsilon}(f)$ and so for some $\varepsilon > 0$, $\rho_\Omega(a) < \lambda_{\Omega_\varepsilon}(f)$. Then there is a σ with $\sigma < \lambda_{\Omega_\varepsilon}(f)$ such that $\mathfrak{n}_{\alpha,\beta}(r, f = a) < K_1 r^\sigma$ for $r \geqslant 1$. If $\sigma \leqslant \omega$, then we have

$$\mathfrak{N}_{\alpha,\beta}\left(r,\frac{1}{f-a}\right) = \omega\int_1^r \frac{\mathfrak{n}_{\alpha,\beta}(t,f=a)}{t^{\omega+1}}dt \leqslant \omega K_1 \log r.$$

This implies that

$$\mathfrak{T}_{\alpha,\beta}(r,f) = \mathfrak{T}_{\alpha,\beta}\left(r,\frac{1}{f-a}\right) + O(1) = Q_{\alpha,\beta}(r,f)$$

and so a contradiction to (2.9.1) is derived. Therefore, we have $\omega < \sigma$ and

$$\begin{aligned} \mathfrak{N}_{\alpha,\beta}\left(r,\frac{1}{f-a}\right) &= \omega\int_1^r \frac{\mathfrak{n}_{\alpha,\beta}(t,f=a)}{t^{\omega+1}}dt \\ &< \omega K_1 \int_1^r t^{\sigma-\omega-1}dt \\ &< K_1 \frac{\omega}{\sigma-\omega} r^{\sigma-\omega}. \end{aligned}$$

This together with (2.9.11) yield that

$$\mathfrak{T}_{\alpha,\beta}(r,f) = \mathfrak{T}_{\alpha,\beta}\left(r,\frac{1}{f-a}\right) + O(1) \leqslant K_1\frac{\omega}{\sigma-\omega}r^{\sigma-\omega} + Q_{\alpha,\beta}(r,f).$$

Thus application of the Tsuji first fundamental theorem and Lemma 2.5.4 yields

$$\mathfrak{T}_{\alpha,\beta}\left(r,\frac{1}{f-b}\right) \leqslant K_2 r^{\sigma-\omega} + O(1)$$

for each $b \in \widehat{\mathbb{C}}$ and a positive constant K_2. In virtue of Lemma 2.3.3, the following implication is clear:

$$\mathfrak{T}_{\alpha,\beta}\left(r,\frac{1}{f-b}\right) \geqslant \mathfrak{N}_{\alpha,\beta}\left(r,\frac{1}{f-b}\right) \geqslant \omega c^{\omega}\frac{N(cr,\Omega_{\varepsilon/2},f=b)}{r^{\omega}}$$

for some $0<c<1$. This implies that

$$N(r,\Omega_{\varepsilon/2},f=b) \leqslant Kr^{\sigma},$$

for a positive constant K, and so $\rho_{\Omega_{\varepsilon/2}}(b) \leqslant \sigma$. In view of Theorem 2.7.1, $\lambda_{\Omega_\varepsilon}(f) \leqslant \sigma$, a contradiction is derived. Theorem 2.9.3 follows. □

We remark that Theorem 2.9.3 straightly comes from Theorem 2.9.2 if "IM" is replaced by "CM". If the Nevanlinna deficient value is taken into account, then we have the following theorem.

Theorem 2.9.4. *Let $f(z)$ and $g(z)$ be both transcendental meromorphic functions and let $f(z)$ be of the finite lower order μ and for some $a \in \widehat{\mathbb{C}}$, $\delta = \delta(a,f) > 0$.*
Given one angular domain $\Omega = \{z : \alpha < \arg z < \beta\}$ with $0 \leqslant \alpha < \beta \leqslant 2\pi$ and

$$\beta - \alpha > \max\left\{\frac{\pi}{\sigma},\ 2\pi - \frac{4}{\sigma}\arcsin\sqrt{\frac{\delta}{2}}\right\}, \tag{2.9.12}$$

where $\mu \leqslant \sigma \leqslant \lambda$ and $\sigma < \infty$ (When $\sigma \leqslant \frac{1}{2}$, set $\Omega = \mathbb{C}$), we assume that $f(z)$ and $g(z)$ have four distinct IM shared values a_j $(j = 1,2,3,4)$ in Ω and $a_j \neq a$ $(j = 1,2,3,4)$. Then $f(z) \equiv g(z)$.

Proof. It suffices to consider the case when $\sigma > \frac{1}{2}$. First of all we want to prove that $f(z)$ satisfies (2.9.1) in Ω_ε for some $\varepsilon > 0$ which will be determined later. Take a sequence of Pólya peak $\{r_n\}$ of order σ. Since $2\pi + \alpha - \beta < \frac{4}{\sigma}\arcsin\sqrt{\frac{\delta}{2}}$, in view of Theorem 2.8.1 we have

$$\operatorname{mes}(D_\Lambda(r_n,a)\cap[\alpha,\beta]) \geqslant d > 0,$$

where $\Lambda(r) = 1/\log r$, for some d and all r_n, which we assume without any loss of generalities. Choose a $\varepsilon > 0$ such that $\varepsilon \leqslant d/8$ and $\sigma > \frac{\pi}{\beta-\alpha-6\varepsilon}$. From the definition of $B(r,f)$ it follows that

$$\begin{aligned}
B_{\alpha+2\varepsilon,\beta-2\varepsilon}\left(r_n,\frac{1}{f-a}\right) &> \frac{2\widetilde{\omega}}{\pi r_n^{\widetilde{\omega}}}\sin(\widetilde{\omega}\varepsilon)\int_{\alpha+3\varepsilon}^{\beta-3\varepsilon}\log^+|f(r_n e^{i\theta})|d\theta \\
&\geqslant \frac{2\widetilde{\omega}}{\pi r_n^{\widetilde{\omega}}}\sin(\widetilde{\omega}\varepsilon)\frac{T(r_n,f)}{\log r_n}(d-6\varepsilon),
\end{aligned}$$

where $\widetilde{\omega} = \frac{\pi}{\beta-\alpha-4\varepsilon}$. Since $\sigma > \frac{\pi}{\beta-\alpha-4\varepsilon}$, one easily gets

$$\frac{B_{\alpha+2\varepsilon,\beta-2\varepsilon}\left(r_n,\frac{1}{f-a}\right)}{\log r_n} \longrightarrow \infty \text{ and so } \frac{S_{\alpha+2\varepsilon,\beta-2\varepsilon}(r_n,f)}{\log r_n} \longrightarrow \infty,$$

as $n \to \infty$. Thus using Lemma 2.2.1, Lemma 2.3.2 and Theorem 2.3.3 implies that $f(z)$ satisfies (2.9.1) in Ω_ε.

Suppose that $f(z) \not\equiv g(z)$. In view of (2.9.10) we have

$$(1-o(1))\mathfrak{T}_{\alpha,\beta}(r,g) \leqslant \mathfrak{T}_{\alpha,\beta}(r,f)+O(\log r), \ r \notin E$$

and therefore

$$\begin{aligned} R_{\alpha+\varepsilon,\beta-\varepsilon}(r_n,g) &= O(\log^+ S_{\alpha+\varepsilon/2,\beta-\varepsilon/2}(r_n,g)+\log r_n) \ \text{(by Theorem 2.5.1)} \\ &= O(\log^+ \mathfrak{T}_{\alpha,\beta}(cr_n,g)+\log r_n) \ \text{(by Theorem 2.3.3)} \\ &= O(\log^+ \mathfrak{T}_{\alpha,\beta}(cr_n,f)+\log r_n) \\ &= O(\log^+ S_{\alpha,\beta}(cr_n,f)+\log r_n) \ \text{(by Theorem 2.3.3)} \\ &= O(\log^+ T(cr_n,f)+\log r_n) \\ &= O(\log^+ T(r_n,f)+\log r_n), \end{aligned}$$

where c is a suitable constant greater than one and that "$r_n \notin E$" has be assumed (otherwise, we choose $R_n \in [r_n, 2r_n] \setminus E$ such that the above inequality holds, while the below implication is still valid for these R_n).

In what follows, we proceed with our discussion in Ω_ε. For the sake of simplification, we omit the subscript of associated notations, for example, write $S(r,f)$ for $S_{\alpha+\varepsilon,\beta-\varepsilon}(r,f)$. In view of the Nevanlinna second fundamental theorem on Ω_ε, as did in (2.9.8), (2.9.9) and (2.9.10), we have

$$S(r_n,f) = S(r_n,g)+O(\log r_n T(r_n,f))$$

and

$$\sum_{j=1}^{4} \overline{C}\left(r_n, \frac{1}{f-a_j}\right) = 2S(r_n,f)+O(\log r_n T(r_n,f))$$

and, therefore, noting that

$$\begin{aligned} 3S(r_n,f) &\leqslant \sum_{j=1}^{4} \overline{C}\left(r_n, \frac{1}{f-a_j}\right)+\overline{C}\left(r_n, \frac{1}{f-a}\right)+O(\log r_n T(r_n,f)) \\ &= 2S(r_n,f)+\overline{C}\left(r_n, \frac{1}{f-a}\right)+O(\log r_n T(r_n,f)), \end{aligned}$$

we easily get

$$B\left(r_n, \frac{1}{f-a}\right) = O(\log r_n T(r_n,f)).$$

On the other hand, we have

$$B\left(r_n, \frac{1}{f-a}\right) \geqslant \frac{2\widehat{\omega}}{\pi r_n^{\widehat{\omega}}} \sin(\widehat{\omega}\varepsilon) \frac{T(r_n,f)}{\log r_n}(d-4\varepsilon),$$

$\widehat{\omega} = \frac{\pi}{\beta-\alpha-2\varepsilon}$. Therefore this yields that $\sigma \leqslant \frac{\pi}{\beta-\alpha-2\varepsilon}$, a contradiction is derived and so we complete the proof of Theorem 2.9.4. □

The inequality (2.9.12) cannot be replaced by "=", that means (2.9.12) is best possible. This is shown by observing the functions e^z and e^{2z}: they cannot assume in the left-half plane any value whose modulus is greater than one.

Theorem 2.9.4 was given by Zheng [43] only by means of the Nevanlinna second fundamental theorem on an angular domain. However, the proof there is incomplete, since the equality "$R_{\alpha,\beta}(r,g) = O(\log r S_{\alpha,\beta}(r,g))$" was used, but the equality does not always hold. Now we raise the following question:

Does Theorem 2.9.4 still hold provided that (2.9.12) is replaced by (2.9.1) and without the assumption of the finite lower order?

The question is closely related to question of whether a Nevanlinna deficient value is a Tsuji deficient value in an angle when the function in question is transcendental in the angle. In view of the similar methods to those in Gundersen[12], the following result is immediately deduced.

Theorem 2.9.5. *Let $f(z)$ and $g(z)$ be both transcendental meromorphic functions(in the Tsuji's sense) in an angular domain $\Omega = \{z : \alpha < \arg z < \beta\}$ with $0 \leqslant \alpha < \beta \leqslant 2\pi$. Assume that $f(z)$ and $g(z)$ have three distinct CM shared values a_j $(j = 1,2,3)$ and one IM shared value a_4 in Ω. Then a_4 is also one CM shared value in Ω of $f(z)$ and $g(z)$.*

If we denote the above result by a simple notation "CM+1IM = 4CM", then we raise a problem of whether "2CM+2IM=4CM" holds.

The four shared values in Theorem 2.9.3 and Theorem 2.9.4 cannot reduce to three CM shared values. Then for a meromorphic function $f(z)$ which satisfies (2.9.1) in an angular domain Ω, how many meromorphic functions do there exist to share $0, 1, \infty$ CM with $f(z)$ in Ω. This is an interesting problem.

It is obvious that there are many questions on uniqueness of meromorphic functions dealing with shared values in one angular domain, which are worthwhile to take into account. Actually, we wish to extend the known results for the complex plane to an angular domain and it is easy to see that some of them can be done by the same idea. The following is an example (for its background, see Yi and Yang [38]).

Theorem 2.9.6. *Let $f(z)$ be a transcendental meromorphic function in an angular domain $\Omega = \{z : \alpha < \arg z < \beta\}$ with $0 \leqslant \alpha < \beta \leqslant 2\pi$. Assume that $f(z)$ and $f^{(k)}(z)$ have three finite distinct IM shared values a_j $(j = 1,2,3)$ in Ω. Then $f(z) \equiv f^{(k)}(z)$.*

Proof. Here we only prove the case when $k = 1$. Suppose on the contrary that $f(z) \not\equiv f'(z)$. It is obvious that ∞ is a IM shared value of $f(z)$ and $f'(z)$ and thus they have four distinct IM shared values in Ω. In view of the same argument as in the proof of Theorem 2.9.3, we have

$$\mathfrak{T}_{\alpha,\beta}(r,f') = \mathfrak{T}_{\alpha,\beta}(r,f) + Q_{\alpha,\beta}(r,f).$$

By noting that (2.1.11) with $\mathfrak{m}_{\alpha,\beta}(r,*)$ in the place of $m(r,*)$ is also available, we have

$$\begin{aligned}\sum_{j=1}^{3}\mathfrak{m}_{\alpha,\beta}\left(r,\frac{1}{f-a_j}\right) &\leqslant \mathfrak{m}_{\alpha,\beta}\left(r,\sum_{j=1}^{3}\frac{1}{f-a_j}\right)+O(1)\\ &\leqslant \mathfrak{m}_{\alpha,\beta}\left(r,\sum_{j=1}^{3}\frac{f'}{f-a_j}\right)+\mathfrak{m}_{\alpha,\beta}\left(r,\frac{1}{f'}\right)+O(1)\\ &\leqslant \mathfrak{m}_{\alpha,\beta}\left(r,\frac{1}{f'}\right)+Q_{\alpha,\beta}(r,f).\end{aligned}$$

The following implication is obvious:

$$\begin{aligned}\sum_{j=1}^{3}\mathfrak{N}_{\alpha,\beta}\left(r,\frac{1}{f-a_j}\right) &\leqslant \sum_{j=1}^{3}\overline{\mathfrak{N}}_{\alpha,\beta}\left(r,\frac{1}{f-a_j}\right)+\mathfrak{N}_{\alpha,\beta}\left(r,\frac{1}{f'}\right)\\ &\leqslant \overline{\mathfrak{N}}_{\alpha,\beta}\left(r,\frac{1}{f-f'}\right)+\mathfrak{N}_{\alpha,\beta}\left(r,\frac{1}{f'}\right)\\ &\leqslant \mathfrak{T}_{\alpha,\beta}(r,f-f')+O(1)+\mathfrak{N}_{\alpha,\beta}\left(r,\frac{1}{f'}\right)\\ &= \mathfrak{N}_{\alpha,\beta}(r,f')+\mathfrak{m}_{\alpha,\beta}\left(r,f(1-\frac{f'}{f})\right)+O(1)+\mathfrak{N}_{\alpha,\beta}\left(r,\frac{1}{f'}\right)\\ &\leqslant \mathfrak{T}_{\alpha,\beta}(r,f)+\overline{\mathfrak{N}}_{\alpha,\beta}(r,f)+\mathfrak{N}_{\alpha,\beta}\left(r,\frac{1}{f'}\right)+Q_{\alpha,\beta}(r,f).\end{aligned}$$

Thus combining the above two inequalities yields

$$\begin{aligned}3\mathfrak{T}_{\alpha,\beta}(r,f) &\leqslant \mathfrak{T}_{\alpha,\beta}(r,f)+\overline{\mathfrak{N}}_{\alpha,\beta}(r,f)+\mathfrak{T}_{\alpha,\beta}(r,f')+Q_{\alpha,\beta}(r,f)\\ &= 2\mathfrak{T}_{\alpha,\beta}(r,f)+\overline{\mathfrak{N}}_{\alpha,\beta}(r,f)+Q_{\alpha,\beta}(r,f)\end{aligned}$$

and then

$$\begin{aligned}\mathfrak{T}_{\alpha,\beta}(r,f) &\leqslant \overline{\mathfrak{N}}_{\alpha,\beta}(r,f)+Q_{\alpha,\beta}(r,f)\\ &\leqslant \frac{1}{2}\mathfrak{T}_{\alpha,\beta}(r,f')+Q_{\alpha,\beta}(r,f)\\ &= \frac{1}{2}\mathfrak{T}_{\alpha,\beta}(r,f)+Q_{\alpha,\beta}(r,f).\end{aligned}$$

This implies that

$$\mathfrak{T}_{\alpha,\beta}(r,f)=Q_{\alpha,\beta}(r,f)$$

and it contradicts our assumption (2.9.1). □

Theorem 2.9.6 for the case of the complex plane was proved by Mues and Steinmetz [24] and Gundersen [13] for $k=1$ and Frank and Schwick [9] for $k\geqslant 2$. Modifying the proof of Frank and Schwick's result in [9] deduces Theorem 2.9.6

for $k \geqslant 2$. The reader may complete its proof as an exercise. We can also extend the results on the complex plane dealing with shared values of a function and its differential polynomial to those in an angular domain. Noting that a transcendental meromorphic function $f(z)$ in $\mathbb{C}$ satisfying $f(z) = f^{(k)}(z)$ has order $\lambda(f) = 1$, therefore a meromorphic function with order greater than one is not transcendental in the Tsuji's sense in an angular domain where it shares three finite values IM with the k-order derivative of it.

Finally, we claim the following result, which was proved by Nevanlinna [27] for the case of the complex plane and which plays a crucial role in discussion of the uniqueness of meromorphic functions.

Theorem 2.9.7. *Let $f_j(z)$ $(j = 1, 2, \cdots, q)$ be meromorphic functions in an angular domain $\Omega = \{z : \alpha < \arg z < \beta\}$ with $0 \leqslant \alpha < \beta \leqslant 2\pi$ each of which satisfies (2.9.1). Assume that for each j, $f_j(z) \not\equiv 0$ and $f_j(z)/f_i(z)$ is not a constant for each pair of j and i with $j \neq i$ and*

$$\sum_{j=1}^{q} \left(\mathfrak{N}_{\alpha,\beta}(r, f_j) + \mathfrak{N}_{\alpha,\beta}\left(r, \frac{1}{f_j}\right) \right) = o(\mathfrak{T}_{\alpha,\beta}(r)),$$

where $\mathfrak{T}_{\alpha,\beta}(r) = \min\left\{\mathfrak{T}_{\alpha,\beta}(r, f_j/f_i) : 1 \leqslant j < i \leqslant q\right\}$. Then for q complex numbers C_j $(j = 1, 2, \cdots, q)$, $\sum_{j=1}^{q} C_j f_j(z) \equiv 0$ if and only if each $C_j = 0$.

That this result holds tell us once more that a great number of unique theorems dealing with the value-points for the complex plane can be extended to the case of an angular domain. The reader is referred to Yi and Yang's book [38] for a complete collection of unique theorems for the complex plane .

We emphasize again that as in Corollary 2.9.1, the transcendental assumption can be replaced with (2.9.3) in all of above Theorems in which it appears.

References

1. Anderson, J. and Baernstein II, A., The size of the set on which a meromorphic function is large, Proc. London Math. Soc., **36**(1978), 518-539.
2. Baernstein II, A., Proof of Edrei's spread conjecture, Proc. London Math. Soc., **26**(1973), 418-434.
3. Cherry, W. and Ye, Z., Nevanlinna's theory of value distribution: the second main theorem and its erro terms, Springer-Verlag, 2001.
4. Chuang, C. T., Une généralisation d'une inégalité de Nevanlinna, Sci. Sinica, **13**(1964), 887-895.
5. Conway, J. B., Functions of one complex variable, Springer-Verlag, 1978.
6. Edrei, A., Meromorphic functions with three radially distributed values, Trans. Amer. Math. Soc., **78**(1955), 276-293.
7. Edrei, A., Sums of deficiencies of meromorphic functions, J. Analyse Math., I. **14**(1965), 79-107;II. **19**(1967), 53-74.
8. Edrei, A. and Fuchs, W. H. J., Bounds for the number of deficient values of certain classes of meromorphic functions, Proc. London Math. Soc., **12** (1962), 315-344.

9. Frank, G. and Schwick, W., Meromorphe funktionen die mit einer Ableitung drei Werteteilen, Results in Math., **22**(1992), 679-684.
10. Frank, G. and Weissenborn, G., Rational deficient functions of meromorphic functions, Bull. London Math. Soc., **18**(1986), 29-33.
11. Goldberg, A. A. and Ostrovskii, I. V., The distribution of values of meromorphic functions (in Russian), Izdat. Nauk. Moscow, 1970.
12. Gundersen, G. G., Meromorphic functions that share three or four values, J. London Math. Soc., **20**(1979), 457-466.
13. Gundersen, G. G., Meromorphic functions that share two finite values with their derivatives, Pacific J. Math., **105**(1983), 299-309.
14. Guo, H., Zheng, J. H. and Ng, T., On a new singular direction of meromorphic functions, Bull. Austral. Math. Soc., **69**(2004), 277-287.
15. Hall and Ruscheweyh, On transforms of functions with bounded boundary rotation, Indian J. Pure Appl. Math., **16**(1985), 1317-1325.
16. Hayman, W. K., Meromorphic functions, Oxford, 1964.
17. Hayman, W. K. and Miles, J., On the growth of a meromorphic function and its derivatives, Complex Variables, **12**(1989), 245-260.
18. Hayman, W. K. and Rossi, J. F., Characteristic, maximum modulus, and value distribution, Trans. Amer. Math. Soc., **284**(1984),651-664.
19. Hayman, W. K. and Yang, L., Growth and values of functions regular in an angle, Proc. London Math. Soc., (3)**44**(1982), 193-214.
20. Hiong King-Lai, Sur les fonctions holomorphes dont les dérivées admettent une valeur exce-tionnelle, Ann. École Norm. Sup., **72**(1955), 165-197.
21. Hu P. C., Li P. and Yang C. C., Unicity of meromorphic mappings, Kluwer Academic Publishers, 2003.
22. Ishizaki, K. and Yanagihara, N., Singular directions of meromorphic solutions of some non-autonomous Schröder euqations, Potential theory in Matsue, Adv. Stud. Pure Math., **44**, Math. Soc. Japan, Tokyo, 2006, 155-166.
23. Li, P. and Yang, C. C., Meromorphic functions that share four small functions, J. Math. Anal. Appl., **263**(2001), 316-326.
24. Mues, E. and Steinmetz, N., Meromorphe funktionen, die mit ihrer ableitung werte teilen, Manuscripta Math., **29**(1997), 195-206.
25. Nevanlinna, R., Untersuchungen über den Picardschen Satz, ActaSoc. Sci. Fenn., (6)**50**(1924).
26. Nevanlinna, R., Analytic functions, Springer, Berlin, Heidelberg, New York, 1970.
27. Nevanlinna, R., Le Théréme de Picard-Borel et la théorie des fonctions méromorphes, Paris, 1929.
28. Osgood, C. F., Sometimes effective Thue-Siegel-Roth-Schmidt-Nevanlinna bounds or better, J. Number Theory, **21**(1985), 347-389.
29. Ostrovskii, A., Deficiencies of meromorphic functions of order less than one, Dokl. Akad. Nauk SSSR, **150**(1963), 32-35.
30. Steinmetz, N., Eine Verallgemeimerung des zweiten Nevanlinnaschen Hauptsatzes, J. Reine Angew Math., **386**(1986), 134-141.
31. Tsuji, M., Potential theory in modern function theory, Maruzen Co. LTD. Tokyo, 1959.
32. Valiron, G., Recherches sur le théoréme de M. Borel dans la théorie des fonctions méromorphes, Acta Math., **52**(1928), 67-92.
33. Valiron, G., Directions de Borel des fonctions méromorphes, Mémor.Sci. Math., fasc., **89**, Paris, 1938.
34. Yamanoi, K., The second main theorem for small functions and related problems, Acta Math., (2)**192**(2004), 225-294.
35. Yamanoi, K., Defect relation for rational functions as targets, Forum Math., **17**(2005), 169-189.
36. Yang, L., New researches on value distribution, Science Press of China, 1982.
37. Yang, L. and Zhang, G. H., Recherches sur le nombre des valeurs déficientes et le nombre des directions de Borel des fonctions méromorphes, Sci. Sinica, **18**(1975), 23-37.

38. Yi, H. X. and Yang, C. C., Uniqueness theory of meromorphic functions, Mathematics and Its Applications, Science Press/Kluwer Academic Publishers, 2003.
39. Zhang, G. H., Theory of entire and meromorphic functions (in Chinese) , Science Press Sinica, 1986.
40. Zhang, X. L., A fundamental inequality for meromorphic functions in an angular domain and its application, Acta Math. Sinica, New Series, (3)**10**(1994), 308-314.
41. Zheng, J. H., On value distribution of meromorphic functions with respect to argument I, Complex Variables and Elliptic Equations, 2008.
42. Zheng, J. H., On uniqueness of meromorphic functions with shared values in some angular domains, Canadian J. Math., (1)**47**(2004), 152-160.
43. Zheng, J. H., On uniqueness of meromorphic functions with shared values in one angular domain, Complex Variables Theory Appl., (9)**48**(2003), 777-785.

Chapter 3
T Directions of a Meromorphic Function

Jianhua Zheng
Department of Mathematical Sciences, Tsinghua University, Beijing 100084, P. R. China
jzheng@math.tsinghua.edu.cn

Abstract: A transcendental meromorphic function has a singular property in any neighborhood of its essential singular point, for example, it assumes there infinitely often all but at most two values on the extended complex plane. This property is preserved in any angular domain containing some fixed ray. Such ray is termed as the singular direction of the function considered. In this chapter, we mainly discuss T directions of meromorphic functions, which was introduced by the author in 2003. First of all we consider the existence of T directions including T directions with small functions as targets. Next we consider connections among T directions and other directions such as Julia directions and Borel directions and mainly introduce a result of Zhang Q. D. which proves that a T (resp. Borel) direction may not be a Borel (resp. T) direction. We list conditions for the existence of singular directions dealing with derivatives of the functions, that is, the Hayman T directions and for the existence of common T directions of a function and its derivatives. We present a simple discussion of distribution of the Julia, Borel and T directions. In terms of their asymptotic form, through the Stokes rays we investigate singular directions of meromorphic solutions of a linear differential equation with rational coefficients. In the case of at least one of the coefficients being transcendental, we use the Nevanlinna's fundamental theorems for an angle to attain the aim of our researches. We conclude this chapter with a simple survey on value distribution of algebroid functions including the Nevanlinna first and second fundamental theorems for a disk and unique theorems and the singular directions.

Key words: T Directions, Hayman T directions, Singular directions, Meromorphic solution, Algebroid functions

The study of singular directions of meromorphic functions is one of important subjects in the theory of value distribution of meromorphic functions, and most of previous attentions were put on the Julia and Borel directions. As we know, the Picard big theorem tells us that a transcendental meromorphic function assumes infinitely often any values possibly except at most two values in any neighborhood of the infinity. This result is refined to be possible for any sector which contains some fixed direction, and this direction is singular and known as Julia direction after G. Julia, for it is G. Julia who found in 1924 the existence of such a direction for all entire and most of meromorphic functions, in particular, for function with at least one asymptotic value by using the Montel Theorem for the normal family. Along this way, in view of the Borel Theorem for a transcendental meromorphic function, a direction corresponding to the Borel Theorem was considered, that is so-called Borel direction. The existence of Borel directions for a meromorphic function with the finite positive order was proved by G. Valiron in 1928 by using the Nevanlinna theory. Corresponding to the Nevanlinna second fundamental theorem, a new singular direction, called T direction, was recently introduced in Zheng [58]. Main purpose of this chapter is to make a careful discussion of this singular direction.

3.1 Notation and Existence of T Directions

We extend the definition of T directions posed in Zheng [58] to a meromorphic function in an angular domain.

Definition 3.1.1. *Let $f(z)$ be a meromorphic function in an angular domain $\overline{\Omega}(\alpha,\beta)$.*

A radial $\arg z = \theta$ contained in Ω is called a T direction of $f(z)$ with respect to Ω, provided that given arbitrary small $\varepsilon > 0$, for any $b \in \widehat{\mathbb{C}}$, possibly with the exception of at most two values of b, we have

$$\limsup_{r\to\infty} \frac{N(r, Z_\varepsilon(\theta), f=b)}{\mathscr{T}(r,\Omega,f)} > 0, \tag{3.1.1}$$

where $Z_\varepsilon(\theta) = \{z:\ \theta-\varepsilon < \arg z < \theta+\varepsilon\}$;

A radial $\arg z = \theta$ is called a precise T direction of $f(z)$ with respect to Ω, if in (3.1.1), $\overline{N}(r, Z_\varepsilon(\theta), f=b)$ is in the place of $N(r, Z_\varepsilon(\theta), f=b)$.

If $f(z)$ is a meromorphic function in the whole complex plane, then a T direction of $f(z)$ with respect to $\mathbb{C}$ is briefly called a T-direction of $f(z)$.

In view of the inequality (2.4.2), it follows that Ahlfors-Shimizu characteristic $\mathscr{T}(r,\Omega,f)$ in the above Definition 3.1.1 can be replaced by $T(r,f)$ for a T direction of $f(z)$ with respect to the whole complex plane, which is the definition of a T direction given in Zheng [58].

The following result characterizes the existence of T direction of $f(z)$ in an angular domain, which will be verified in terms of Theorem 2.4.3 and Theorem 2.4.4, as we did in [10].

Theorem 3.1.1. *Let $f(z)$ be a meromorphic function in $\overline{\Omega}(\alpha,\beta)$ such that*

$$\lim_{r\to\infty} \frac{\mathscr{T}(r,\Omega,f)}{(\log r)^2} = \infty.$$

If for a fixed $\phi \in (\alpha,\beta)$ and arbitrary $\varepsilon > 0$, we have

$$\limsup_{r\notin F\to\infty}\frac{\mathscr{T}(r,Z_\varepsilon(\phi),f))}{\mathscr{T}(r,\Omega,f)}>0, \tag{3.1.2}$$

where $F=F(\Omega)$ is a set with finite logarithmic measure appeared in Lemma 2.4.2, then $\arg z=\phi$ *is a (precise) T direction of $f(z)$ with respect to Ω.*

Proof. Suppose the theorem fails, that is, $\arg z=\phi$ is not a precise T direction. Then we have a fixed $\varepsilon>0$ and three distinct points a, b and c in $\hat{\mathbb{C}}$ such that

$$\begin{aligned}\overline{N}(r,Z_{2\varepsilon}(\phi),f=a)+\overline{N}(r,Z_{2\varepsilon}(\phi),f=b)+\overline{N}(r,Z_{2\varepsilon}(\phi),f=c)\\ =o(\mathscr{T}(r,\Omega)),\end{aligned} \tag{3.1.3}$$

where $Z_\varepsilon(\phi)=\{z:\phi-\varepsilon<\arg z<\phi+\varepsilon\}$. In view of (2.4.8) with (2) we have

$$\mathscr{T}(r,Z_\varepsilon(\phi),f)=o(\mathscr{T}(r,\Omega)),\ r\notin F.$$

This contradicts our assumption (3.1.2).

We complete the proof of Theorem 3.1.1. □

Analyzing the proof of Theorem 3.1.1, we do not know if the theorem is still true under the assumption

$$\limsup_{r\to\infty}\frac{\mathscr{T}(r,\Omega)}{(\log r)^2}=\infty. \tag{3.1.4}$$

Theorem 3.1.2. *Let $f(z)$ satisfy the assumption of Theorem 3.1.1. If for certain $\varepsilon>0$*

$$\limsup_{r\notin F\to\infty}\frac{\mathscr{T}(r,\Omega_\varepsilon,f)}{\mathscr{T}(r,\Omega,f)}>0, \tag{3.1.5}$$

where $F=F(\Omega)$ is a set with finite logarithmic measure appeared in Lemma 2.4.2, then Ω contains a (precise) T direction of $f(z)$ with respect to Ω.

Proof. According to (3.1.5), there exist a sequence of positive numbers $\{r_n\}$ outside F such that (3.1.5) holds for this sequence $\{r_n\}$. Now we divide Ω_ε into two equal angular domains and then in view of (3.1.5) for at least one of these two domains, denoted by Ω_1, we have

$$\limsup_{n\to\infty}\frac{\mathscr{T}(r_n,\Omega_1,f)}{\mathscr{T}(r_n,\Omega,f)}>0. \tag{3.1.6}$$

In this way, we can obtain a sequence of angular domains $\{\Omega_j\}$ such that $\Omega_{j+1}\subset\Omega_j$ and the opening of Ω_j tends to zeros as $j\to\infty$ and for each Ω_j in the place of Ω_1, (3.1.6) holds. There exists a unique direction $\arg z=\phi$ in $\bigcap_{j=1}^{\infty}\overline{\Omega}_j$. From the proof of Theorem 3.1.1 it is easy to see that $\arg z=\phi$ is a T direction of $f(z)$ with respect to Ω. □

Clearly we cannot assert that $\arg z=\theta$ is a T direction of $f(z)$ with respect to Ω if (3.1.1) is satisfied only for one value of b. However, it is true under some additional assumption in view of Corollary 2.7.1 and Theorem 2.7.2.

Theorem 3.1.3. *Let $f(z)$ be a meromorphic function in $\overline{\Omega}(\alpha,\beta)$ and $\phi \in (\alpha,\beta)$. If for arbitrarily small $\varepsilon > 0$ and a $d > 1$,*

$$\liminf_{r\to\infty} \frac{\mathscr{T}(dr, Z_\varepsilon(\phi), f)}{\mathscr{T}(r, Z_\varepsilon(\phi), f)} > d^{\pi/(2\varepsilon)} \tag{3.1.7}$$

and

$$\limsup_{r\notin F\to\infty} \frac{N(r, Z_\varepsilon(\phi), f=0)}{\mathscr{T}(r,\Omega,f)} > 0,$$

where $Z_\varepsilon(\phi) = \{z : \phi - \varepsilon < \arg z < \phi + \varepsilon\}$, then $\arg z = \phi$ is a T direction of $f(z)$ with respect to Ω.

Proof. In view of (3.1.7), for some $\varepsilon > 0$ and for $r \geqslant r_0$ we have $\mathscr{T}(dr) > d^2\mathscr{T}(r)$, $\mathscr{T}(r) = \mathscr{T}(r, Z_\varepsilon(\phi), f)$. For $r > d^{n_0}$ with $d^{n_0} \geqslant r_0$, we have $d^n \leqslant r < d^{n+1}$ for a $n \geqslant n_0$ and therefore

$$\mathscr{T}(r) \geqslant \mathscr{T}(d^n) \geqslant (d^2)^{n-n_0}\mathscr{T}(r_0) > rd^{-n_0}T(r_0).$$

This together with $\mathscr{T}(r) \leqslant \mathscr{T}(r,\Omega,f)$ yields $\lim_{r\to\infty} \frac{\mathscr{T}(r,\Omega)}{(\log r)^2} = \infty$. It follows from Corollary 2.7.1 in terms of (3.1.7) that for some $K > 0$

$$N(r, Z_\varepsilon(\phi), f=0) \leqslant K\mathscr{T}(r, Z_{2\varepsilon}(\phi), f).$$

Thus under the assumption of Theorem 3.1.3, (3.1.2) holds for $Z_{2\varepsilon}(\phi)$ and hence Theorem 3.1.1 asserts that $\arg z = \phi$ is a T direction of $f(z)$ with respect to Ω. □

Put

$$\rho_\phi(a) = \lim_{\varepsilon\to 0^+} \limsup_{r\to\infty} \frac{\log N(r, Z_\varepsilon(\phi), f=a)}{\log r}.$$

$\rho_\phi(a)$ is called convergent exponent of a-value points of $f(z)$ for the radial $\arg z = \phi$.

Theorem 3.1.4. *Let $f(z)$ be a meromorphic function in $\overline{\Omega}(\alpha,\beta)$ and $\phi \in (\alpha,\beta)$. If $\rho_\phi(0) = \infty$ and for arbitrary $\varepsilon > 0$,*

$$\liminf_{r\notin F\to\infty} \frac{N(r, Z_\varepsilon(\phi), f=0)}{\mathscr{T}(r,\Omega,f)} > 0, \tag{3.1.8}$$

$F = F(f,\Omega)$, then $\arg z = \phi$ is a T direction of $f(z)$ with respect to Ω.

Proof. Suppose that $\arg z = \phi$ is not a T direction of $f(z)$ with respect to Ω. Then for some $\varepsilon > 0$ and three distinct complex points a, b and c on $\widehat{\mathbb{C}}$

$$N(r, Z_{2\varepsilon}(\phi), f=a) + N(r, Z_{2\varepsilon}(\phi), f=b) + N(r, Z_{2\varepsilon}(\phi), f=c)$$
$$= o(\mathscr{T}(r,\Omega,f)). \tag{3.1.9}$$

It follows from $\rho_\phi(0) = \infty$ and employing Theorem 2.7.2 (consult the proof of Theorem 2.7.3) that $\mathscr{T}(r, Z_\varepsilon(\phi), f)$ is of infinite order and hence in view of Lemma

1.1.3 there exist a sequence of positive numbers $\{r_n\}$ outside $F(f,\Omega)$ such that for $1 \leqslant t \leqslant r_n$,

$$t^{-\omega-1}\mathscr{T}(t,Z_\varepsilon(\phi),f) \leqslant \mathrm{e}r_n^{-\omega-1}\mathscr{T}(r_n,Z_\varepsilon(\phi),f),$$

where $\omega = \pi/(2\varepsilon)$. Thus

$$\mathscr{T}(r_n,Z_\varepsilon(\phi),f) \geqslant \mathrm{e}^{-1}\mathscr{T}(1,Z_\varepsilon(\phi),f)r_n^{\omega+1}.$$

In terms of the first inequality in Theorem 2.7.2, using these two inequalities produces

$$\begin{aligned}
N(r_n,Z_{\varepsilon/2}(\phi),f=0) &\leqslant K_1\mathscr{T}(r_n,Z_\varepsilon(\phi),f)+O(r_n^\omega)\\
&\quad +K_1r_n^\omega\int_1^{r_n}\frac{\mathscr{T}(t,Z_\varepsilon(\phi),f)}{t^{\omega+1}}\mathrm{d}t\\
&\leqslant (K_1+O(r_n^{-1}))\mathscr{T}(r_n,Z_\varepsilon(\phi),f)\\
&\quad +K_1r_n^\omega\int_1^{r_n}\mathrm{e}\frac{\mathscr{T}(r_n,Z_\varepsilon(\phi),f)}{r_n^{\omega+1}}\mathrm{d}t\\
&\leqslant (K_1+O(r_n^{-1})+\mathrm{e}K_1)\mathscr{T}(r_n,Z_\varepsilon(\phi),f).
\end{aligned}$$

However, in view of (2.4.8) with (2) and (3.1.9) we have

$$\mathscr{T}(r,Z_\varepsilon(\phi),f) = o(\mathscr{T}(r,\Omega,f)),\ r=r_n \notin F(f,\Omega).$$

This contradicts (3.1.8) and so Theorem 3.1.4 follows. □

Now we come to the case of a transcendental meromorphic function in $\mathbb{C}$.

Theorem 3.1.5. *Let $f(z)$ be a transcendental meromorphic function. Then for any unbounded sequence $\{r_n\}$ of positive real numbers outside $F(f)$ such that*

$$\lim_{n\to\infty}\frac{T(r_n,f)}{(\log r_n)^2}=\infty,$$

there exist a direction $\arg z=\theta$ such that for arbitrary small $\varepsilon>0$ we have

$$\limsup_{n\to\infty}\frac{\overline{N}(r_n,Z_\varepsilon(\theta),f=b)}{T(r_n,f)}>0,$$

possibly except at most two values of b.

The radial in Theorem 3.1.5 is actually a (precise) T direction of $f(z)$. We will below call such a radial satisfying (3.1.1) for a sequence $\{r_n\}$ T direction for $\{r_n\}$.

Proof. Suppose on contrary that $f(z)$ has no precise T-direction for $\{r_n\}$. Then there exist a finite number of the radials $\arg z=\theta_j$ $(j=1,2,\cdots,m)$ and a $\varepsilon>0$ such that $\{Z_\varepsilon(\theta_j):1\leqslant j\leqslant m\}$ is a covering of $\mathbb{C}\setminus\{0\}$ and for each j, we have three extended complex numbers a_j, b_j and c_j satisfying

$$\overline{N}(r_n,Z_{2\varepsilon}(\theta_j),f=a_j)+\overline{N}(r_n,Z_{2\varepsilon}(\theta_j),f=b_j)+\overline{N}(r_n,Z_{2\varepsilon}(\theta_j),f=c_j)$$

$$= o(T(r_n, f)), \text{ as } n \to \infty.$$

From Corollary 2.4.1 it follows that

$$\mathscr{T}(r_n, Z_\varepsilon(\theta_j), f) = o(T(r_n, f)).$$

However this together with the fact that

$$\sum_{j=1}^{m} \mathscr{T}(r_n, Z_\varepsilon(\theta_j), f) \geqslant \mathscr{T}(r_n, \mathbb{C}, f) = T(r_n, f) + O(1)$$

yields $T(r_n, f) = o(T(r_n, f))$, a contradiction has been derived. We complete the proof of Theorem 3.1.5. □

Corollary 3.1.1. *Let $f(z)$ be a transcendental meromorphic function with (3.1.4) with $T(r,f)$ in the place of $\mathscr{T}(r,\Omega,f)$. Then $f(z)$ has at least a (precise) T direction.*

The result of Corollary 3.1.1 was conjectured in [58] and later proved by Guo, Zheng and Ng [10] by using Ahlfors-Shimizu characteristic $\mathscr{T}(r,\Omega)$ of a meromorphic function in an angular domain Ω. Actually, this result for transcendental meromorphic functions of the finite positive order can be attained through the existence of the Borel directions of maximal kind proved by G. Valiron (for the detail see Section 3.3 below) and for ones of zero order was proved by Tsuji [32] in 1935 and for ones of infinite order can be also verified through the Nevanlinna second fundamental theorem in an angle (the reader is referred to the proof of Theorem 3.2.2 below).

In Section 3.3, we shall make a remark on that the condition (3.1.4) in those results above is necessary, that is, there exists a transcendental meromorphic function which does not satisfy (3.1.4) and has no T directions at all.

The following is an analogy of Theorem of Cartwrighte and Valiron (cf. [36]) for entire function and Yang [44] for meromorphic function concerning Borel directions.

Theorem 3.1.6. *Let $f(z)$ be a transcendental meromorphic function with finite lower order μ and non-zero order λ and have a Nevanlinna deficient value $a \in \widehat{\mathbb{C}}$ with $\delta = \delta(a,f) > 0$. For any positive and finite τ with $\mu \leqslant \tau \leqslant \lambda$, consider the angular domain $\Omega(\alpha,\beta)$ with*

$$\beta - \alpha > \max\left\{\frac{\pi}{\tau}, 2\pi - \frac{4}{\tau}\arcsin\sqrt{\frac{\delta}{2}}\right\}. \tag{3.1.10}$$

Then $f(z)$ has a (precise) T direction in Ω.

Proof. Suppose on contrary that $f(z)$ has no precise T direction in Ω. Then as in the proof of Theorem 3.1.5, employing Theorem 2.4.4 yields

$$\mathscr{T}(t,\Omega) = o(T(2t,f)) + O((\log t)^2) \text{ as } t \to \infty.$$

Take a sequence of Pólya peak $\{r_n\}$ of $f(z)$ of order τ. Then we have

$$\begin{aligned}\int_1^{r_n} \frac{\mathscr{T}(t,\Omega)}{t^{\omega+1}}\mathrm{d}t &= o\left(\int_1^{r_n} \frac{T(2t,f)}{t^{\omega+1}}\mathrm{d}t\right) + \int_1^{r_n} \frac{O((\log t)^2)}{t^{\omega+1}}\mathrm{d}t \\ &\leqslant o\left(\int_1^{r_n} \frac{T(r_n,f)}{t^{\omega+1}}\left(\frac{2t}{r_n}\right)^{\tau}\mathrm{d}t\right) + O((\log r_n)^2) \\ &= o\left(\frac{T(r_n,f)}{r_n^{\omega}}\right) + O((\log r_n)^2),\end{aligned}$$

and hence by noting $\tau > \omega = \frac{\pi}{\beta-\alpha}$, we attain

$$\frac{r_n^{\omega}}{T(r_n,f)}\int_1^{r_n} \frac{\mathscr{T}(t,\Omega)}{t^{\omega+1}}\mathrm{d}t \to 0 \text{ as } n \to \infty.$$

We define the real function $\Lambda(r)$ by

$$\Lambda(r)^2 = \max\left\{\frac{\mathscr{T}(r_n,\Omega)}{T(r_n,f)}, \frac{r_n^{\omega}}{T(r_n,f)}\int_1^{r_n} \frac{\mathscr{T}(t,\Omega)}{t^{\omega+1}}\mathrm{d}t\right\}, \text{ for } r_n \leqslant r < r_{n+1}.$$

Obviously $\Lambda(r) \to 0$ as $r \to \infty$. In light of Theorem 2.8.1 for all sufficiently large n and a small $\varepsilon > 0$ with $\beta - \alpha > 2\pi - \frac{4}{\tau}\arcsin\sqrt{\frac{\delta}{2}} + 4\varepsilon$, we have

$$\mathrm{mes}D_{\Lambda}(r_n) \geqslant \min\{2\pi, \frac{4}{\tau}\arcsin\sqrt{\frac{\delta}{2}}\} - \varepsilon,$$

where

$$\mathrm{mes}D_{\Lambda}(r_n) = \{\theta \in [0,2\pi) : \log^+ \frac{1}{|f(r_n\mathrm{e}^{\mathrm{i}\theta}) - a|} > \Lambda(r_n)T(r_n,f)\}.$$

From (3.1.10) we easily see

$$\mathrm{mes}(D_{\Lambda}(r_n) \cap [\alpha+\varepsilon, \beta-\varepsilon]) \geqslant \mathrm{mes}(D_{\Lambda}(r_n)) - \mathrm{mes}([\beta-\varepsilon, 2\pi+\alpha+\varepsilon]) > \varepsilon$$

and hence

$$\begin{aligned}B_{\alpha,\beta}\left(r_n, \frac{1}{f-a}\right) &\geqslant \frac{2\omega}{\pi r_n^{\omega}}\sin(\omega\varepsilon)\int_{\alpha+\varepsilon}^{\beta-\varepsilon} \log^+ \frac{1}{|f(r_n\mathrm{e}^{\mathrm{i}\theta}) - a|}\mathrm{d}\theta \\ &\geqslant \frac{2\varepsilon\omega}{\pi r_n^{\omega}}\sin(\omega\varepsilon)\Lambda(r_n)T(r_n,f).\end{aligned}$$

Below we estimate $B_{\alpha,\beta}\left(r_n, \frac{1}{f-a}\right)$ from above. In light of the first fundamental theorem for an angle, Lemma 2.2.1 and Theorem 2.4.7 we have

$$\begin{aligned}
B_{\alpha,\beta}\left(r_n,\frac{1}{f-a}\right) &\leqslant S_{\alpha,\beta}\left(r_n,\frac{1}{f-a}\right)\\
&= S_{\alpha,\beta}(r_n,f)+O(1)\\
&\leqslant 2\omega\frac{\mathscr{T}(r_n,\Omega)}{r_n{}^{\omega}}+\omega^2\int_1^{r_n}\frac{\mathscr{T}(t,\Omega)}{t^{\omega+1}}\mathrm{d}t+O(1)\\
&= \frac{2\omega\Lambda(r_n)^2}{r_n{}^{\omega}}T(r_n,f)+\omega^2\frac{\Lambda(r_n)^2}{r_n{}^{\omega}}T(r_n,f)+O(1).
\end{aligned}$$

These inequalities imply that

$$\Lambda(r_n)\leqslant\frac{(2+\omega)\pi}{2\varepsilon\sin(\omega\varepsilon)}\Lambda(r_n)^2+O\left(\frac{r_n^{\omega}}{T(r_n,f)}\right)\leqslant O(\Lambda(r_n)^2),$$

a contradiction is derived for $\Lambda(r_n)\to 0$ as $n\to\infty$, and hence Theorem 3.1.6 follows. □

Finally, we conclude this section with an extension of concept of T direction.

Definition 3.1.2. *Let $f(z)$ be a transcendental meromorphic function. A curve $L: z=t\mathrm{e}^{\mathrm{i}\theta(t)}$ $(0\leqslant t<\infty)$ is called B-regular T curve of $f(z)$, provided that L is B-regular and for arbitrary small $\varepsilon>0$ we have (3.1.1) with $T(r,f)$ in the place of $\mathscr{T}(r,\Omega,f)$ for $Z_\varepsilon(L)=\{z:\ \theta(t)-\varepsilon<\arg z<\theta(t)+\varepsilon\}$ and*

$$N(r,Z_\varepsilon(L),f=b)=\int_1^r\frac{n(t,Z_\varepsilon(L),f=b)}{t}\mathrm{d}t$$

for all but at most two values of $b\in\widehat{\mathbb{C}}$;

A curve L is called precise B-regular T curve of $f(z)$, if in (3.1.1), $T(r,f)$ is in the place of $\mathscr{T}(r,\Omega,f)$ and $\overline{N}(r,Z_\varepsilon(L),f=b)$ in the place of $N(r,Z_\varepsilon(L),f=b)$.

A continuous path $L: z=t\mathrm{e}^{\mathrm{i}\alpha(t)}$ $(0\leqslant t_0\leqslant t)$ in $\mathbb{C}$ is called B-regular if for any pair (t_1,t_2), the portion of L which lies in $t_1\leqslant|z|\leqslant t_2$ is of length $\leqslant B(t_2-t_1)$.

The existence of a B-regular T curve is guaranteed by Theorem 3.1.5, since a T direction must be a B-regular T curve. However, we ask if a transcendental meromorphic function $f(z)$ with $T(r,f)=O((\log r)^2)$ must have a B-regular T curve.

3.2 T Directions Dealing with Small Functions

This section is devoted to discussing the existence of T directions of a meromorphic function concerning not only complex values but also small functions as his targets. Let $f(z)$ and $a(z)$ be two meromorphic functions. We recall that $a(z)$ is called small function of $f(z)$ if $T(r,a)=o(T(r,f))$ as $r\to\infty$ possibly outside a set with finite measure (here a is allowed to be a constant) and absolutely small if $T(r,a)=o(T(r,f))$ as $r\to\infty$ without except set.

Theorem 3.2.1. *Let $f(z)$ be a transcendental meromorphic function with the non-zero order and the finite lower order. Then there exists a direction* $\arg z = \theta$ *such that for arbitrary small $\varepsilon > 0$ we have*

$$\limsup_{r\longrightarrow\infty} \frac{N(r, Z_\varepsilon(\theta), f = a)}{T(r, f)} > 0,$$

for all small function $a(z)$ with possible exception of at most two functions of $a(z)$.

Proof. Since $f(z)$ is of finite lower order, in view of Theorem 1.1.3, then $T(r) = T(r, f)$ has a sequence of Pólya peaks $\{r_n\}$ outside $F(f)$ with the finite positive order σ between $\mu(f)$ and $\lambda(f)$. Set

$$\psi(r) = \int_1^r \frac{T(256t)}{t} \mathrm{d}t.$$

Since $T(256t) \leqslant K\left(\frac{256t}{r_n}\right)^\sigma T(r_n)$ for $1 \leqslant t \leqslant r_n$, we then have

$$\begin{aligned}\int_1^{r_n} \frac{\psi(t)}{t} \mathrm{d}t &= \psi(r_n)\log r_n - \int_1^{r_n} \frac{T(256t)}{t} \log t \ \mathrm{d}t \\ &= \int_1^{r_n} \frac{T(256t)}{t} (\log r_n - \log t)\mathrm{d}t \\ &\leqslant \int_1^{r_n} K\left(\frac{256t}{r_n}\right)^\sigma \frac{T(r_n)}{t} \log \frac{r_n}{t} \mathrm{d}t \\ &= K(256)^\sigma \frac{T(r_n)}{r_n^\sigma} \int_1^{r_n} \log \frac{r_n}{t}\, t^{\sigma-1} \mathrm{d}t. \end{aligned} \tag{3.2.1}$$

A straightforward calculation deduces that

$$\int_1^r \log \frac{r}{t}\, t^{\sigma-1}\mathrm{d}t = -\frac{1}{\sigma}\log r + \frac{1}{\sigma^2}(r^\sigma - 1) < \frac{1}{\sigma^2} r^\sigma.$$

Substituting the above inequality into (3.2.1) gets

$$\int_1^{r_n} \frac{\psi(t)}{t} \mathrm{d}t \leqslant K(256)^\sigma \sigma^{-2} T(r_n, f). \tag{3.2.2}$$

There exists a direction $\arg z = \theta_0$, as did in the proof of Theorem 3.1.2, such that for arbitrarily small $\varepsilon > 0$, we have

$$\lim_{n\longrightarrow\infty} \frac{\mathscr{T}(r_n, Z_\varepsilon(\theta_0))}{T(r_n, f)} > 0,$$

where $Z_\varepsilon(\theta_0) = \{z : \theta_0 - \varepsilon < \arg z < \theta_0 + \varepsilon\}$. For any three small functions $a_1(z)$, $a_2(z)$ and $a_3(z)$ with respect to $f(z)$, set

$$g(z) = \frac{f(z) - a_1(z)}{f(z) - a_2(z)} \frac{a_3(z) - a_2(z)}{a_3(z) - a_1(z)}.$$

Since $T(r,a) = \sum_{j=1}^{3} T(r,a_j) = o(T(r,f))$ for all r possibly outside a set E of finite measure, in view of Theorem 2.4.5 and Lemma 1.1.11 it follows therefore that

$$\begin{aligned}\mathscr{A}(r, Z_\varepsilon(\theta_0), f) &\leqslant 27\mathscr{A}(64r, Z_{2\varepsilon}(\theta_0), g) + O\left(\int_1^{128r} \frac{T(t,a)}{t} \mathrm{d}t\right) \\ &= 27\mathscr{A}(64r, Z_{2\varepsilon}(\theta_0), g) + O\left(\int_1^{r/2} \frac{T(256t,a)}{t} \mathrm{d}t\right) \\ &= 27\mathscr{A}(64r, Z_{2\varepsilon}(\theta_0), g) + o(\psi(r/2)), \ r \notin E^*,\end{aligned}$$

where E^* is a set of finite measure. From Lemma 1.1.7 we have for all sufficiently large r

$$\mathscr{A}(r, Z_\varepsilon(\theta_0), f) \leqslant 27\mathscr{A}(128r, Z_{2\varepsilon}(\theta_0), g) + o(\psi(r))$$

and hence, in view of Theorem 2.4.4 and (3.2.2), we have

$$\begin{aligned}\mathscr{T}(r_n, Z_\varepsilon(\theta_0), f) &\leqslant 27\mathscr{T}(128r_n, Z_{2\varepsilon}(\theta_0), g) + o\left(\int_1^{r_n} \frac{\psi(t)}{t} \mathrm{d}t\right) \\ &\leqslant 81[\overline{N}(256r_n, Z_{3\varepsilon}(\theta_0), g = 0) + \overline{N}(256r_n, Z_{3\varepsilon}(\theta_0), g = \infty) \\ &\quad + \overline{N}(256r_n, Z_{3\varepsilon}(\theta_0), g = 1)] + O((\log r_n)^2) + o(T(r_n, f)) \\ &\leqslant 81 \sum_{j=1}^{3} \overline{N}(256r_n, Z_{3\varepsilon}(\theta_0), f = a_j) + o(T(r_n, f)).\end{aligned}$$

By noting that $\{r_n\}$ is a sequence of Pólya peak, we have $T(256r_n, f) \leqslant K_0 T(r_n, f)$ and this implies that

$$\lim_{n\to\infty} \frac{\sum_{j=1}^{3} \overline{N}(256r_n, Z_{3\varepsilon}(\theta_0), f = a_j)}{T(256r_n, f)} > 0.$$

We complete the proof of Theorem 3.2.1. □

Theorem 3.2.2. *Let $f(z)$ be a transcendental meromorphic function with the infinite lower order. Then there exists a direction $\arg z = \theta$ such that for arbitrary small $\varepsilon > 0$ we have*

$$\limsup_{r\to\infty} \frac{N(r, Z_\varepsilon(\theta), f = a)}{T(r, f)} > 0,$$

for all absolutely small function $a(z)$ with possible exception of at most two functions of $a(z)$.

Proof. Take a sequence of increasing positive numbers $\{s_n\}$ which tends to infinity and then since $T(r,f)$ is of infinite lower order, by means of Lemma 1.1.3 we can therefore find a sequence of positive numbers $\{r_n\}$ outside $F(f)$ tending to the infinity such that

$$T(t,f) \leqslant \mathrm{e}\left(\frac{t}{r_n}\right)^{s_n} T(r_n,f) \tag{3.2.3}$$

for $1 \leqslant t \leqslant r_n$.

By Theorem 3.1.5 there exists a T direction $\arg z = \theta_0$ of $f(z)$ for $\{r_n\}$. Suppose on the contrary that there exist three absolutely small functions $a_1(z)$, $a_2(z)$ and $a_3(z)$ with respect to $f(z)$ such that

$$\sum_{j=1}^{3} \overline{N}(r, Z_{2\varepsilon}(\theta_0), f = a_j) = o(T(r,f)).$$

Then in terms of the first and second fundamental theorem of Nevanlinna in an angle and (2.2.14), we have

$$\begin{aligned} S_{Z_{2\varepsilon}}(r,f) &\leqslant S_{Z_{2\varepsilon}}(r,g) + L \\ &\leqslant \sum_{j=1}^{3} \overline{C}_{Z_{2\varepsilon}}(r, f = a_j) + O(\log rT(r,f)) + L \\ &\leqslant 4\omega \frac{\overline{N}(r)}{r^{\omega}} + 2\omega^2 \int_1^r \frac{\overline{N}(t)}{t^{\omega+1}} \mathrm{d}t + O(\log rT(r,f)) + L, \end{aligned}$$

$r \notin F(f)$, where $\omega = \frac{\pi}{4\varepsilon}$, $L = O(\sum_{j=1}^{3} S_{Z_{2\varepsilon}}(r,a_j) + 1)$ and

$$\overline{N}(r) = \sum_{j=1}^{3} \overline{N}(r, Z_{2\varepsilon}(\theta_0), f = a_j).$$

Applying Theorem 2.4.7 to a_j yields

$$S_{Z_{2\varepsilon}}(r,a_j) \leqslant 2\omega \frac{T(r,a_j)}{r^{\omega}} + \omega^2 \int_1^r \frac{T(t,a_j)}{t^{\omega+1}} \mathrm{d}t + O(1)$$

so that

$$L \leqslant o\left(\frac{T(r,f)}{r^{\omega}}\right) + o\left(\int_1^r \frac{T(t,f)}{t^{\omega+1}} \mathrm{d}t\right) + O(1).$$

Thus we have

$$S_{Z_{2\varepsilon}}(r,f) \leqslant o\left(\frac{T(r,f)}{r^{\omega}}\right) + o\left(\int_1^r \frac{T(t,f)}{t^{\omega+1}} \mathrm{d}t\right) + O(\log rT(r,f)),\ r \notin F(f).$$

On the other hand, in view of Lemma 2.2.2 for any $c \in \hat{\mathbb{C}}$ we have

$$\begin{aligned} S_{Z_{2\varepsilon}}(r,f) &\geqslant C_{Z_{2\varepsilon}}(r, f = c) + O(1) \\ &\geqslant 2\omega \sin(\omega\varepsilon) \frac{N(r, Z_{\varepsilon}(\theta_0), f = c)}{r^{\omega}} + O(1). \end{aligned}$$

Thus

$$\begin{aligned} N(r,Z_\varepsilon(\theta_0),f=c) &\leqslant o(T(r,f))+o\left(r^\omega\int_1^r\frac{T(t,f)}{t^{\omega+1}}\mathrm{d}t\right)+O(r^\omega\log rT(r,f))\\ &=o(T(r,f))+o\left(r^\omega\int_1^r\frac{T(t,f)}{t^{\omega+1}}\mathrm{d}t\right), \end{aligned}\tag{3.2.4}$$

by noting that the lower order of $f(z)$ is infinite. In view of (3.2.3), for $s_n>\omega$ we have

$$\begin{aligned} r_n^\omega\int_1^{r_n}\frac{T(t,f)}{t^{\omega+1}}\mathrm{d}t &\leqslant \mathrm{e}r_n^\omega\int_1^{r_n}\left(\frac{t}{r_n}\right)^{s_n}\frac{T(r_n,f)}{t^{\omega+1}}\mathrm{d}t\\ &\leqslant \mathrm{e}r_n^{\omega-s_n}T(r_n,f)\int_1^{r_n}t^{s_n-\omega-1}\mathrm{d}t\\ &\leqslant \mathrm{e}r_n^{\omega-s_n}T(r_n,f)\frac{1}{s_n-\omega}(r_n^{s_n-\omega}-1)\\ &<\frac{\mathrm{e}}{s_n-\omega}T(r_n,f). \end{aligned}$$

Then for $s_n>\omega$, substituting the above inequality into (3.2.4) yields

$$N(r_n,Z_\varepsilon(\theta_0),f=c)=o(T(r_n,f)).$$

A contradiction is derived, because $\arg z=\theta_0$ is a T-direction. From this Theorem 3.2.2 follows. □

It is natural to ask whether the assumption that $f(z)$ is not of zero order is necessary and analyzing the proof of Theorem 3.2.1, we also raise a question: must a T direction be also a T direction dealing with small functions? Indeed, we can prove that each T direction of a meromorphic function must be a T direction with absolutely small functions as targets if $\lim_{r\to\infty}\frac{T(dr,f)}{T(r,f)}=\infty$ in view of Lemma 1.1.2 (Note: This condition implies that the function is of infinite lower order). Finally, we remark that we do not know if Theorem 3.2.2 holds for small functions.

3.3 Connection Among T Directions and Other Directions

In this section, our main purpose is to take into account connection of T directions with the Julia and Borel directions. We also consider the latter in an angular domain. Let $f(z)$ be a meromorphic function in an angular domain $\overline{\Omega}(\alpha,\beta)$. A direction in Ω is called a Julia direction of $f(z)$ if in any angular domain containing it $f(z)$ can assume infinitely often any values, possibly with the exception of at most two values. It is obvious that a T direction must be a Julia direction if

$$\mathscr{T}(r,\Omega,f)/\log r\to\infty \text{ as } r\to\infty. \tag{3.3.1}$$

Therefore a T direction of a transcendental meromorphic function must be a Julia direction, since (3.3.1) always holds for $\mathscr{T}(r,\mathbb{C},f)=T(r,f)+O(1)$. By means of this fact, we make a remark on that the condition (3.1.4) for the existence of T directions and the corresponding condition in Theorem 3.1.5 are necessary, since Ostrowski[22] gave a simple example of transcendental meromorphic function with $T(r,f)=O((\log r)^2)$ which has no Julia directions and so no T directions.

However, conversely a Julia direction may not be a T direction. Indeed, observe the function $e^{z^2}+e^z$ and the positive and negative imaginary axes are its two Julia directions but not T directions at all. However, the positive and negative imaginary axes are in fact its two Borel directions of order 1.

We have another example with a Julia direction which is not a T direction. To the end, we first show the following

Theorem 3.3.1. *Let $f(z)$ be analytic in $\overline{\Omega}$ and let $\widetilde{\Omega}$ be an angle contained in Ω. Assume that for all $r>0$, on the part of $|z|=r$ lying in $\widetilde{\Omega}$ we have*

$$\log|f(z)|\leqslant o\left(\frac{\mathscr{T}(r/2,\Omega,f)}{(\log r)^{2+\alpha}}\right)+M,$$

where M and α are two positive constants. If

$$\lim_{r\to\infty}\frac{\mathscr{T}(r,\Omega,f)}{(\log r)^2\log\log r}=\infty, \tag{3.3.2}$$

then $f(z)$ has no T directions in $\widetilde{\Omega}$ with respect to Ω.

Proof. It suffices to verify that for any fixed $\varepsilon>0$, $\widetilde{\Omega}_\varepsilon$ contains no T direction of $f(z)$ with respect to Ω.

As in the proof of Lemma 2.7.1, we put the same definitions on r_n, A_n, A_{jn} and B_{jn}. We denote by z_{jn} the center of A_{jn} and B_{jn}. Then

$$\begin{aligned}
n(A_{jn},f=a) &\leqslant (\log 5)^{-1}N(B_{jn},f=a)\\
&\leqslant (\log 5)^{-1}T\left(B_{jn},\frac{1}{f-a}\right)\\
&\leqslant (\log 5)^{-1}\left(T(B_{jn},f)+\log 2+\log^+|a|+\log\frac{1}{|f(z_{jn})-a|}\right)\\
&\leqslant (\log 5)^{-1}\left(m(B_{jn},f)+\log 2+\log\frac{1}{|f(z_{jn}),a|}\right)\\
&\leqslant (\log 5)^{-1}(o(\mathscr{T}(r_n,\Omega,f)/(\log r_n)^{2+\alpha})+M+\log 2+2\log n)\\
&\leqslant (\log 5)^{-1}(o(\mathscr{A}(r_n,\Omega,f)/n^{1+\alpha})+M+\log 2+2\log n),
\end{aligned}$$

for $a\in\widehat{\mathbb{C}}$ possibly outside a disk E_{jn} with the sphere radius $e^{-2\log n}$. As in the proof of Lemma 2.7.1, we can find three distinct complex numbers a_ν $(\nu=1,2,3)$ such that the above inequality with $a=a_\nu$ holds for all large n. Thus for sufficiently large r we have

$$N(r,\widetilde{\Omega}_\varepsilon,f=a_v)\leqslant o(\mathscr{T}(r,\Omega,f))+O((\log r)^2\log\log r)=o(\mathscr{T}(r,\Omega,f)).$$

This implies that $\widetilde{\Omega}_\varepsilon$ contains no T directions of $f(z)$ with respect to Ω. □

We say that an analytic function $f(z)$ has polynomial growth in an angular domain V with the vertex at the origin, provided that $|f(z)|\leqslant |z|^d+M, \forall z\in V$ for a positive number d and M. Then Theorem 3.3.1 is available for an analytic function of polynomial growth under (3.3.2).

Now we observe the following function

$$g(z)=\prod_{n=0}^{\infty}\left(\frac{e^{\sqrt{n}}-z}{e^{\sqrt{n}}+z}\right). \tag{3.3.3}$$

Rossi [24] proved that $T(r,g)=(1/3+o(1))(\log r)^3$ and $g(z)$ has exactly two Julia directions, the negative and positive imaginary axes and it is of polynomial growth in some angular domain containing the positive imaginary axis. These two Julia directions are also T directions of $g(z)$ in view of Corollary 3.1.1 and $\overline{g(z)}=g(\overline{z})$, for no radial other than the negative and positive imaginary axes is a Julia direction and so a T direction of $g(z)$, but $g(z)$ must have at least one T direction. Sauer[25] considered the function $f(z)=(e^{iz}+1)g(z)$ and proved that the positive imaginary axis is a Julia direction of $f(z)$ but not of $f'(z)$. On the other hand, since $f(z)$ is of polynomial growth in some sector containing the positive imaginary axis and $T(r,f)\sim r\ (r\to\infty)$, in view of Theorem 3.3.1 the positive imaginary axis therefore is neither a T direction of $f(z)$ nor a Borel direction of positive order.

A direction $\arg z=\theta$ in Ω is called a Borel direction of order ρ, $0<\rho$, of a function f meromorphic in Ω, if for arbitrary $\varepsilon>0$ and any $a\in\hat{\mathbb{C}}$, possibly except at most two values of a, we have

$$\limsup_{r\to\infty}\frac{\log^+ n(r,Z_\varepsilon(\theta),f=a)}{\log r}\geqslant\rho. \tag{3.3.4}$$

In the case when $\rho=\lambda_\Omega(f)$, we simply call the Borel direction of order $\lambda_\Omega(f)$ the Borel direction (with respect to Ω).

When $0<\lambda(f)<\infty$, a routine calculation deduces a T direction for $\{r_n\}$ such that $\lim\limits_{n\to\infty}\frac{\log T(r_n,f)}{\log r_n}=\lambda(f)$ is also a Borel direction, and so from Theorem 3.1.5 this gives a proof of the existence of the Borel direction.

G. Valiron is the first one who introduced the concept of a proximate order $\lambda(r)$ for a meromorphic function f with finite positive order and the type function $U(r)=r^{\lambda(r)}$ of f or $T(r,f)$ (for detail the reader is referred to Theorem 1.1.1). In 1932, G. Valiron raised in terms of his type function the concept of one Borel direction of maximal kind, that is, a direction such that for arbitrary $\varepsilon>0$ and any $a\in\hat{\mathbb{C}}$, possibly except at most two values of a, we have

$$\limsup_{r\to\infty}\frac{n(r,Z_\varepsilon(\theta),f=a)}{U(r)}>0, \tag{3.3.5}$$

and proved the existence of such a direction for a meromorphic function of finite positive order. Certainly, the Borel direction of maximal kind can be considered in an angular domain in terms of the type function of $\mathscr{T}(r,\Omega,f)$. It is clear that one Borel direction of maximal kind must be a Borel direction as well as a T direction. However, a T direction for $\{r_n\}$ such that $\lim\limits_{n\to\infty}\frac{T(r_n,f)}{U(r_n)}=1$ is also a Borel direction of maximal kind. Below we verify this result. Assume that for $a\in\widehat{\mathbb{C}}$ and $\varepsilon>0$

$$\limsup_{n\to\infty}\frac{N(r_n,Z_\varepsilon(\theta),f=a)}{T(r_n,f)}>0 \text{ and so } \limsup_{n\to\infty}\frac{N(r_n,Z_\varepsilon(\theta),f=a)}{U(r_n)}>0.$$

If (3.3.5) fails, that is, $n(r,Z_\varepsilon(\theta),f=a)=o(U(r))$, then in view of Lemma 1.1.2 we have

$$N(r,Z_\varepsilon(\theta),f=a)=\int_1^r\frac{n(t,Z_\varepsilon(\theta),f=a)}{t}\mathrm{d}t=o\left(\int_1^r\frac{U(t)}{t}\mathrm{d}t\right)=o(U(r)),$$

a contradiction is derived and hence (3.3.5) holds. This shows $\arg z=\theta$ is a Borel direction of $f(z)$ of maximal kind. From Theorem 3.1.5 this gives a proof of the existence of the Borel direction of maximal kind.

Of course, the Borel directions rely heavily on the order of meromorphic functions. Obviously, (3.3.4) makes no sense for the case $\rho=0$ and hence we cannot consider Borel directions of a meromorphic function with zero order like that. However, some mathematicians used (3.3.4) with $\log\log r$ in the place of $\log r$ for this case. It is rude that (3.3.4) is used for the case $\rho=\infty$ and so more log is considered to put on the numerator in (3.3.4).

In order to treat the case of a meromorphic function with the infinite order, Hiong K. L. [14] introduced for a meromorphic function f a continuous, non-decreasing real positive function $\rho(r)$ which tends to infinity as $r\to\infty$, called an infinite order of f, such that

$$\lim_{r\to\infty}\frac{\rho(R)}{\rho(r)}=1,\ R=r+\frac{r}{\rho(r)\log r} \text{ and } \limsup_{r\to\infty}\frac{\log T(r,f)}{\rho(r)\log r}=1.$$

Then a Borel direction of $\rho(r)$ order can be defined to be a direction $\arg z=\theta$ such that for arbitrary $\varepsilon>0$ and any $a\in\widehat{\mathbb{C}}$, possibly except at most two values of a, we have

$$\limsup_{r\to\infty}\frac{\log^+ n(r,Z_\varepsilon(\theta),f=a)}{\rho(r)\log r}=1. \tag{3.3.6}$$

When f is of finite positive order, $\rho(r)\equiv\lambda(f)$ and a Borel direction of $\rho(r)$ order is a Borel direction; Hiong in the same paper proved the existence of a Borel direction of $\rho(r)$ order for a meromorphic function with the infinite order. We may also introduce proximate order and type function for a meromorphic function with the infinite order, provided that the condition (1.1.1) is suitably weaken (see Theorem 1.1.2).

Obviously, for a Borel direction of $\rho(r)$ order, the growth of the number of a-value points in the corresponding angle is characterized in terms of order $\rho(r)$. However, the order $\rho(r)$ or the proximate order $\lambda(r)$ are essentially neither clear nor explicit for a meromorphic function with the infinite or finite order. In some sense, $\rho(r)$ seems to be $\frac{\log T(r,f)}{\log r}$, but it has certain better regular property than that the latter does. Thus it seems that (3.3.6) would be equivalent to

$$\limsup_{r\to\infty} \frac{\log^+ N(r,\Omega(\theta-\varepsilon,\theta+\varepsilon), f=a)}{\log T(r,f)} = 1.$$

Therefore, it does not seem to be satisfactory that we use the order $\rho(r)$ to characterize the growth of the number of a-value points of such a meromorphic function.

From the previous observation, the author believes that it would be a good direct way to characterize the growth of the number of a-value points by comparing it to the Nevanlinna's characteristic $T(r,f)$, which explains the significance of the concept of the T directions.

It is easy to see that a T direction for an angle Ω must be a Borel direction of order $\mu_\Omega(f)$ if $\mu_\Omega(f) > 0$. For a meromorphic function of the regular growth with finite positive order, that is, its lower order equals to its order, its T direction must be also its Borel direction, but we do not know if the converse is true. Then it is natural to ask whether T directions have to be the Borel directions. However, it is well-known that one Borel exceptional value may not be a Nevanlinna deficient value and neither is the converse, then we wonder that a T direction might not be a Borel direction of $\rho(r)$ order.

Previously, we have pointed out that a meromorphic function with the finite positive order has at least a direction which is a T direction and a Borel direction as well. We have not discovered further connection between T direction and Borel direction of $\rho(r)$ order until the recent work of Zhang Qingde[56]. He proved the following

Theorem 3.3.2. *For arbitrary positive number λ, positive integers p_1 and p_2, let D_1 and D_2 be the systems of p_1 and p_2 radials, respectively, such that $D_1 \cap D_2 = \varnothing$. Then there exist two meromorphic functions $f(z)$ and $g(z)$ with the order λ such that every direction in $D_1 \cup D_2$ is a T direction of $f(z)$ without other T directions and a Borel direction of $g(z)$ without other Borel directions, while the Borel directions of $f(z)$ and the T directions of $g(z)$ are exactly radials in D_2.*

Following Zhang [56] we construct the meromorphic functions $f(z)$ and $g(z)$ which satisfy the requirements of Theorem 3.3.2, some of whose ideas come from Yang and Zhang [48].

Write

$$D_1 = \bigcup_{j=1}^{p_1} \{z : \arg z = \theta_j\} \text{ and } D_2 = \bigcup_{k=1}^{p_2} \{z : \arg z = \phi_k\}.$$

For a positive number λ, let q be the greatest integer with $q < \lambda$. Take a τ with $\max\{q, \lambda/2\} < \tau < \lambda$ and set

$$a_{nj} = \mathrm{e}^n \mathrm{e}^{\mathrm{i}\theta_j} \text{ and } A_{nk} = \mathrm{e}^{\mathrm{e}^n} \mathrm{e}^{\mathrm{i}\phi_k} \tag{3.3.7}$$

$$b_{nj}=\mathrm{e}^{n}\mathrm{e}^{\mathrm{i}(\theta_j+\mathrm{e}^{-(\tau+1)n})} \text{ and } B_{nk}=\mathrm{e}^{\mathrm{e}^{n}}\mathrm{e}^{\mathrm{i}(\phi_k+\mathrm{e}^{-(\lambda+1)\mathrm{e}^{n}})}$$

$$m_n=\left[\frac{\mathrm{e}^{\tau n}}{n^s}\right] \text{ and } M_n=\left[\frac{\mathrm{e}^{\lambda \mathrm{e}^{n}}}{n^s}\right]$$

$$n=1,2,\cdots;\ j=1,2,\cdots,p_1;\ k=1,2,\cdots,p_2,$$

where s is a number greater than 1. Consider the canonical products

$$h_1(z)=\prod_{n=1}^{\infty}\prod_{j=1}^{p_1}\left(E(q,\frac{z}{a_{nj}})\right)^{m_n},\quad h_2(z)=\prod_{n=1}^{\infty}\prod_{j=1}^{p_1}\left(E(q,\frac{z}{b_{nj}})\right)^{m_n}$$

and

$$H_1(z)=\prod_{n=1}^{\infty}\prod_{j=1}^{p_2}\left(E(q,\frac{z}{A_{nj}})\right)^{M_n},\quad H_2(z)=\prod_{n=1}^{\infty}\prod_{j=1}^{p_2}\left(E(q,\frac{z}{B_{nj}})\right)^{M_n},$$

where $E(q,z)$ is the Weierstrass factor defined by $E(q,z)=(1-z)\exp(z+z^2/2+\cdots+z^q/q)$ for $q\neq 0$ and $E(0,z)=1-z$. Since for arbitrarily small $\varepsilon>0$

$$\sum_{n=1}^{\infty}\sum_{j=1}^{p_1}\frac{m_n}{|a_{nj}|^{\tau}}=p_1\sum_{n=1}^{\infty}\mathrm{e}^{-\tau n}\left[\frac{\mathrm{e}^{\tau n}}{n^s}\right]<+\infty$$

and

$$\sum_{n=1}^{\infty}\sum_{j=1}^{p_1}\frac{m_n}{|a_{nj}|^{\tau-\varepsilon}}=p_1\sum_{n=1}^{\infty}\mathrm{e}^{-(\tau-\varepsilon)n}\left[\frac{\mathrm{e}^{\tau n}}{n^s}\right]=+\infty,$$

it follows that $h_i(z)$ $(i=1,2)$ has the order τ. And the same argument implies that $H_i(z)$ $(i=1,2)$ is of order λ and its exponent of convergence of zeros is also λ.

Set

$$f(z)=\frac{h_1(z)H_1(z)}{h_2(z)H_2(z)}. \tag{3.3.8}$$

Then $f(z)$ is of order λ and its exponent of convergence of zeros is also λ. Below we denote positive constants by notations $K,K_1,K_2,\cdots$ which are not the same at each occurrence.

Lemma 3.3.1.

$$K_1\frac{r^{\tau}}{\log^s r}\leqslant N\left(r,\frac{1}{h_i}\right)\leqslant K_2\frac{r^{\tau}}{\log^s r}$$

and for $\mathrm{e}^{\mathrm{e}^{n}+1}\leqslant r<\mathrm{e}^{\mathrm{e}^{n}+2}$,

$$N\left(r,\frac{1}{H_i}\right)\leqslant K_2\frac{r^{\lambda}}{(\log\log r)^s}$$

and for $\mathrm{e}^{\mathrm{e}^{n+1}-2}\leqslant r<\mathrm{e}^{\mathrm{e}^{n+1}}$,

$$N\left(r,\frac{1}{H_i}\right) \leqslant K_2 r^{\lambda/\mathrm{e}}(\log r)^2,\ i=1,2,$$

where K_1 and K_2 are positive constants independent of n and r and not same at each appearance.

Proof. Here we only consider $i=1$. Given r, we have $\mathrm{e}^n \leqslant r < \mathrm{e}^{n+1}$ for some n and equivalently $\log r - 1 < n \leqslant \log r$. Then

$$n\left(r,\frac{1}{h_1}\right) \geqslant p_1 m_n \geqslant p_1\left(\frac{\mathrm{e}^{\tau n}}{n^s}-1\right) \geqslant K_1\frac{r^\tau}{\log^s r}$$

and

$$n\left(r,\frac{1}{h_1}\right) = p_1\sum_{k=1}^{n} m_k \leqslant p_1\sum_{k=1}^{n}\frac{\mathrm{e}^{\tau k}}{k^s} \leqslant p_1\frac{\mathrm{e}^{\tau n}}{n^s}\sum_{k=1}^{n}\left(\frac{n}{k}\right)^s \mathrm{e}^{-\tau(n-k)}.$$

We estimate the sum in the above inequality by dividing it into two parts:

$$\sum_{1\leqslant k\leqslant n/2}^{n}\left(\frac{n}{k}\right)^s \mathrm{e}^{-\tau(n-k)} \leqslant \frac{n}{2}n^s\mathrm{e}^{-\tau n/2} \to 0\ (n\to+\infty)$$

and

$$\sum_{n/2<k\leqslant n}^{n}\left(\frac{n}{k}\right)^s \mathrm{e}^{-\tau(n-k)} \leqslant 2^s\sum_{k=0}^{\infty}\mathrm{e}^{-\tau k} < +\infty.$$

A straightforward calculation implies that

$$\int_{\mathrm{e}}^{r}\frac{t^\tau}{\log^s t}\frac{\mathrm{d}t}{t} \sim \frac{1}{\tau}\frac{r^\tau}{\log^s r}.$$

Combining these inequalities yields the first inequality desired.

Now we want to establish the second inequality. Under the equivalent condition that $\log(\log r-2) < n \leqslant \log(\log r - 1)$, we have

$$\begin{aligned} N\left(r,\frac{1}{H_1}\right) &= p_2\sum_{k=1}^{n} M_k\log\frac{r}{|A_{k1}|} \\ &\leqslant p_2\sum_{k=1}^{n}\frac{\mathrm{e}^{\lambda\mathrm{e}^k}}{k^s}(\mathrm{e}^n+2-\mathrm{e}^k) \\ &= p_2\frac{\mathrm{e}^{\lambda\mathrm{e}^n}}{n^s}\sum_{k=1}^{n}\left(\frac{n}{k}\right)^s\frac{\mathrm{e}^n+2-\mathrm{e}^k}{\mathrm{e}^{\lambda(\mathrm{e}^n-\mathrm{e}^k)}} \\ &\leqslant K_2\frac{\mathrm{e}^{\lambda\mathrm{e}^n}}{n^s} \leqslant K_2\frac{r^\lambda}{(\log\log r)^s}. \end{aligned}$$

The final inequality follows from the following implication by noting $\log\log r - 1 \leqslant n < \log(\log r + 1) - 1$

$$\begin{aligned} N\left(r,\frac{1}{H_1}\right) &= p_2\sum_{k=1}^{n} M_k \log\frac{r}{|A_{k1}|}\\ &\leqslant p_2\sum_{k=1}^{n}\frac{\mathrm{e}^{\lambda \mathrm{e}^k}}{k^s}(\mathrm{e}^{n+1}-\mathrm{e}^k)\\ &= p_2 n \mathrm{e}^{\lambda \mathrm{e}^n}\mathrm{e}^{n+1} \leqslant K_2 r^{\lambda/\mathrm{e}}(\log r)^2. \end{aligned}$$

□

Lemma 3.3.2. *For z with $|z| = r \geqslant 2$, if the distance of z from all zeros and poles of $f(z)$ is greater than dr with $d > 0$, then*

$$\left|\frac{f'(z)}{f(z)}\right| \leqslant Kr^{q-1}.$$

Proof. Let us begin with the estimation of logarithmic derivative of $\frac{E(q,\frac{z}{a_{nj}})}{E(q,\frac{z}{b_{nj}})}$. We have

$$\begin{aligned} &\left|\frac{E'(q,\frac{z}{a_{nj}})}{E(q,\frac{z}{a_{nj}})} - \frac{E'(q,\frac{z}{b_{nj}})}{E(q,\frac{z}{b_{nj}})}\right|\\ &= \left(\frac{1}{z-a_{nj}} - \frac{1}{z-b_{nj}}\right) + \left(\frac{1}{a_{nj}} - \frac{1}{b_{nj}}\right) + \cdots + z^{q-1}\left(\frac{1}{a_{nj}^q} - \frac{1}{b_{nj}^q}\right)\\ &= (a_{nj}-b_{nj})\left\{\frac{1}{(z-a_{nj})(z-b_{nj})} + \frac{1}{a_{nj}b_{nj}}\left[1 + z\left(\frac{1}{a_{nj}} + \frac{1}{b_{nj}}\right)\right.\right.\\ &\quad \left.\left. + \cdots + z^{q-1}\left(\frac{1}{a_{nj}^{q-1}} + \frac{1}{a_{nj}^{q-2}}\frac{1}{b_{nj}} + \cdots + \frac{1}{b_{nj}^{q-1}}\right)\right]\right\}\\ &\leqslant \mathrm{e}^{-\tau n}\left(\frac{1}{(dr)^2} + 1 + 2\mathrm{e}^{-n}r + \cdots + q\mathrm{e}^{-(q-1)n}r^{q-1}\right)\\ &\leqslant \mathrm{e}^{-\tau n}\left(\frac{1}{4d^2} + 1 + r + \cdots + r^{q-1}\right)\\ &\leqslant \mathrm{e}^{-\tau n}\left(\frac{1}{4d^2} + q\right)r^{q-1},\ r \geqslant 2, \end{aligned}$$

where we used the inequality

$$|a_{nj} - b_{nj}| = \mathrm{e}^n|1 - \mathrm{e}^{\mathrm{i}\mathrm{e}^{-(\tau+1)n}}| = \mathrm{e}^n 2\sin\left(\frac{1}{2}\mathrm{e}^{-(\tau+1)n}\right) < \mathrm{e}^n\mathrm{e}^{-(\tau+1)n} = \mathrm{e}^{-\tau n}.$$

Thus

$$\left|\frac{h_1'}{h_1}-\frac{h_2'}{h_2}\right| \leqslant \sum_{n=1}^{\infty}\sum_{j=1}^{p_1} m_n \left|\frac{E'(q,\frac{z}{a_{nj}})}{E(q,\frac{z}{a_{nj}})}-\frac{E'(q,\frac{z}{b_{nj}})}{E(q,\frac{z}{b_{nj}})}\right|$$
$$\leqslant (\frac{1}{4d^2}+q)r^{q-1}p_1\sum_{n=1}^{\infty} m_n \mathrm{e}^{-\tau n}$$
$$\leqslant (\frac{1}{4d^2}+q)p_1\left(\sum_{n=1}^{\infty} n^{-s}\right)r^{q-1}.$$

The same argument implies that

$$\left|\frac{H_1'}{H_1}-\frac{H_2'}{H_2}\right| \leqslant (\frac{1}{4d^2}+q)p_2\left(\sum_{n=1}^{\infty} n^{-s}\right)r^{q-1}.$$

Clearly, we actually complete the proof of Lemma 3.3.2. □

Lemma 3.3.3. *Let Γ be a Jordan curve such that the distance of Γ from zeros and poles of $f(z)$ is at least dr for some $d>0$. Then either*

$$\log^+\frac{1}{|f(z)|}=0,\ \forall z\in\Gamma \text{ or } \log^+|f(z)|\leqslant K|\Gamma|\max_{z\in\Gamma}|z|^{q-1},\ \forall z\in\Gamma,$$

where $|\Gamma|$ is the length of Γ.

Proof. We can assume without any loss of generalities that there exists a point $z_0\in\Gamma$ with $|f(z_0)|=1$. Then in view of Lemma 3.3.2, for each $z\in\Gamma$ we have

$$\log^+|f(z)| \leqslant |\log|f(z)|| \leqslant |\log f(z)|$$
$$= \left|\int_{\Gamma_{z_0z}}\frac{f'(\zeta)}{f(\zeta)}\mathrm{d}\zeta\right| \leqslant |\Gamma|\max_{z\in\Gamma}\left|\frac{f'(z)}{f(z)}\right|$$
$$\leqslant K|\Gamma|\max_{z\in\Gamma}|z|^{q-1},$$

where Γ_{z_0z} is a sub-arc of Γ between z_0 and z. □

Lemma 3.3.4. *For $\frac{2}{3}\mathrm{e}^{\mathrm{e}^n+2}\leqslant r<\frac{5}{6}\mathrm{e}^{\mathrm{e}^n+2}$, we have*

$$T(r,f)<K\frac{r^{\lambda}}{(\log\log r)^s}$$

and for $\frac{2}{3}\mathrm{e}^{[\mathrm{e}^{n+1}]}\leqslant r<\frac{5}{6}\mathrm{e}^{[\mathrm{e}^{n+1}]}$

$$T(r,f)<K\frac{r^{\tau}}{(\log r)^s},$$

where $[x]$ denotes the greatest integer less than x.

Proof. In the first case, by noting $\frac{2}{3}\mathrm{e}^{\mathrm{e}^n+2} - \mathrm{e}^{\mathrm{e}^n} > \left(\frac{2}{3} - \mathrm{e}^{-2}\right)\mathrm{e}^{\mathrm{e}^n+2} > \frac{2}{5}r$ and $\mathrm{e}^{\mathrm{e}^{n+1}} - \frac{5}{6}\mathrm{e}^{\mathrm{e}^n+2} > r$, therefore the distance of the circle $|z| = r$ from zeros and poles of $H(z) = H_1(z)/H_2(z)$ is greater than $\frac{2}{5}r$. Employing Lemma 3.3.3 to $H(z)$ yields either

$$m(r,H) \leqslant K(r^q+1) \text{ or } m\left(r,\frac{1}{H}\right) = 0. \tag{3.3.9}$$

From Lemma 3.3.1 it follows that

$$N\left(r,\frac{1}{H_1}\right) + N\left(r,\frac{1}{H_2}\right) \leqslant K_2\frac{r^\lambda}{(\log\log r)^s}.$$

The first inequality is derived by the two above inequalities together with

$$T(r,f) \leqslant T(r,H) + T(r,h_1/h_2)$$

and by noting that h_i is of order τ and $q < \tau < \lambda$.

In the second case, it is also easy to see that the distance of the circle $|z| = r$ from zeros and poles of $H(z)$ is greater than $\frac{1}{6}r$. Then we still have (3.3.9). Since $\mathrm{e}^{[\mathrm{e}^{n+1}]} - \frac{5}{6}\mathrm{e}^{[\mathrm{e}^{n+1}]} > \frac{1}{6}r$ and $\frac{2}{3}\mathrm{e}^{[\mathrm{e}^{n+1}]} - \mathrm{e}^{[\mathrm{e}^{n+1}]-1} > \left(\frac{2}{3} - \mathrm{e}^{-1}\right)r$, the distance of the circle $|z| = r$ from zeros and poles of $h(z) = h_1(z)/h_2(z)$ is greater than $\frac{1}{6}r$. Thus we have (3.3.9) for $h(z)$ in the place of $H(z)$.

Recalling $\tau > \lambda/2$, it follows from Lemma 3.3.1 that

$$N\left(r,\frac{1}{h_i}\right) + N\left(r,\frac{1}{H_i}\right) \leqslant K_2\frac{r^\tau}{(\log r)^s},\ i = 1,2.$$

This deduces the second inequality of Lemma 3.3.4. □

Now we are in position to prove Theorem 3.3.2, that is, that $f(z)$ satisfies the requirement of Theorem 3.3.2.

(I) We first of all prove that each radial line $\arg z = \phi \in D_2$ is a T-direction as well as Borel direction of $f(z)$.

Given a sufficiently small $\varepsilon > 0$, we consider disk $U_n = \{z:\ |z - z_n| < d\mathrm{e}^{\mathrm{e}^n}\}$, $z_n = \mathrm{e}^{\mathrm{e}^n}\mathrm{e}^{\mathrm{i}\phi}$, where $0 < d < 1$ is chosen such that U_n is contained in the angular domain $Z_\varepsilon(\phi) = \{z:\ |\arg z - \phi| < \varepsilon\}$. In view of Lemma 3.3.3, we have either

$$m\left(\frac{1}{2}d\mathrm{e}^{\mathrm{e}^n}, z_n, f\right) \leqslant K_1(\mathrm{e}^{q\mathrm{e}^n}+1) \text{ or } m\left(\frac{1}{2}d\mathrm{e}^{\mathrm{e}^n}, z_n, \frac{1}{f}\right) = 0. \tag{3.3.10}$$

Then by noting $\mathscr{A}(U_n,f) = \mathscr{A}(U_n,\frac{1}{f})$ the following implication always holds

$$\begin{aligned}
\mathscr{A}(U_n,f) &\geqslant \int_{\frac{1}{2}de^{e^n}}^{de^{e^n}} \frac{\mathscr{A}(t,z_n,f)}{t}\,\mathrm{d}t \\
&= \mathscr{T}(de^{e^n},z_n,f) - \mathscr{T}(\frac{1}{2}de^{e^n},z_n,f) \\
&\geqslant T(de^{e^n},z_n,f) - T(\frac{1}{2}de^{e^n},z_n,f) - \log 2 \\
&\geqslant \int_{\frac{1}{2}de^{e^n}}^{de^{e^n}} \frac{n(t,z_n,f)}{t}\,\mathrm{d}t - m(\frac{1}{2}de^{e^n},z_n,f) - \log 2 \\
&\geqslant (\log 2)\, n(\frac{1}{2}de^{e^n},z_n,f) - K_1(e^{qe^n}+1) \\
&\geqslant (\log 2)\, M_n - K_1(e^{qe^n}+1) \\
&\geqslant K_2 \frac{e^{\lambda e^n}}{n^s}.
\end{aligned}$$

Therefore for $\frac{2}{3}e^{e^n+2} \leqslant r < \frac{5}{6}e^{e^n+2}$, we have

$$\begin{aligned}
\mathscr{T}(r,Z_\varepsilon(\phi),f) &\geqslant \int_{\frac{2}{3}r}^{r} \frac{\mathscr{A}(t,Z_\varepsilon(\phi),f)}{t}\,\mathrm{d}t \\
&\geqslant \mathscr{A}(U_n,f) \\
&\geqslant K_2 \frac{r^\lambda}{(\log\log r)^s} \\
&\geqslant KT(r,f).
\end{aligned}$$

By noting that $\mathscr{A}(r,f)\log r$ is of order λ, in view of (2.4.11) in the proof of Theorem 2.4.3 we obtain for arbitrary three distinct complex numbers a_j $(j=1,2,3)$

$$\mathscr{T}(r,Z_\varepsilon(\phi),f) \leqslant \sum_{j=1}^{3} \overline{N}(r,Z_{2\varepsilon}(\phi),f=a_j) + O(r^{\hat{\tau}}),\ \frac{\lambda}{2} < \hat{\tau} < \tau. \tag{3.3.11}$$

This immediately yields the desired result.

(II) For the radial line $\arg z = \theta \in D_1$, the same argument as above for $V_n = \{z : |z-z_n| < de^{[e^{n+1}]-1}\}$, $z_n = e^{[e^{n+1}]-1}e^{i\theta}$, implies that for $\frac{2}{3}e^{[e^{n+1}]} \leqslant r < \frac{5}{6}e^{[e^{n+1}]}$

$$T(r,Z_\varepsilon(\theta),f) \geqslant K_2 \frac{r^\tau}{(\log r)^s} \geqslant KT(r,f).$$

This together with (3.3.11) demonstrates that $\arg z = \theta$ is a T-direction of $f(z)$.

(III) Next we want to prove that $\arg z = \theta \in D_1$ is not a Borel direction of $f(z)$. Given a small $\varepsilon > 0$, by $z = h_n(\zeta)$ we mean a Riemann mapping from the unit disk Δ onto

$$B_n = \{z:\ |\arg z - \theta| < 2\varepsilon, \frac{3}{2}e^{n-2} < |z| < \frac{3}{4}e^{n+1}\}$$

with $h_n(0) = z_n = \frac{1}{2}(e^{n-2} + e^{n+1})e^{i\theta}$. A computation asserts the existence of two constants b and c with $0 < b < c < 1$ such that

$$C_n = \{z : |\arg z - \theta| < \varepsilon, \frac{1}{2}(e^{n-2} + e^{n-1}) < |z| < \frac{1}{2}(e^n + e^{n+1})\}$$

is mapped by $h_n^{-1}(z)$ into disk $\{\zeta : |\zeta| < c\}$ and all zeros and poles of $f(z)$ in B_n go into $\{\zeta : b < |\zeta| < c\}$ under the mapping $h_n^{-1}(z)$. Employing Lemma 3.3.2 gets either

$$\max_{z\in\partial B_n} \log^+ |f(z)| \leqslant K_2(e^{nq} + 1) \text{ or } \max_{z\in\partial B_n} \log^+ \frac{1}{|f(z)|} = 0.$$

Set $F_n(\zeta) = f(h_n(\zeta))$ and $\gamma_n = \{a : |a, F_n(0)| \leqslant e^{-n}\}$. For $a \notin \gamma_n$ and $a \neq 0, \infty$, we have

$$\begin{aligned}
n(C_n, f = a) &\leqslant n(c, F_n = a)\\
&\leqslant \log\frac{1}{c}\int_c^1 \frac{n(t, F_n = a)}{t}\mathrm{d}t\\
&\leqslant \log\frac{1}{c} T\left(1, \frac{1}{F_n - a}\right)\\
&\leqslant \log\frac{1}{c}\left(T(1, F_n) + \log\frac{1}{|a, F_n(0)|} + \log 2\right)\\
&\leqslant \log\frac{1}{c}\left(K_2(e^{nq} + 1) + n(1, F_n)\log\frac{1}{b} + n + \log 2\right)\\
&\leqslant K_2\left(\frac{e^{\tau n}}{n^s} + n\right).
\end{aligned}$$

Taking a sufficiently large n_0 such that $\sum_{n=n_0}^{\infty} e^{-n} < 1/2$, then we can find three distinct complex numbers a outside $\cup_{n=n_0}^{\infty}\gamma_n \cup \{F_n(0)\}_{n=1}^{\infty} \cup \{0, \infty\}$. For these three a and $e^n \leqslant r < e^{n+1}$ with $n > n_0$, we have

$$\begin{aligned}
n(r, Z_\varepsilon(\theta), f = a) &\leqslant \sum_{k=n_0}^{n} n(C_k, f = a) + O(1)\\
&\leqslant K_2 \sum_{k=n_0}^{n}\left(\frac{e^{\tau k}}{k^s} + k\right) + O(1)\\
&\leqslant K_2 \frac{r^\tau}{(\log r)^s},
\end{aligned}$$

so that

$$\limsup \frac{\log^+ n(r, Z_\varepsilon(\theta), f = a)}{\log r} \leqslant \tau < \lambda.$$

Finally, it follows from the similar argument to above context that each $\arg z = \varphi \in [0, 2\pi) \setminus (D_1 \cup D_2)$ is neither a T-direction nor a Borel direction of $f(z)$.

Thus we have shown that $f(z)$ is our desired function.

The $g(z)$ in Theorem 3.3.2 is constructed in the same form as that of $f(z)$ with a_{nj}, b_{nj}, A_{nj}, B_{nj} and m_n and M_n defined by

$$a_{nj} = e^n e^{i\theta_j} \text{ and } A_{nk} = e^n e^{i\phi_k}$$

$$b_{nj} = e^n e^{i(\theta_j + e^{-(\lambda+1)n})} \text{ and } B_{nk} = e^n e^{i(\phi_k + e^{-(\lambda+1)e^n})}$$

$$m_n = \left[\frac{e^{\lambda n}}{n^{s+1}}\right] \text{ and } M_n = \left[\frac{e^{\lambda n}}{n^s}\right]$$

$$n = 1,2,\cdots; \ j = 1,2,\cdots,p_1; \ k = 1,2,\cdots,p_2,$$

where s is a number greater than 1.

We leave to the reader the proof of that $g(z)$ satisfies the requirement of Theorem 3.3.2.

□

Finally, we consider the Nevanlinna direction, which was introduced by Lü and Zhang [19] in 1983 according to the Nevanlinna deficiency relation (2.8.1). They first defined the deficiency and deficient value for an angle and then for a direction and finally by a Nevanlinna direction meant a direction for which the total sum of deficiencies with respect to this direction does not exceed 2. Obverse the exponential function $f(z) = e^z$. A simple calculation implies that for any angle Ω which does not contain the positive or negative imaginary axis, we have $m\log r \leqslant \mathscr{T}(r,\Omega) \leqslant M\log r$ for two suitable positive constants m and M. On the other hand, it is also clear that e^z has exactly two Nevanlinna directions, that is, the positive and negative imaginary axes. Therefore for the exponential function, the Julia, Borel, Nevanlinna and T directions coincide.

The Nevanlinna direction must be a Julia direction and the Borel direction of $\rho(r)$ order for the case of $0 < \lambda(f) \leqslant +\infty$ must be a Nevanlinna direction, which was proved by Zhang Q. D. [55] in 1986. Then we ask if a T direction is a Nevanlinna direction.

3.4 Singular Directions Dealing with Derivatives

The Hayman inequality reveals that the characteristic of a meromorphic function can be controlled in terms of an integrated counting function of its a-value points and an integrated counting function of b-value points with $b \neq 0$ of its k order derivative. According to the Hayman inequality, Yang [46] posed the existence of a singular direction which is named nowadays Hayman direction after W. Hayman. For a transcendental meromorphic function $f(z)$ with order $0 < \lambda(f) < \infty$, a radial $\arg z = \theta$ is called Hayman direction if for arbitrary $\varepsilon > 0$, arbitrary positive integer k and arbitrary two complex numbers a and $b(\neq 0)$, we have

$$\limsup_{r\to\infty} \frac{\log\{n(r,Z_\varepsilon(\theta),f=a)+n(r,Z_\varepsilon(\theta),f^{(k)}=b)\}}{\log r} = \lambda(f).$$

The existence of Hayman directions was proved later by Yang and Zhang [49] and Chen [2]. Actually, their result is that each Borel direction must be a Hayman direction.

It is natural to raise a singular direction in view of $\mathscr{T}(r,\Omega)$ corresponding to the Hayman inequality as follows.

Definition 3.4.1. *Let $f(z)$ be a meromorphic function in an angle Ω. A radial $\arg z = \theta \in \Omega$ is called Hayman T direction of $f(z)$ with respect to Ω if for arbitrary $\varepsilon > 0$, arbitrary positive integer k and arbitrary two complex numbers a and $b(\neq 0)$, we have*

$$\limsup_{r\to\infty} \frac{N(r,Z_\varepsilon(\theta),f=a)+N(r,Z_\varepsilon(\theta),f^{(k)}=b)}{\mathscr{T}(r,\Omega)} > 0;$$

If $f(z)$ is a transcendental meromorphic function, then we define Hayman T direction of $f(z)$ by using the above inequality with $T(r,f)$ replacing $\mathscr{T}(r,\Omega)$.

In order to prove the existence of Hayman T direction, we need the following result of Valiron type established by Chen [2].

Lemma 3.4.1. *Let $f(z)$ be a meromorphic function in $\{z : |z| < R\}$ $(0 < R < +\infty)$ and let*

$$N = n(R, f = a) + n(R, f^{(k)} = b)$$

for $k > 0$ and two complex numbers a and $b(\neq 0)$. Then for each $\alpha \in \widehat{\mathbb{C}}$, we have

$$n\left(\frac{R}{32}, f=\alpha\right) < C\left(N+1+\log^+ R+\log^+\frac{R}{R-r}+\log^+\log^+|f(z_0)|\right.$$
$$\left.+\log\frac{1}{|f(z^*),\alpha|}\right), \tag{3.4.1}$$

for each $z_0 \in \{z : |z| < \frac{R}{32}\}\setminus(\gamma')\cup(\gamma'')$, where (γ') and (γ'') are two sets of Boutroux-Cartan exceptional disks with $h = \frac{R}{512e}$ respectively corresponding to the a-points of $f(z)$ and b-points of $f^{(k)}(z)$ and to the poles of $f(z)$, and C is a constant only depending on a and b, and z^* is a point in $\{z : |z| < \frac{R}{8}\}$.

We remark that when $z^* = z_0$, the term $\log^+\log^+|f(z_0)|$ in (3.4.1) can be removed, that is to say, we can establish

$$n\left(\frac{R}{128}, f=\alpha\right) < C\left(N+1+\log^+ R+\log^+\frac{R}{R-r}+\log\frac{1}{|f(z^*),\alpha|}\right) \tag{3.4.2}$$

for some fixed point $z^* \in \{z : |z| < \frac{R}{32}\}\setminus(\gamma')\cup(\gamma'')$. We prove this result. If for some $z_0 \in \{z : |z| < \frac{R}{32}\}\setminus(\gamma')\cup(\gamma'')$, $|f(z_0)-a| \leqslant 1$, then the result follows from (3.4.1). Now assume that $|f(z)-a| > 1$ in $\{z : |z| < \frac{R}{32}\}\setminus(\gamma')\cup(\gamma'')$. Take a point $z_0 \in \{z : |z| < \frac{R}{128}\}\setminus(\gamma')\cup(\gamma'')$ and a real number $\frac{5}{2}\frac{R}{128} < \rho < \frac{3R}{128}$ such that $\{z : |z-z_0| = \rho\}\cap((\gamma')\cup(\gamma'')) = \varnothing$. Obviously, $\{z : |z| < R/128\} \subset \{z : |z-z_0| < R/64\} \subset \{z : |z-z_0| < \rho\} \subset \{z : |z| < R/32\}$. Then we have

$$m\left(\rho, z_0, \frac{1}{f-a}\right) = 0$$

and

$$\begin{aligned} n\left(\frac{R}{128}, f=\alpha\right) &\leqslant n\left(\frac{R}{64}, z_0, f=\alpha\right) \\ &\leqslant N(\rho, z_0, f=\alpha)\left(\log\frac{5}{4}\right)^{-1} \\ &\leqslant T\left(\rho, z_0, \frac{1}{f-\alpha}\right)\left(\log\frac{5}{4}\right)^{-1} \\ &\leqslant \Big(T(\rho, z_0, f-a) + \log^+|a| + \log^+|\alpha| \\ &\quad + \log\frac{1}{|f(z_0)-\alpha|} + 2\log 2\Big)\left(\log\frac{5}{4}\right)^{-1} \\ &\leqslant \left(N\left(\rho, z_0, \frac{1}{f-a}\right) + 2\log^+|a| + 3\log 2\right. \\ &\quad \left. + \log^+|f(z_0)| + \log^+|\alpha| + \log\frac{1}{|f(z_0)-\alpha|}\right)\left(\log\frac{5}{4}\right)^{-1} \\ &\leqslant C\left(N + 1 + \log\frac{1}{|f(z_0),\alpha|}\right). \end{aligned}$$

Lemma 3.4.1 improves the corresponding result of Zhang [54] (see Theorem 6.5 of Yang [46]). Yang and Zhang [49] also established such a theorem of Valiron type, that is, their Theorem 2, with $N\log N$ in the place of N, but without the term $\log^+\log^+|f(z_0)|$ in the equality (3.4.1). Lemma 3.4.1 is improved in Chen and Guo [4] concerning a meromorphic function as target and hence using their result replacing Lemma 3.4.1, we can discuss the existence of Hayman T directions dealing with small functions as targets. As in the proof of Lemma 2.7.1 using Lemma 3.4.1 and (3.4.2) we can get

Lemma 3.4.2. *Let $f(z)$ be a meromorphic function in $\overline{\Omega}(\alpha,\beta)$ and a and $b(\neq 0)$ are two complex numbers and $\varepsilon > 0$. Then we have*

$$\begin{aligned} N(r,\Omega_\varepsilon, f=c) \leqslant K\{&N(2r,\Omega, f=a) + N(2r,\Omega, f^{(k)}=b)\} \\ &+ O((\log r)^3) \end{aligned} \tag{3.4.3}$$

for all $c \in \widehat{\mathbb{C}}$ possibly outside a set with measure zero, where K is a constant only depends on ε, a and b. We have (3.4.3) with N and $(\log r)^3$ replaced by n and $(\log r)^2$.

The term $O((\log r)^3)$ in (3.4.3) comes from the term $\log R$ in (3.4.2). Since

$$\mathscr{T}(r,\Omega_\varepsilon, f) = \int_{\widehat{\mathbb{C}}} N(r,\Omega_\varepsilon, f=c)\mathrm{d}m(c),$$

finding the integration of two side of (3.4.3) under the sphere metric, we have

$$\mathscr{T}(r,\Omega_\varepsilon,f) \leqslant K\{N(2r,\Omega,f=a)+N(2r,\Omega,f^{(k)}=b)\} \\ +O((\log r)^3). \tag{3.4.4}$$

Since (3.4.3) holds for n in the place of N with $(\log r)^3$ replaced by $(\log r)^2$, we therefore have

$$\mathscr{A}(r,\Omega_\varepsilon,f) \leqslant K\{n(2r,\Omega,f=a)+n(2r,\Omega,f^{(k)}=b)\} \\ +O((\log r)^2). \tag{3.4.5}$$

(3.4.4) and (3.4.5) are deduced in [60] in a different way. In terms of the Rossi's example, we pointed out in [60] that the term $O((\log r)^3)$ in (3.4.4) cannot be replaced by a quantity $\phi(r)$ such that $\liminf_{r\to\infty}\phi(r)(\log r)^{-3}=0$. In order to treat the case of infinite lower order, we need the following

Lemma 3.4.3. *Let $f(z)$ be a meromorphic function in $\overline{\Omega}(\alpha,\beta)$ and a, c and $b(\neq 0)$ are three complex numbers and $\varepsilon>0$. Then we have*

$$N(r,\Omega_\varepsilon,f=c) \leqslant K\left(N(r,\Omega)+r^\omega\int_1^r \frac{N(t,\Omega)}{t^{\omega+1}}\mathrm{d}t\right) \\ +O(r^\omega\log r\mathscr{T}(r,\Omega)),\ r\notin F_\Omega(f), \tag{3.4.6}$$

where $N(r,\Omega)=N(r,\Omega,f=a)+N(r,\Omega,f^{(k)}=b)$ and $\omega=\frac{\pi}{\beta-\alpha-\varepsilon}$.

Proof. The inequality (3.4.6) results from the following implication

$$\begin{aligned}
&2\omega\sin(\varepsilon\omega/2)\frac{N(r,\Omega_\varepsilon,f=c)}{r^\omega}\\
&\leqslant C_{\Omega_{\varepsilon/2}}\left(r,\frac{1}{f-c}\right)\\
&\leqslant S_{\Omega_{\varepsilon/2}}\left(r,\frac{1}{f-c}\right)\\
&= S_{\Omega_{\varepsilon/2}}(r,f)+O(1)\\
&\leqslant \left(2+\frac{1}{k}\right)C_{\Omega_{\varepsilon/2}}\left(r,\frac{1}{f-a}\right)+\left(2+\frac{2}{k}\right)C_{\Omega_{\varepsilon/2}}\left(r,\frac{1}{f^{(k)}-b}\right)\\
&\quad +O(\log rS_\Omega(r,f))\\
&\leqslant \left(2+\frac{1}{k}\right)\left(4\omega\frac{N(r,\Omega,f=a)}{r^\omega}+2\omega^2\int_1^r\frac{N(t,\Omega,f=a)}{t^{\omega+1}}\mathrm{d}t\right)\\
&\quad +\left(2+\frac{2}{k}\right)\left(4\omega\frac{N(r,\Omega,f^{(k)}=b)}{r^\omega}+2\omega^2\int_1^r\frac{N(t,\Omega,f^{(k)}=b)}{t^{\omega+1}}\mathrm{d}t\right)\\
&\quad +O(\log r\mathscr{T}(r,\Omega)),\ r\notin F_\Omega(f),\ \omega=\frac{\pi}{\beta-\alpha-\varepsilon},
\end{aligned}$$

where we used the inequalities (2.2.15), (2.2.10)(Hayman inequality for an angle), Theorem 2.5.1, (2.2.14) and finally Theorem 2.4.7. □

Now we can establish the main result of this section, which shows the existence of Hayman T directions (See [60]).

Theorem 3.4.1. *Let $f(z)$ be a transcendental meromorphic function such that*

$$\limsup_{r\to\infty}\frac{T(r,f)}{(\log r)^3}=\infty. \tag{3.4.7}$$

Then $f(z)$ has at least a Hayman T direction which is also a T direction.

Proof. We treat two cases below.

(I) $f(z)$ is of finite lower order. If the order of $f(z)$ is positive or infinite, then Theorem 1.1.3 asserts the existence of a sequence of Pólya peaks $\{r_n\}$ with order $\sigma>0$ such that

$$T(2r_n,f)\leqslant 2^{\sigma+1}T(r_n,f) \text{ and } \lim_{n\to+\infty}\frac{\log T(r_n,f)}{\log r_n}>0.$$

It is easy to see, from the latter inequality, that $(\log r_n)^3=o(T(r_n,f))$. In view of Theorem 3.1.5 there exists a T direction $\arg z=\theta$ of $f(z)$ for $\{r_n\}$. From Lemma 3.4.2 it easily follows that $\arg z=\theta$ is also a Hayman T direction of $f(z)$.

Now assume that the order of $f(z)$ vanishes. In view of Lemma 1.1.8, $\overline{\log\operatorname{dens}}W=1$, where $W=\{r>0:T(2r,f)\leqslant 2T(r,f)\}$. From the condition (3.4.7), we can take a sequence of positive numbers $\{r_n\}$ such that $(\log r_n)^3=o(T(r_n,f))$. It is easy to see that for sufficiently large n, $W\cap(r_n,r_n^2)\neq\varnothing$. Take an $r_n'\in W\cap(r_n,r_n^2)$ and then we have $T(2r_n',f)\leqslant 2T(r_n',f)$ and

$$(\log r_n')^3\leqslant 8(\log r_n)^3=o(T(r_n,f))=o(T(r_n',f)).$$

Employing Theorem 3.1.5 and Lemma 3.4.2 again yields that there exists a T direction of $f(z)$ for $\{r_n'\}$ which is also a Hayman T direction of $f(z)$.

Thus we have shown the existence of a Hayman T direction for $f(z)$ being of finite lower order.

(II) $f(z)$ is of infinite lower order. In view of Lemma 3.4.3 it is not difficult to complete our proof for this case in question by consulting the proof of Theorem 3.2.2 in view of Lemma 3.4.3. The proof is left to the reader. □

From the above proof, we cannot confirm that each T direction must be a Hayman T direction and so this is a problem. However, we can show that a Hayman T direction may not be a T direction by using the example which Yang and Zhang [49] used to describe that a Hayman direction may not be a Borel direction. In fact, the relation between Hayman T direction and T direction has something to do with the existence of common T direction of $f(z)$ and its derivatives $f^{(p)}(z)$. The latter will be carefully discussed in the next section.

For a meromorphic function of zero order, Yang [47] gave out a singular direction similar to the Julia direction corresponding to the Hayman inequality and his result was later reinforced by Chen [3] who proved that if

$$\limsup_{r\to\infty}\frac{T(r,f)}{(\log r)^3}=\infty,$$

then there exists a radial $\arg z=\theta$ such that for arbitrary $\varepsilon>0$, arbitrary positive integer k and arbitrary two complex numbers a and $b(\neq 0)$, we have

$$\limsup_{r\to\infty}\frac{n(r,Z_\varepsilon(\theta),f=a)+n(r,Z_\varepsilon(\theta),f^{(k)}=b)}{(\log r)^2}=\infty. \tag{3.4.8}$$

It is obvious that Theorem 3.4.1 is an improvement of Chen's above result. Actually, from the inequality $\frac{1}{2}n(r^{\frac{1}{2}},*)\log r\leqslant N(r,*)\leqslant n(r,*)\log r$ it follows that (3.4.8) is equivalent to the following equation

$$\limsup_{r\to\infty}\frac{N(r,Z_\varepsilon(\theta),f=a)+N(r,Z_\varepsilon(\theta),f^{(k)}=b)}{(\log r)^3}=\infty.$$

Certainly, we can also determine condition of growth for the existence of Hayman T direction with respect to an angle in terms of Lemma 3.4.2 and Lemma 3.4.3.

3.5 The Common T Directions of a Meromorphic Function and Its Derivatives

From this section on, we shall discuss several topics on T directions which should attract many interests and in the final section, we shall suggest the investigation of other functions than meromorphic functions.

In 1928, G. Valiron [33] asked if there exists a common Borel direction of a meromorphic function and its derivative. This problem was investigated by many mathematicians such as Valiron [33], Rauch [23], Chuang [6], Milloux [20], Yang [45] and Zhang [54]. In the end of this section, we shall introduce some of their works. However, the Valiron problem is still open. Here we consider a problem of Valiron type for T directions, that is,

Question 3.5.1. *Does there exist a common T direction of a transcendental meromorphic function and its derivative?*

A Borel direction of derivative of a meromorphic function may not be itself Borel direction. This can be described by observing the following examples of Steinmetz's:

(1) $f(z)=\mathrm{e}^{-\mathrm{i}z}/(\mathrm{e}^z+1)$ has Borel directions $\arg z=\pi/4,\pi,3\pi/2$, while $f'(z)=\mathrm{e}^{-\mathrm{i}z}(\mathrm{i}\mathrm{e}^z-\mathrm{e}^z+\mathrm{i})/(\mathrm{e}^z+1)^2$ has in addition the Borel direction $\arg z=\pi/2$;

(2) $f(z)=\cos(z)/\cosh(z)$ has Borel directions $\arg z=\pi/4,3\pi/4,5\pi/4,7\pi/4$, while $f'(z)$ has in addition the Borel directions $\arg z=\pi/2$ and $3\pi/2$.
(which Prof. Steinmetz was very kind to send to the author by e-mail.) We can also give out more examples. The assertions about the above examples can be demon-

strated in terms of the Stokes rays of exponential functions (Please consult Section 3.7).

It is obvious that each Borel direction of Steinmetz examples above is also their T direction and vice versa and so is it for their derivatives. Therefore, a T direction of derivative of a meromorphic function may not be itself T direction.

Furthermore, let $g(z)$ be defined in (3.3.3). The two Julia directions of $g(z)$, the positive and negative imaginary axes, are not Julia directions of $g'(z)$ (see Rossi[24]) and hence are not T directions of $g'(z)$. According to the growth of $g'(z)$ and Theorem 3.1.5, $g'(z)$ has to have T directions and therefore, we obtain the following

Theorem 3.5.1. *Let $g(z)$ be the function with the form in (3.3.3). Then $g(z)$ and $g'(z)$ have no common T directions.*

In the following we introduce some notations for simplicity. Let $f(z)$ be a transcendental meromorphic function. Set

$$JD(f) = \{\theta \in [0, 2\pi) : \arg z = \theta \text{ is a Julia direction of } f\}$$

and $BD(f)$ and $TD(f)$ are defined in the same way for Borel directions and T directions, respectively.

Theorem 3.5.2. *Let $f(z)$ be a transcendental meromorphic function with*

$$\liminf_{r\to\infty} \frac{T(dr, f)}{T(r, f)} = \infty \tag{3.5.1}$$

for some $d > 1$. Then there exists at least a radial which is one common T direction of $f(z)$ and $f^{(p)}(z)$ for each p, that is, $\bigcap_{j=1}^{\infty} TD(f^{(j)}) \neq \varnothing$.

Proof. In view of (2.2.5) and (2.2.15), we have

$$S_{\alpha,\beta}(r, f^{(p)}) \geqslant C_{\alpha,\beta}(r, f^{(p)} = a) + O(1) \geqslant 2\omega \sin(\omega\delta) \frac{N(r, \Omega_\delta, f^{(p)} = a)}{r^\omega} + O(1),$$

where $\Omega = \Omega(\alpha, \beta)$.

On the other hand, it follows from the Nevanlinna second fundamental theorem for an angle and (2.2.14) that

$$\begin{aligned}
S_{\alpha,\beta}(r, f^{(p)}) &\leqslant (p+1) S_{\alpha,\beta}(r, f) + O(\log rT(r, f)) \\
&\leqslant (p+1) \sum_{v=1}^{3} C_{\alpha,\beta}(r, f = a_v) + O(\log rT(r, f)) \\
&\leqslant 4(p+1)\omega \frac{N(r, \Omega)}{r^\omega} + 2(p+1)\omega^2 \int_1^r \frac{N(t, \Omega)}{t^{\omega+1}} dt \\
&\quad + O(\log rT(r, f)), \ r \notin E(f),
\end{aligned}$$

where $N(r, \Omega) = \sum_{v=1}^{3} N(r, \Omega, f = a_v)$.

Under the assumption (3.5.1) employing Lemma 1.1.2 yields that for arbitrary $\sigma > 0$ and sufficiently large r

$$\int_1^r \frac{T(t,f)}{t^{\sigma+1}}\mathrm{d}t \leqslant K\frac{T(r,f)}{r^\sigma}+O(1).$$

Let $M(\mathrm{e})$ be the set $M(K)$ for $K=\mathrm{e}$ in Theorem 2.6.4. Take a sequence of positive number $\{r_n\}$ tending to ∞ outside $M(\mathrm{e})\cup E(f)$ and so $T(r_n,f)\leqslant 4\mathrm{e}^2T(r_n,f^{(p)})$. It is easy from (3.5.1) to see that $\lim_{n\to\infty}\frac{\log T(r_n,f)}{\log r_n}=\infty$.

Let $\arg z=\theta_0$ be a T direction of $f^{(p)}$ for $\{r_n\}$. Suppose that $\arg z=\theta_0$ is not a T direction of $f(z)$. Then for some $\varepsilon>0$ and three distinct points a_ν in $\hat{\mathbb{C}}$,

$$\sum_{\nu=1}^{3} N(r,Z_{2\varepsilon}(\theta_0),f=a_\nu)=o(T(r,f)).$$

As before, we denote the sum on the left side of above equality by $N(r,Z_{2\varepsilon}(\theta_0))$.

Combining the above inequalities together yields that

$$\begin{aligned}
N(r_n,Z_\varepsilon(\theta_0),f^{(p)}=a) &\leqslant \frac{2(p+1)}{\sin(\omega\varepsilon)}N(r_n,Z_{2\varepsilon}(\theta_0))\\
&\quad+\frac{(p+1)\omega}{\sin(\omega\varepsilon)}r_n^\omega\int_1^{r_n}\frac{N(t,Z_{2\varepsilon}(\theta_0))}{t^{\omega+1}}\mathrm{d}t\\
&\quad+O(r_n^\omega \log r_n T(r_n,f))\\
&=o(T(r_n,f))+o\left(r_n^\omega\int_1^{r_n}\frac{T(t,f)}{t^{\omega+1}}\mathrm{d}t\right)\\
&=o(T(r_n,f)).
\end{aligned}$$

On the other hand, we have

$$\limsup_{n\to\infty}\frac{N(r_n,Z_\varepsilon(\theta_0),f^{(p)}=a)}{T(r_n,f)}\geqslant\frac{1}{4\mathrm{e}^2}\limsup_{n\to\infty}\frac{N(r_n,Z_\varepsilon(\theta_0),f^{(p)}=a)}{T(r_n,f^{(p)})}>0.$$

A contradiction is derived, from which we have proved that at least one of T directions of $f^{(p)}(z)$ must be a T direction of $f(z)$ and actually a common T direction of $f^{(j)}(z)$ $(j=1,2,\cdots,p)$. This is because

$$T(r_n,f^{(j)})\leqslant 2(j+1)T(r_n,f) \text{ and } T(r_n,f^{(j)})\leqslant 8\mathrm{e}^2(j+1)T(r_n,f^{(p)})$$

and in view of Theorem 2.6.1 and (2.6.1), we have

$$\liminf_{r\to\infty}\frac{T(d^2r,f^{(j)})}{T(r,f^{(j)})}=\liminf_{r\to\infty}\frac{C_d^{-1}T(dr,f)}{(j+1)T(r,f)+o(T(dr,f))}=\infty.$$

The above argument is also available for the case when $f(z)$ is replaced by $f^{(j)}(z)$.

Thus it is clear that Theorem 3.5.2 follows. □

For the case of the infinite order, we can prove

Theorem 3.5.3. *Let $f(z)$ be of infinite order. Then $TD(f^{(p)})\cap TD(f^{(q)})\neq\varnothing$ for each pair of non-negative integers p and q in $\mathbb{N}$.*

Proof. Suppose that Theorem 3.5.3 does not hold and so we assume without any loss of generalities that $TD(f)\cap TD(f^{(p)})=\varnothing$ for some positive integer p. Then for each $\theta\in TD(f^{(p)})$ there exist $\varepsilon(\theta)>0$ and three distinct points $a_\nu(\theta)$ $(\nu=1,2,3)$ in $\hat{\mathbb{C}}$ such that

$$\sum_{\nu=1}^{3}N(r,Z_{2\varepsilon(\theta)}(\theta),f=a_\nu)=o(T(r,f)).$$

Clearly, $\{(\theta-\varepsilon(\theta),\theta+\varepsilon(\theta)):\ \theta\in TD(f^{(p)})\}$ is a covering of $TD(f^{(p)})$..Let ε is the Lebesgue number of this covering and $\varepsilon>0$ exists as $TD(f^{(p)})$ is compact in $[0,2\pi]$.

Given sufficiently large K, let $M(K)$ and $\delta(K)$ be the set and the constant in Theorem 2.6.4. By induction, we construct a non-decreasing sequence $\{\tilde{r}_n\}$ such that $\tilde{r}_n\in[1,K^n]$ and

$$\frac{T(\tilde{r}_n,f)}{\tilde{r}_n^{\omega+1}}=\max_{1\leqslant t\leqslant K^n}\left\{\frac{T(t,f)}{t^{\omega+1}}\right\}$$

with $\omega=\frac{\pi}{2\varepsilon}$. Assume that we have obtained $\tilde{r}_n$ satisfying the above requirement. Now take a $\tilde{r}_{n+1}\in[\tilde{r}_n,K^{n+1}]$ such that

$$\frac{T(\tilde{r}_{n+1},f)}{\tilde{r}_{n+1}^{\omega+1}}=\max_{\tilde{r}_n\leqslant t\leqslant K^{n+1}}\left\{\frac{T(t,f)}{t^{\omega+1}}\right\}.$$

It is obvious that $\tilde{r}_{n+1}$ satisfies our requirement and hence the desired sequence $\{\tilde{r}_n\}$ is attained.

Since $f(z)$ is of infinite order, we thus have that $\{\tilde{r}_n\}\uparrow\infty$ as $n\to\infty$,

$$\int_1^{\tilde{r}_n}\frac{T(t,f)}{t^{\omega+1}}\mathrm{d}t\leqslant\frac{T(\tilde{r}_n,f)}{\tilde{r}_n^{\omega}}$$

and $\liminf_{n\to\infty}\frac{\log T(\tilde{r}_n,f)}{\log\tilde{r}_n}\geqslant\omega+1.$

Since

$$(\log K)\delta(K)\leqslant\frac{\log K}{2\mathrm{e}^{K-1}-1}\to0$$

as $K\to\infty$, we therefore find two fixed C and K such that $2C<K$ and $\log C>(\log K)\delta(K)$. Below we need to treat two cases.

(I) For each n, $C\tilde{r}_n<\tilde{r}_{n+1}$. Set $I=\bigcup_{n=1}^{\infty}[\tilde{r}_n,C\tilde{r}_n]$ and then since

$$\int_{I(C\tilde{r}_n)}\frac{\mathrm{d}t}{t}=\sum_{k=1}^{n}\int_{\tilde{r}_k}^{C\tilde{r}_k}\frac{\mathrm{d}t}{t}=n\log C\text{ and }\log C\tilde{r}_n\leqslant\log C+n\log K,$$

I has the upper logarithmic density at least $\frac{\log C}{\log K}>\delta(K)$ and hence there exist a sequence $\{r_{n_k}\}$ such that $r_{n_k}\in[\tilde{r}_{n_k},C\tilde{r}_{n_k}]\setminus M(K)$. Thus, we can obtain a sequence $\{r_n\}$ such that $r_n\in[\tilde{r}_n,C\tilde{r}_n]$ which contains a subsequence outside $M(K)$.

Let $\arg z = \theta_0$ be a T direction of $f^{(p)}$ for $\{r_n\}$ such that

$$\limsup_{k\to\infty} \frac{N(r_{n_k}, Z_\varepsilon(\theta_0), f^{(p)} = a)}{T(r_{n_k}, f^{(p)})} > 0.$$

This follows from Theorem 3.1.5, for the argument in the proof of Theorem 3.1.5 is available for any subsequence of $\{r_n\}$. By noting that ε is the Lebesgue number, we have

$$\sum_{v=1}^{3} N(r, Z_{2\varepsilon}(\theta_0), f = a_v) = o(T(r,f)).$$

For these r_n we have

$$\begin{aligned}\int_1^{r_n} \frac{T(t,f)}{t^{\omega+1}}\,dt &= \int_1^{\tilde r_n} \frac{T(t,f)}{t^{\omega+1}}\,dt + \int_{\tilde r_n}^{r_n} \frac{T(t,f)}{t^{\omega+1}}\,dt \\ &\leqslant \frac{T(\tilde r_n, f)}{\tilde r_n^{\omega}} + T(r_n, f)\frac{C^\omega - 1}{\omega C^\omega}\frac{1}{\tilde r_n^{\omega}} \\ &\leqslant \left(C^\omega + \frac{C^\omega - 1}{\omega}\right)\frac{T(r_n, f)}{r_n^\omega}.\end{aligned}$$

As in the proof of Theorem 3.5.2, we have

$$N(r_n, Z_\varepsilon(\theta_0), f^{(p)} = a) = o(T(r_n, f))$$

and since $\{r_n\}$ contains a subsequence $\{r_{n_k}\}$ outside $M(K)$, a contradiction can be derived by the same argument as in the proof of Theorem 3.5.2.

(II) There exist a sequence $\{n_k\}$ such that $\tilde r_{n_k+1} \leqslant C\tilde r_{n_k}$. For these n_k, we have $2C\tilde r_{n_k} < K^{n_k+1}$ and hence

$$T(2C\tilde r_{n_k}, f) \leqslant \left(\frac{2C\tilde r_{n_k}}{\tilde r_{n_k+1}}\right)^{\omega+1} T(\tilde r_{n_k+1}, f) \leqslant (2C)^{\omega+1} T(C\tilde r_{n_k}, f).$$

Set $r_n = C\tilde r_n$ and then for a subsequence of $\{r_n\}$, $T(2r_n, f) \leqslant (2C)^{\omega+1} T(r_n, f)$. For each n we have

$$\int_1^{2r_n} \frac{T(t,f)}{t^{\omega+1}}\,dt \leqslant \left(C^\omega + \frac{(2C)^\omega - 1}{\omega 2^\omega}\right)\frac{T(2r_n, f)}{r_n^\omega}.$$

Let $\arg z = \theta_0$ be a T direction of $f^{(p)}$ for $\{2r_n\}$ such that

$$d = \liminf_{k\to\infty} \frac{N(2r_{n_k}, Z_\varepsilon(\theta_0), f^{(p)} = a)}{T(2r_{n_k}, f^{(p)})} > 0.$$

Actually, it follows from Theorem 3.1.5 that the above lim inf is positive for a subsequence of $\{r_{n_k}\}$ which is still denoted by the same notations. As in the proof of Theorem 3.5.2, we have

$$N(2r_n,Z_\varepsilon(\theta_0),f^{(p)}=a)=o(T(2r_n,f))$$

and hence for n_k, $N(2r_{n_k},Z_\varepsilon(\theta_0),f^{(p)}=a)=o(T(r_{n_k},f))$. Then for sufficiently large k in view of Chuang's inequality (2.6.2) we have

$$T(r_{n_k},f)\leqslant K_0T(2r_{n_k},f^{(p)})\leqslant \frac{2}{d}K_0N(2r_{n_k},Z_\varepsilon(\theta_0),f^{(p)}=a)=o(T(r_{n_k},f)).$$

A contradiction is derived.

Consequently Theorem 3.5.3 follows. □

We should mention that under the assumption of Theorem 3.5.3, we don't know if $\bigcap_{j=0}^{\infty} TD(f^{(j)})\neq\varnothing$. However, the author guesses it would be true.

Up to now, to solve Question 3.5.1, we are only required to answer problem of whether there exist a common T direction of a meromorphic function $f(z)$ and its derivative if $f(z)$ is of finite order and satisfies (3.4.7) in view of Theorems 3.5.1 and 3.5.3.

The Hayman T directions of a meromorphic function have something to do with common T directions of itself and its derivatives. From Theorem 3.4.1 we immediately deduce the following

Theorem 3.5.4. *Let $f(z)$ be a transcendental meromorphic function such that*

$$\limsup_{r\to\infty}\frac{T(r,f)}{(\log r)^3}=\infty.$$

If for some $a\in\mathbb{C}$, $\delta(a,f)=1$, then $\bigcap_{j=0}^{\infty} TD(f^{(j)})\neq\varnothing$.

Proof. As in the proof of Theorem 3.4.1, we can find a sequence of positive numbers $\{r_n\}$ outside $F(f)$ such that the T direction $\arg z=\theta$ of $f(z)$ for $\{r_n\}$ must be also a Hayman T direction of $f(z)$. Since $\delta(a,f)=1$, that is $N(r,f=a)=o(T(r,f))$, this implies that for each $b\neq 0$

$$\limsup_{n\to\infty}\frac{N(r_n,Z\varepsilon(\theta),f^{(k)}=b)}{T(r_n,f)}>0.$$

Noting that $T(r_n,f^{(k)})\leqslant(k+1)T(r_n,f)+O(\log r_nT(r_n,f))$, we have

$$\limsup_{n\to\infty}\frac{N(r_n,Z\varepsilon(\theta),f^{(k)}=b)}{T(r_n,f^{(k)})}>0$$

and hence this Hayman T direction $\arg z=\theta$ must be a T direction of $f^{(k)}(z)$ for each k. Theorem 3.5.4 has been proved. □

In the following we consider the case when the function in question has few poles. For this, we first establish a preliminary result.

Given an angular domain $\Omega(\alpha,\beta)$, consider the conformal transformation of $w(z)$, then there exists a sector S containing the ray such that $n(r,S,w=0) = O(\log r)$. Therefore, (3.7.4) confirms that the ray $\arg z = \theta$ is a Stokes ray of $w(z)$ of order λ. Conversely, (3.7.4) follows from the asymptotic representation of $w(z)$ in the boundary of a neighborhood of the Stokes ray given in Theorem 3.7.1.

The following is a consequence of Theorem 3.7.1.

Theorem 3.7.2. *Let $w(z)$ be given as in Theorem 3.7.1. Then there exist finitely many rays $\arg z = \theta_j$ $(j = 1,2,\cdots,N)$ such that the number of a-points of $w(z)$ for every $a \in \mathbb{C}$ in $|z| \leqslant r$ but outside the logarithmic strips*

$$|\arg z - \theta_j| < K\frac{\log^+ |z|}{|z|^{1/p}} \tag{3.7.6}$$

is $O(\log r)$ for two positive constants K and p.

Thus Stokes rays of $w(z) - a$ lie only on the rays in Theorem 3.7.2. It is easy to see that $w(z) - a$ has the equal Stokes rays of the order $\lambda(w)$ for all complex number a except possibly at most one of a.

Define the indicator function $h_w(\theta)$ of $w(z)$ by

$$h_w(\theta) = \limsup_{r\to\infty} \frac{\log|w(re^{i\theta})|}{r^{\lambda(w)}}$$

for $0 < \lambda(w) < \infty$. If $w(z)$ is a meromorphic solution of the equation (3.7.1) with rational coefficients, then in view of Theorem 3.7.1 we have

$$\log|w(re^{i\theta})| = h_w(\theta)r^{\lambda} + O(r^{\lambda-\varepsilon}),$$

uniformly as $r \to \infty$, possibly outside a set $E(r) \subset [0,2\pi)$ of measure $\operatorname{mes}(E(r)) = O(r^{-\varepsilon})$, where $\lambda = \lambda(w)$. Since $w(z)$ has at most finitely many poles, we easily get

$$T(r,w) = m(r,w) + O(\log r) = r^{\lambda}\frac{1}{2\pi}\int_0^{2\pi} h_w^+(\theta)\mathrm{d}\theta + O(r^{\lambda-\varepsilon/2})$$

and

$$\begin{aligned} N\left(r,\frac{1}{w}\right) &= N(r,w) + \frac{1}{2\pi}\int_0^{2\pi} \log|w(re^{i\theta})|\mathrm{d}\theta - \log|c(0)| \\ &= r^{\lambda}\frac{1}{2\pi}\int_0^{2\pi} h_w(\theta)\mathrm{d}\theta + O(r^{\lambda-\varepsilon/2}), \end{aligned}$$

where $h_w^+(\theta) = h_w(\theta)$, when $h_w(\theta) \geqslant 0$, and otherwise, $h_w^+(\theta) = 0$ and $c(0)$ is the coefficient of the first term in Laurant series of $w(z)$ at 0. Thus

$$\delta(0,w) = \Delta(0,w) = 1 - \frac{\int_0^{2\pi} h_w(\theta)\mathrm{d}\theta}{\int_0^{2\pi} h_w^+(\theta)\mathrm{d}\theta},$$

$$\zeta = \frac{u^{\omega} - R^{\omega}}{u^{\omega} + R^{\omega}} = \phi^{-1}(u), \tag{3.5.2}$$

where R is a positive number, $\omega = \pi/(\beta - \alpha)$ and then

$$z = \mathrm{e}^{\mathrm{i}\xi}\phi(\zeta) = R\mathrm{e}^{\mathrm{i}\xi}\left(\frac{1+\zeta}{1-\zeta}\right)^{1/\omega}, \quad \xi = \frac{\alpha+\beta}{2},$$

conformally maps the unit disk $|\zeta| < 1$ onto Ω and sends 0 to $R\mathrm{e}^{\mathrm{i}\xi}$. Putting $z = r\mathrm{e}^{\mathrm{i}\theta} \in \Omega$ and $u = \mathrm{e}^{-\mathrm{i}\xi}z$, from (3.5.2) we have

$$1 - |\zeta|^2 = \frac{4R^{\omega}r^{\omega}\cos(\omega[\theta - \xi])}{R^{2\omega} + r^{2\omega} + 2R^{\omega}r^{\omega}\cos(\omega[\theta - \xi])} \tag{3.5.3}$$

and by noting $\frac{1}{2}(1 - |\zeta|^2) \leqslant 1 - |\zeta| \leqslant 1 - |\zeta|^2$ and $-\frac{\pi}{2} < \omega[\theta - \xi] < \frac{\pi}{2}$ we have

$$1 - |\zeta| \leqslant 4\left(\frac{R}{r}\right)^{\omega}$$

and

$$1 - |\zeta| \geqslant \frac{2R^{\omega}r^{\omega}}{(R^{\omega} + r^{\omega})^2}\cos(\omega[\theta - \xi]).$$

Therefore $\phi^{-1}(\mathrm{e}^{-\mathrm{i}\xi}z)$ maps the domain

$$\Omega_{\varepsilon}[M,R] = \{z : M^{-1}R \leqslant |z| \leqslant MR, \alpha + \varepsilon \leqslant \arg z \leqslant \beta - \varepsilon\}$$

into the disk $\{\zeta : |\zeta| < 1 - \eta\}$, $\eta = \frac{2M^{\omega}}{(M^{\omega}+1)^2}\sin(\omega\varepsilon)$ and $z = \mathrm{e}^{\mathrm{i}\xi}\phi(\zeta)$ maps the disk $\{\zeta : |\zeta| < 1 - \tau\}$ into

$$|z| \leqslant R\left|\frac{1+\zeta}{1-\zeta}\right|^{1/\omega} \leqslant 2^{1/\omega}R\tau^{-1/\omega}.$$

It is easy to deduce that

$$\phi'(\zeta) = \frac{2}{\omega}\phi(\zeta)\frac{1}{1-\zeta^2}$$

so that

$$|\phi'(\zeta)| \leqslant \frac{2r}{\omega}\frac{1}{1-|\zeta|} \quad \text{and} \quad \frac{1}{|\phi'(\zeta)|} = \frac{\omega}{2r}|1 - \zeta^2| \leqslant \frac{\omega}{r},$$

where $r = |\phi(\zeta)|$.

Lemma 3.5.1. *Let $f(z)$ be a transcendental meromorphic function. Consider an angular domain $\Omega(\alpha,\beta)$. Let M and R be two positive numbers. Then for a fixed positive number $\delta \leqslant \frac{\beta-\alpha-2\varepsilon}{10}$, there exists a positive constant K only depending on ω, δ and ε such that if*

$$K(M^{3\omega}\log T(dMR,f)+\log MR)+NM^{2\omega}\log M+1)<n(\Omega_\varepsilon[M,R],f'=0),\tag{3.5.4}$$

where $d=6\times 8^{1/\omega}(\sin(\omega\varepsilon))^{-1/\omega}$ *and* $N=n(dMR,\Omega,f=0)+n(dMR,\Omega,f=\infty)$, *then for* $z,z_0\in\Omega_\varepsilon[M,R]\setminus(\gamma)$, (γ) *is the set of disks the total sum of whose diameters is not greater than* $\delta M^{-1}R$, *we have*

$$\log^+|f(z)|\leqslant\log^+|f(z_0)|+\pi\tag{3.5.5}$$

and furthermore

$$\log^+|f^{(-p)}(z)|\leqslant\sum_{j=0}^{p}\log^+|f^{(-j)}(z_0)|+K_p\log(MR)\tag{3.5.6}$$

for some positive constant K_p *and each positive integer* p, *where* $f^{(-p)}$ *is the pth primitive of* $f(z)$ *(if exists), that is,* $(f^{(-p)})^{(p)}=f$.

Proof. Set

$$\tau=\frac{2(2M)^\omega}{((2M)^\omega+1)^2}\sin(\omega\varepsilon)\text{ and }\nu=\frac{2(3M)^\omega}{((3M)^\omega+1)^2}\sin(\omega\varepsilon).$$

A routine calculation confirms the existence of ρ between τ and ν such that each point z whose corresponding point ζ lies on the circle $|\zeta|=1-\rho$ by $z=\mathrm{e}^{\mathrm{i}\xi}\phi(\zeta)$ has the distance $\delta(z)\geqslant(MR\,T(dMR,f))^{-1}$ from zeros and poles of $f(z)$. Since

$$\begin{aligned}\rho^{-1/\omega}\leqslant\nu^{-1/\omega}&=\frac{((3M)^\omega+1)^{2/\omega}}{(2(3M)^\omega)^{1/\omega}}(\sin(\omega\varepsilon))^{-1/\omega}\\&\leqslant(2(3M)^\omega)^{1/\omega}(\sin(\omega\varepsilon))^{-1/\omega}\\&=3\times 2^{1/\omega}(\sin(\omega\varepsilon))^{-1/\omega}M,\end{aligned}$$

the disk $|\zeta|\leqslant 1-\rho$ is mapped by $z=\mathrm{e}^{\mathrm{i}\xi}\phi(\zeta)$ into the disk $\{z:|z|\leqslant\frac{d}{2}MR\}$ where $d=6\times 8^{1/\omega}(\sin(\omega\varepsilon))^{-1/\omega}$.

Define $F(\zeta)=f(\mathrm{e}^{\mathrm{i}\xi}\phi(\zeta))=f(z)$. We come to estimate $\log^+\left|\frac{F'(\zeta)}{F(\zeta)}\right|$ in $|\zeta|<1-\eta$. Employing Lemma 2.5.2 for $R=1-\rho$ and $r=1-\eta$ yields that in $|\zeta|<1-\eta$

$$\begin{aligned}\log\left|\frac{F'(\zeta)}{F(\zeta)}\right|\leqslant&\frac{2}{\eta-\tau}m\left(1-\rho,\frac{F'}{F}\right)\\&+(n(1-\rho,F=0)+n(1-\rho,F=\infty))\log\frac{2}{H}\\&-\frac{(1-\rho-t)^2}{2(1-\rho)^2}n(t,F'=0),\end{aligned}\tag{3.5.7}$$

where $1-\eta\leqslant t<1-\rho$ and $\zeta\notin(\gamma)_\zeta$, $(\gamma)_\zeta$ is the set of Boutroux-Cartan exceptional disks for zeros and poles of F in $|\zeta|<1-\rho$ and H.

In the following, let us estimate each term in the right side of (3.5.7). In view of Lemma 2.5.1 on $|\zeta| = 1-\rho$ we have

$$\begin{aligned}\log^+\left|\frac{F'(\zeta)}{F(\zeta)}\right| &\leqslant \log^+\left|\frac{f'(z)}{f(z)}\right| + \log^+|\phi'(\zeta)| \\ &\leqslant K_1(\log T(dMR,f)+\log MR),\end{aligned}$$

K_1 is a constant only depending on ε, so that

$$m\left(1-\rho,\frac{F'}{F}\right) \leqslant K_1(\log T(dMR,f)+\log MR).$$

In view of the formulae of η and τ, we have

$$\begin{aligned}\eta-\tau &= 2\sin(\omega\varepsilon)M^{\omega}\left(\frac{1}{(M^{\omega}+1)^2}-\frac{2^{\omega}}{((2M)^{\omega}+1)^2}\right)\\ &= 2(2^{\omega}-1)\sin(\omega\varepsilon)\frac{M^{\omega}(2^{\omega}M^{2\omega}-1)}{(M^{\omega}+1)^2((2M)^{\omega}+1)^2}\\ &\geqslant \frac{(2^{\omega}-1)^2\sin(\omega\varepsilon)}{2(2^{\omega}+1)^2}\frac{1}{M^{\omega}}.\end{aligned}$$

Noting that $\Omega_{\varepsilon}[M,R]$ is mapped into $\{\zeta:|\zeta|\leqslant 1-\eta\}$ and $\{\zeta:|\zeta|\leqslant 1-\rho\}$ into $\{z:|z|\leqslant \frac{d}{2}MR\}$, we have

$$n(1-\eta,F'=0)\geqslant n(\Omega_{\varepsilon}[M,R],f'=0)$$

and

$$n(1-\rho,F=0)+n(1-\rho,F=\infty)\leqslant N.$$

Thus in virtue of (3.5.7), putting $t=1-\eta$ and $H=cM^{-\omega-2}$, $c\leqslant \frac{2^{\omega}\omega\sin(\omega\varepsilon)}{de(2^{\omega}+1)^2}\delta$ such that $2eH\pi<1$, we obtain

$$\begin{aligned}\log\left|\frac{F'(\zeta)}{F(\zeta)}\right| &\leqslant K_2M^{\omega}(\log T(dMR,f)+\log MR)\\ &\quad +N((\omega+2)\log M+\log(2/c))\\ &\quad -K_3M^{-2\omega}n(\Omega_{\varepsilon}[M,R],f'=0),\end{aligned}$$

for $\zeta\notin(\gamma)_{\zeta}$ and $|\zeta|<1-\eta$, where we used $(1-\rho-t)^2=(\eta-\rho)^2\geqslant(\eta-\tau)^2$. Then this can assert the existence of K such that if (3.5.4) holds, then $\left|\frac{F'(\zeta)}{F(\zeta)}\right|\leqslant 1$ for such ζ. Hence for this K we have

$$|\log|F(\zeta)|-\log|F(\zeta_0)||\leqslant\int_{L(\zeta,\zeta_0)}\left|\frac{F'(\zeta)}{F(\zeta)}\right||d\zeta|\leqslant\pi,$$

where $\zeta,\zeta_0\notin(\gamma)_\zeta$ and $L(\zeta,\zeta_0)$ is the shortest curve in $\{\zeta:|\zeta|<1\}$ outside $(\gamma)_\zeta$ connecting ζ and ζ_0 with the length less than π. This implies that

$$\log^+|F(\zeta)|\leqslant\log^+\left|\frac{F(\zeta)}{F(\zeta_0)}\right|+\log^+|F(\zeta_0)|\leqslant\log^+|F(\zeta_0)|+\pi,$$

where $\zeta,\zeta_0\notin(\gamma)_\zeta$. Thus (3.5.5) follows. We denote by (γ) the set of disks which just cover the images of elements in $(\gamma)_\zeta$ via $z=e^{i\xi}\phi(\zeta)$ and then the total sum of diameters of disks in (γ) does not exceed

$$\begin{aligned}(\max_{|\zeta|<1-\rho}|\phi'(\zeta)|)2eH&=(\max_{|\zeta|=1-\rho}|\phi'(\zeta)|)2eH\\&\leqslant\frac{dMR}{\omega\rho}2eH\leqslant\frac{dMR}{\omega\tau}2eH\\&=\frac{de(2^\omega+1)^2}{2^\omega\omega\sin(\omega\varepsilon)}M^{\omega+2}H\frac{R}{M}\\&\leqslant\delta\frac{R}{M}.\end{aligned}$$

(3.5.6) follows from (3.5.5) and the equality (2.6.4). Thus Lemma 3.5.1 follows. □

To treat the derivative, we need the following, which can be immediately deduced in view of Lemma 2.7.1.

Lemma 3.5.2. *Let $f(z)$ be a function meromorphic in an angular domain Ω. If $\arg z=\theta$ in Ω is not a T direction of $f(z)$ for Ω, then for a fixed $a\in\mathbb{C}$, a fixed positive integer m and some small $\varepsilon>0$ we have*

$$N(r,Z_{2\varepsilon}(\theta),f=az^m+b)=o(T(2r,\Omega,f))+O((\log r)^3)$$

for all $b\in\widehat{\mathbb{C}}$ possibly outside a set of b with measure zero.

Now let us establish final result of this section dealing with few poles, which supplements the Milloux Theorem mentioned in end of this section in some sense.

Theorem 3.5.5. *Let $f(z)$ be a transcendental meromorphic function with order λ, $0<\lambda<\infty$. If $\delta(\infty,f)=1$, then $\bigcap_{j=0}^{\infty}TD(f^{(j)})\neq\varnothing$*

Proof. Since $f(z)$ has the finite positive order λ, in view of Theorem 2.6.3 we can take a sequence of relaxed Pólya peak $\{r_n\}$ of $f(z)$ of order λ which is also a sequence of relaxed Pólya peak $f^{(m)}(z)$ of order λ for each m, that is to say, for a sequence of positive numbers $\{\eta_n\}$ such that $\eta_n\to 0$ as $n\to\infty$, we have

$$T(t,f)\leqslant K\left(\frac{t}{r_n}\right)^\lambda T(r_n,f)$$

and for each positive integer m we have some positive constant $\hat{K}_m$ such that

$$T(t,f^{(m)}) \leqslant \hat{K}_m \left(\frac{t}{r_n}\right)^{\lambda} T(r_n,f^{(m)})$$

for $\eta_n r_n \leqslant t \leqslant r_n/\eta_n$. We can assume $K = \hat{K}_1 = 1$ and $r_n \geqslant 2^n$ in the below discussion.

Assume that $\arg z = \theta_0$ is a T direction of $f'(z)$ for $\{r_n\}$, but not of $f(z)$. We can take a complex number a and a $\varepsilon > 0$ such that

$$N(r_n, Z_\varepsilon(\theta_0), f' = a) > C_0 T(r_n, f')$$

and find three complex numbers $a_j (j = 1,2,3)$ such that

$$\sum_{j=1}^{3} N(r, Z_{3\varepsilon}(\theta_0), f = a_j) = o(T(r,f)).$$

Then in view of Lemma 3.5.2 and the assumption $\delta(\infty, f) = 1$, we have

$$N(r, Z_{2\varepsilon}(\theta_0), f = az+b) + N(r, Z_{2\varepsilon}(\theta_0), f = \infty) = o(T(2r,f))$$

for some $b \in \mathbb{C}$. Take a number $\tau > 1$ with $\tau^{-\lambda} < C_0$ and it is easy to see that

$$\begin{aligned} n(r_n, Z_\varepsilon(\theta_0), f' = a) &\geqslant (\log\tau)^{-1}(N(r_n, Z_\varepsilon(\theta_0), f' = a) \\ &\quad - N(r_n/\tau, Z_\varepsilon(\theta_0), f' = a)) \\ &\geqslant (\log\tau)^{-1}(C_0 T(r_n, f') - T(r_n/\tau, f')) + O(1) \\ &\geqslant (\log\tau)^{-1}(C_0 T(r_n, f') - \tau^{-\lambda} T(r_n, f')) + O(1) \\ &\geqslant C_1 T(r_n, f'). \end{aligned}$$

Set

$$N(r_n) = N(2dr_n, Z_{2\varepsilon}(\theta_0), f = az+b) + N(2dr_n, Z_{2\varepsilon}(\theta_0), f = \infty),$$

and therefore

$$N(r_n) = o(T(4dr_n, f)) = o(T(r_n, f)),$$

where d is defined in Lemma 3.5.1 with $\omega = \frac{\pi}{4\varepsilon}$. Choose $\delta_n \geqslant \eta_n$ with $\delta_n \to 0$ as $n \to \infty$ and such that letting $M_n = 1/\sqrt{\delta_n}$, we have

$$N(r_n) M_n^{2\omega} \log M_n + M_n^{3\omega} \log r_n = o(T(r_n, f)).$$

Set $R_n = 2\sqrt{\delta_n} r_n$ and so $M_n R_n = 2r_n$ and $M_n^{-1} R_n = 2\delta_n r_n$. Since $f(z)$ is of finite order, we have

$$M_n^{3\omega}(\log T(dM_nR_n, f) + \log M_nR_n) = o(T(r_n, f)).$$

Then

$$\begin{aligned}n(Z_\varepsilon(\theta_0)[M_n,R_n],f'=a) &= n(2r_n,Z_\varepsilon(\theta_0),f'=a)-n(2\delta_n r_n,Z_\varepsilon(\theta_0),f'=a)\\ &\geqslant C_1T(r_n,f')-\frac{1}{\log 2}N(4\delta_n r_n,Z_\varepsilon(\theta_0),f'=a)\\ &\geqslant C_1T(r_n,f')-\frac{1}{\log 2}(4\delta_n)^\lambda T(r_n,f')+O(1)\\ &\geqslant C_2T(r_n,f')+O(1)\\ &\geqslant C_2 2^{-\lambda}T(2r_n,f')+O(1)\\ &\geqslant C_3T(r_n,f),\end{aligned} \tag{3.5.8}$$

and therefore

$$\begin{aligned}N(r_n)M_n^{2\omega}\log M_n+M_n^{3\omega}(\log T(dM_nR_n,f)+\log M_nR_n) &= o(T(r_n,f))\\ &= o(n(Z_\varepsilon(\theta_0)[M_n,R_n],f'=a)).\end{aligned}$$

Using Lemma 3.5.1 for sufficiently large n we deduce that for $z,z_0\in Z_\varepsilon(\theta_0)[M_n,R_n]\setminus(\gamma)_n$, $(\gamma)_n$ is the set of disks the total sum of whose radius is not greater than $\frac{\varepsilon}{5}M_n^{-1}R_n$, with $|z_0|=2R_n/M_n$, we have

$$\begin{aligned}\log^+|f(z)| &\leqslant \log^+|f(z)-az-b|+\log^+|az+b|+\log 2\\ &\leqslant \log^+|f(z_0)-az_0-b|+\pi+\log^+|2ar_n|+\log^+|b|+2\log 2\\ &\leqslant \log^+|f(z_0)|+\pi+2\log^+|2ar_n|+2\log^+|b|+4\log 2\\ &\leqslant C_4T(4R_n/M_n,f)+\pi+2\log^+|2ar_n|+2\log^+|b|+4\log 2\\ &\leqslant C_5((8\delta_n)^\lambda T(r_n,f)+\log r_n),\end{aligned} \tag{3.5.9}$$

where we used Lemma 2.1.3 to estimate $\log^+|f(z_0)|$.

We can use a finite number of disks Γ_j $(j=1,2,\cdots,q_n)$ with center at z_j contained in $Z_{2\varepsilon}(\theta_0)\setminus(\gamma)_n$ to cover the domain $U_n=\{4\delta_n r_n\leqslant|z|\leqslant r_n\}\cap Z_\varepsilon(\theta_0)$ and $q_n=O(\log(\delta_n^{-1}))$. By means of Lemma 2.1.6, Lemma 2.5.1 and (3.5.9) we have

$$\begin{aligned}&n(\Gamma_j,f'=\alpha)\leqslant\frac{1}{\log 2}N(2\Gamma_j,f'=\alpha)\\ &\leqslant\frac{1}{\log 2}\left(T(2\Gamma_j,f')+\log^+|\alpha|+\log 2+\log\frac{1}{|f'(z_j)-\alpha|}\right)\\ &\leqslant\frac{1}{\log 2}\left(2T(2\Gamma_j,f)+m\left(2\Gamma_j,\frac{f'}{f}\right)+\log\frac{1}{|f'(z_j),\alpha|}\right)\\ &\leqslant K_2(\sum_{i=1}^3 n(4\Gamma_j,f=a_i)+\log^+|f(z_j)|+\log(r_nT(2r_n,f))+\log\frac{1}{|f'(z_j),\alpha|})\\ &\leqslant K_2\sum_{i=1}^3 n(4\Gamma_j,f=a_i)+K_3(\delta_n^\lambda T(r_n,f)+\log r_n)+K_2\log\frac{1}{|f'(z_j),\alpha|}.\end{aligned}$$

Set $E_{jn}=\{\alpha:|f'(z_j),\alpha|\leqslant e^{-n}\}$. Then for $\alpha\notin\cup_{n=N}^\infty\cup_{j=1}^{q_n}E_{jn}$ we have for $n\geqslant N$

$$\begin{aligned} n(U_n, f' = \alpha) &\leqslant K_4 \sum_{i=1}^{3} N(4r_n, Z_{3\varepsilon}(\theta_0), f = a_i) \\ &\quad + K_3 q_n(\delta_n^{\lambda} T(r_n, f) + \log r_n) + K_2 q_n n \\ &= o(T(r_n, f)). \end{aligned}$$

On the other hand, using the argument which derived (3.5.8) we can attain that

$$n(U_n, f' = \alpha) \geqslant C(\alpha) T(r_n, f)$$

for all but at most two values of α, for $\arg z = \theta_0$ is a T direction of f' for $\{r_n\}$, where $C(\alpha)$ is a positive constant depending on α.

A contradiction is derived and hence $\arg z = \theta_0$ is also a T direction of $f(z)$. Using the same argument as in above to $f^{(m)}(z)$ and $f^{(j)}(z)$ for $0 < j < m$ with suitable modification implies that for each positive integer m, every T direction of $f^{(m)}(z)$ for $\{r_n\}$ must be a common T direction of $f^{(j)}(z)$ $(j = 0, 1, \cdots, m)$. Thus Theorem 3.5.5 follows. □

Finally, let us recall some of main results about common Borel directions of a meromorphic function and its derivatives. In 1951, H. Milloux [20] proved that for a transcendental entire function $f(z)$ of finite positive order, every Borel direction of its derivative $f'(z)$ must be a Borel direction of itself $f(z)$. We do not know if the Milloux Theorem would hold for T directions, that is to say, whether every T direction of derivative of a meromorphic function with $\delta(\infty, f) = 1$ must be a T direction of this function itself. However, Theorem 3.5.5 only asserts that they have at least one common T direction. Zhang [54] in 1977 and Yang [45] in 1979 extended the Milloux Theorem to the case of meromorphic function with the infinity as a Borel exceptional value and predigested the complicated proof of the Milloux Theorem H. Milloux gave. Chuang [7] in 1951 proved that under some additional assumption imposed on function considered besides it being of finite positive order, a Borel direction of a transcendental meromorphic function must be a Borel direction of its derivative. Chuang's theorem was improved by Zhang [54] who proved Theorem 3.5.4 with Borel directions in the place of T directions and with $\delta(a, f) = 1$ replaced by that a is a Borel exceptional value of $f(z)$.

3.6 Distribution of the Julia, Borel Directions and T Directions

There exists a meromorphic function of infinite order or finite order such that all of $JD(f)$, $BD(f)$ and $TD(f)$ are finite. In 1976, Drasin and Weitsman [9] gave sufficient and necessary conditions for a non-empty closed subset E of $[0, 2\pi)$ such that E is exactly the set of arguments of all Borel directions of an entire function with given order. Yang and Zhang [48] proved that any non-empty closed subset of $[0, 2\pi)$ (without any more requirements) is exactly the set of arguments of all Borel directions of a meromorphic function with given order. In view of the methods of

Yang and Zhang [48] and Zhang Q. D. [56] (the reader is referred to the proof of Theorem 3.3.2), we can prove the following

Theorem 3.6.1. *For $\lambda > 0$ and a closed subset E of $[0,2\pi)$, then there exists a meromorphic function with order λ such that*

$$BD(f) = TD(f) = E.$$

It is, however, obvious that a non-empty closed subset of $[0,2\pi)$ may not be the set of arguments of all Borel (T) directions of any meromorphic function with given order and given $\delta = \delta(\infty, f) > 0$. Given $\lambda > 1/2$ and $0 < \delta \leqslant 1$, let E be a non-empty closed subset of $[0,\beta]$ with

$$0 < \beta < \min\left\{2\pi - \frac{\pi}{\lambda}, \frac{4}{\lambda}\arcsin\sqrt{\frac{\delta}{2}}\right\}.$$

Then for any meromorphic function $f(z)$ with order λ and $\delta(\infty, f) = \delta$, we have

$$BD(f) \neq E \text{ and } TD(f) \neq E.$$

This is because

$$2\pi - \beta > \max\left\{\frac{\pi}{\lambda}, 2\pi - \frac{4}{\lambda}\arcsin\sqrt{\frac{\delta}{2}}\right\}$$

and in view of Theorem 3.1.6 there exists at least one Borel direction and T direction with argument in $(\beta, 2\pi)$.

Then what conditions are sufficient and necessary for this case such that $BD(f) = E$ or/and $TD(f) = E$?

We are given m radials $\arg z = \theta_j$ $(1 \leqslant j \leqslant m)$ from the origin with

$$\theta_1 < \theta_2 < \cdots < \theta_m < \theta_1 + 2\pi.$$

If $\{\theta_j\}_{j=1}^m$ is the set of arguments of all Borel (T-) directions of a meromorphic function $f(z)$ with order $\lambda > 1/2$ and $\delta = \delta(\infty, f)$, Theorem 3.1.6 then shows that

(1) $\theta_{j+1} - \theta_j \leqslant \Theta$, $j = 1, 2, \cdots, m-1$;

(2) $\theta_m - \theta_1 \geqslant 2\pi - \Theta$,

where $\Theta = \Theta(\delta, \lambda) = \max\left\{\frac{\pi}{\lambda}, 2\pi - \frac{4}{\lambda}\arcsin\sqrt{\frac{\delta}{2}}\right\}$.

It is natural to ask whether or not (1) and (2) are sufficient to the existence of a meromorphic function with order $\lambda > 1/2$ and $\delta = \delta(\infty, f)$ such that $BD(f) = \{\theta_j\}_{j=1}^m$ or/and $TD(f) = \{\theta_j\}_{j=1}^m$.

Drasin and Weitsman [9] proved that (1) and (2) with $\frac{\pi}{\lambda}$ in the place of Θ and $2\pi - \Theta$ are sufficient to the existence of such an entire function with order $\lambda > 1/2$. It is easily seen that (1) and (2) given by Drasin and Weitsman imply (1) and (2) stated above by noting $\frac{\pi}{\lambda} \leqslant \Theta$ and $2\pi - \Theta \leqslant \frac{\pi}{\lambda}$ when $\delta(\infty, f) = 1$.

It is interesting to discuss the linear measure of these three sets $JD(f)$, $BD(f)$ and $TD(f)$. That $\text{mes}BD(f) > 0$ and $\text{mes}TD(f) > 0$ follows from Theorem 4.1.4 below provided that $f(z)$ is an entire function of finite positive lower order and has an infinite number of Nevanlinna deficient values. Here we can establish the following

Theorem 3.6.2. *Let $f(z)$ be a transcendental meromorphic function. If there are an unbounded sequence $\{r_n\}$ of positive numbers such that*

$$\log|f(z)| \geqslant T^d(r_n,f), |z| = r_n, d > 0$$

and

$$\lim_{n\to\infty} \frac{\log T(r_n,f)}{\log r_n} = \infty,$$

then

$$BD(f) = JD(f) = [0,2\pi),$$

that is, each ray from the origin is a Borel direction of $f(z)$ with infinite order .

Proof. We are given an arbitrary radial $\arg z = \theta$. According to the definition of $B_{\alpha,\beta}(r,f)$, for $\varepsilon > 0$ we have

$$\begin{aligned} r_n^{\omega} S_{\theta-\varepsilon,\theta+\varepsilon}(r_n,f) &\geqslant r_n^{\omega} B_{\theta-\varepsilon,\theta+\varepsilon}(r_n,f) \\ &\geqslant \frac{2\omega}{\pi} T^d(r_n,f) \int_{\theta-\varepsilon}^{\theta+\varepsilon} \sin(\omega(\phi-\theta+\varepsilon))\mathrm{d}\phi \\ &= \frac{4}{\pi} T^d(r_n,f), \end{aligned}$$

where $\omega = \frac{\pi}{2\varepsilon}$. On the other hand, we use Lemma 2.2.1, Theorem 2.4.7 and Theorem 2.4.4 in turn to obtain that for three distinct complex numbers a_v $(v = 1,2,3)$

$$\begin{aligned} r^{\omega} S_{\theta-\varepsilon,\theta+\varepsilon}(r,f) &= r^{\omega} \dot{S}_{\theta-\varepsilon,\theta+\varepsilon}(r,f) + O(r^{\omega}) \\ &\leqslant 2\omega \mathscr{T}(r,Z_\varepsilon(\theta)) + \omega^2 r^{\omega} \int_1^r \frac{\mathscr{T}(t,Z_\varepsilon(\theta))}{t^{\omega+1}}\mathrm{d}t + O(r^{\omega}) \\ &\leqslant 2\omega \mathscr{T}(r,Z_\varepsilon(\theta)) + \omega r^{\omega} \mathscr{T}(r,Z_\varepsilon(\theta)) + O(r^{\omega}) \\ &\leqslant 6\omega N(2r) + 3\omega r^{\omega} N(2r) + O(r^{\omega}(\log r)^2) \\ &\leqslant (6\omega^2 + 3\omega^2 r^{\omega}) N(2r) + O(r^{\omega}(\log r)^2), \end{aligned}$$

where $N(t) = \sum_{v=1}^{3} N(t, Z_{2\varepsilon}(\theta), f = a_v)$. This yields that

$$\lim_{n\to\infty} \frac{\log N(2r_n)}{\log T(r_n,f)} \geqslant d \tag{3.6.1}$$

and therefore $\arg z = \theta$ is a Borel direction of $f(z)$ with infinite order, from which Theorem 3.6.2 follows. □

Actually, if $r_n \notin E(f)$, then using the Nevanlinna second fundamental theorem on an angle we can get (3.6.1) with $N(r_n)$ in the place of $N(2r_n)$. However, we do not know if the above result is still true for $TD(f)$.

Finally, we conclude this section with the following result, which is a direct consequence of Lemma 2.7.1.

Theorem 3.6.3. *Let $f(z)$ be a transcendental meromorphic function. If $f(z)$ has no T directions in an angular domain $\Omega(\alpha,\beta)$, then for arbitrarily small $\varepsilon>0$, we have*

$$N(r,\Omega_\varepsilon,f=a)=o(T(2r,f))+O((\log r)^2\log\log r) \tag{3.6.2}$$

for all $a\in\mathbb{C}$ possibly outside a set of a with measure zero.

Proof. Since $f(z)$ has no T directions in $\Omega(\alpha,\beta)$, there exist m directions $\arg z=\theta_j(1\leqslant j\leqslant m)$ such that $\Omega_\varepsilon\subset\cup_{j=1}^m V_j(\varepsilon)$, $V_j(\varepsilon)=\{z:|\arg z-\theta_j|<\varepsilon\}$ and for three distinct values a_j,b_j and c_j $(1\leqslant j\leqslant m)$,

$$N(r,V_j(2\varepsilon),f=a_j)+N(r,V_j(2\varepsilon),f=b_j)+N(r,V_j(2\varepsilon),f=c_j)=o(T(r,f)).$$

In view of Lemma 2.7.1, we get

$$N(r,V_j(\varepsilon),f=a)=o(T(2r,f))+O((\log r)^2\log\log r)$$

for all $a\in\mathbb{C}$ possibly outside a set of a with measure zero. Since

$$N(r,\Omega_\varepsilon,f=a)\leqslant\sum_{j=1}^m N(r,V_j(\varepsilon),f=a),$$

(3.6.2) therefore follows. □

3.7 Singular Directions of Meromorphic Solutions of Some Equations

In this section we mainly consider a linear differential equation

$$w^{(n)}+a_{n-1}(z)w^{(n-1)}+\cdots+a_0(z)w=0 \tag{3.7.1}$$

with meromorphic functions $a_j(z)$ $(j=0,1,\cdots,n-1)$. When every $a_j(z)$ is an entire function, each solution of (3.7.1) is an entire function. In this section, we discuss the singular directions of meromorphic solutions of (3.7.1).

Let us begin with the case when the coefficient functions $a_j(z)$ $(j=0,1,\cdots,n-1)$ are rational. It is well-known that the results from the theory of asymptotic integration are important tools in discussion of such an equation. For this reason, we collect some basic concepts and results from the theory of asymptotic integration which will be often used below.

Consider n linearly independent formal functions

$$w_j(z)=\mathrm{e}^{P_j(z)}z^{\lambda_j}[\log z^{1/p}]^{m_j}(1+o(1)),\ 1\leqslant j\leqslant n, \tag{3.7.2}$$

where $P_j(z)$ is a polynomial in $z^{1/p}$ for some $p \in \mathbb{N}\backslash\{0\}$, $\lambda_j \in \mathbb{C}$, $m_j \in \mathbb{N}$.

For this function system $\{w_1, w_2, \cdots, w_n\}$, let us introduce the concept of its Stokes ray.

Definition 3.7.1. *Let a function system $\{w_1, w_2, \cdots, w_n\}$ with the form (3.7.2) be given. A ray $\arg z = \theta \in \mathbb{R}$ is called a Stokes ray of this system, provided that for some $\delta > 0$, there exist two different $P_i(z)$ and $P_j(z)$ such that*

$$\lim_{r\to\infty} \frac{\mathrm{Re}(P_j(re^{i\varphi}) - P_i(re^{i\varphi}))}{r^{\lambda}} \begin{cases} > 0, \ \varphi \in (\theta, \theta+\delta), \\ < 0, \ \varphi \in (\theta-\delta, \theta), \end{cases}$$

where λ is a positive rational number called order of the Stokes ray.

Indeed, λ is the degree of $P_j(z) - P_i(z)$ in $z^{1/p}$ divided by p. Obviously, there exist only a finite number of Stokes rays for a system.

Let $w(z)$ be a meromorphic function such that for arbitrary ray $\arg z = \theta$, it is a linear combination of the system $\{w_1, w_2, \cdots, w_n\}$ in a sector containing this ray, that is,

$$w(z) = c_1 w_{i_1}(z) + c_2 w_{i_2}(z) + \cdots + c_m w_{i_m}(z), \tag{3.7.3}$$

in $|\arg z - \theta| < h$ where $c_j \neq 0, 1 \leqslant i_1 < \cdots < i_m \leqslant n$.

Below we introduce concept of Stokes rays for such a function, which is very useful in studying singular directions of such a function.

Definition 3.7.2. *Let $w(z)$ be given as above. A ray $\arg z = \theta \in \mathbb{R}$ is called a Stokes ray of this function, provided that it is expressed in (3.7.3) and for some δ, $0 < \delta < h$, there exist $P_{i_k}(z)$ and $P_{i_v}(z)$ with $P_{i_k}(z) \not\equiv P_{i_v}(z)$ such that for $\theta < \varphi < \theta + \delta$ and every $P_{i_s}(z) \not\equiv P_{i_k}(z)$, we have*

$$\mathrm{Re}(P_{i_k}(re^{i\varphi}) - P_{i_s}(re^{i\varphi})) \to +\infty,$$

as $r \to \infty$ and $P_{i_v}(z)$ has the same property for $\theta - \delta < \varphi < \theta$. Furthermore, if $P_{i_k}(z) - P_{i_v}(z) \sim az^{\lambda}$ as $|z| \to \infty$, then the ray $\arg z = \theta$ is called a Stokes ray of order λ.

Obviously, a Stokes ray of the function $w(z)$ is also one of the corresponding system $\{w_{i_1}(z), \cdots, w_{i_m}(z)\}$, but the converse may not be correct. We understand by observing a simple example: $w(z) = e^{z^2} + e^{iz} + 1$. The function $w(z)$ has all Stokes rays at $\arg z = \frac{\pi}{4}, \frac{3\pi}{4}, \frac{5\pi}{4}$ and $\frac{7\pi}{4}$. However, the positive and negative real axes are also Stokes rays of the system $\{e^{z^2}, e^{iz}, 1\}$. Fortunately, every Stokes ray of order $\lambda(w)$ of the system $\{w_{i_1}(z), \cdots, w_{i_m}(z)\}$ must be one of $w(z)$ of order $\lambda(w)$.

In view of basic property of value distribution of exponential polynomial and the Rochousé Theorem, we can prove (compare Lemma 1 of [1]) that a ray $\arg z = \theta$ is a Stokes ray of $w(z)$ of order λ if and only if for arbitrary small sector S containing the ray

$$n(r, S, w = 0) = cr^{\lambda}(1 + o(1)), \tag{3.7.4}$$

where c is a positive constant depending on S. This can be shown in view of below Theorem 3.7.1. For instance, consider the function $w(z) = e^{z^2} + e^z + 1$. The ray $\arg z = \frac{\pi}{2}$ is a Stokes ray of $w(z)$ with order 1 and it is easy to see that (3.7.4) holds with $\lambda = 1$.

According to the theory of asymptotic integration (compare Sternberg [30], Wasow [39], Dietrich [8] and Brüggemann[1]), the following is true. The equation (3.7.1) with rational functions $a_j(z)$ $(j = 0, 1, \cdots, n-1)$ has a formal fundamental system (FS for short) $\{w_1, w_2, \cdots, w_n\}$ of solutions with the form

$$w_j(z) = e^{P_j(z)} z^{\lambda_j} [\log z^{1/p}]^{m_j} Q_j(z, \log z),\ 1 \leqslant j \leqslant n, \tag{3.7.5}$$

where $P_j(z)$, λ_j, p and m_j are as in (3.7.2) and $Q_j(z, \log z)$ is a polynomial in $(\log z)^{-1}$ over the field of formal series $\sum_{s=0}^{\infty} c_s z^{-s/p}$, and $Q_j(z, \log z) = 1 + O(1/\log z)$, as $|z| \to \infty$.

The function system $\{w_1, w_2, \cdots, w_n\}$ in (3.7.5) is a system in (3.7.2) and therefore, we can consider its Stokes rays. The system has only finitely many Stokes rays $\arg z = \theta_j$ $(1 \leqslant j \leqslant m)$ with

$$0 \leqslant \theta_1 < \theta_2 < \cdots < \theta_m < 2\pi.$$

Important is that every meromorphic solution of (3.7.1) must be a linear combination of the system $\{w_1, w_2, \cdots, w_n\}$ in (3.7.5) in each sector

$$S_j = \{z \in \mathbb{C} : \theta_{j-1} < \arg z < \theta_{j+1}\}, \quad j = 1, 2, \cdots, m,$$

where $\theta_0 = \theta_m - 2\pi$ and $\theta_{m+1} = \theta_1 + 2\pi$.

In view of the proof of Theorem 1 of Steinmetz [27], we can establish the following result, which is an improvement of his Theorem 1.

Theorem 3.7.1. *Let $w(z)$ be a transcendental meromorphic solution of (3.7.1) with rational coefficients and have the Stokes rays $\arg z = \theta_j$ $(1 \leqslant j \leqslant m)$ as in above. Then for each j, there exists a $P(z) = P_{i_\nu}(z)$ such that*

$$\log|w(z)| = \mathrm{Re}P(z) + O(\log|z|)$$

as $z \to \infty$ in $\theta_j \leqslant \arg z \leqslant \theta_{j+1}$, possibly outside an exceptional set consisting of

(1) countably many disks $\{z : |z - z_n| < |z_n|^{1-\tau}\}$, where τ is positive and the counting function of the sequence $\{z_n\}$ is $O(\log r)$, and

(2) two logarithmic semi-strips

$$0 \leqslant \arg z - \theta_j < C\frac{\log^+|z|}{|z|^{1/p}} \text{ and } 0 \leqslant \theta_{j+1} - \arg z < C\frac{\log^+|z|}{|z|^{1/p}}.$$

Theorem 3.7.1 applies to the case when $w(z)$ is a linear combination of the system (3.7.5) in a sector but may not be a meromorphic solution of an equation (3.7.1). From Theorem 3.7.1, one immediately implies that if $\arg z = \theta$ is not a Stokes ray

and this implies that $\delta(0,w) = \Delta(0,w) = 1$ is equivalent to that 0 is a Borel exceptional value (BEV for brevity) of $w(z)$ (cf. Corollary 2 of [27]).

From the above discussion and Theorem 3.7.2, we immediately obtain the following

Theorem 3.7.3. *Let $w(z)$ be given as in Theorem 3.7.1. Then*

(1) $w(z)$ has only finitely many Borel and T directions;

(2) its Borel and T directions coincide and are Stokes rays of order $\lambda(w)$ of $w(z) - a$ for some $a \in \mathbb{C}$ and vice versa;

(3) a ray $\arg z = \theta$ is a Borel direction of $w(z)$ if and only if for some $a \in \mathbb{C}$, the a-points of $w(z)$ has the exponent $\lambda(w)$ of convergence in the logarithmic strip (3.7.6) for $\arg z = \theta$.

Let us consider a second order linear differential equation

$$w'' + A(z)w = 0, \tag{3.7.7}$$

where $A(z)$ is a rational function with $A(z) \sim cz^n, n \geqslant 0$ as $z \to \infty$. We calculate the leading terms of $P_j(z)$ in the formal solutions (3.7.5) for (3.7.7). To the end, we note that the corresponding algebraic equation

$$y^2 + A(z) = 0$$

has two solutions with the form, near $z = \infty$, $y_j = \pm\sqrt{c}z^{n/2}(1+o(1)), j = 1,2$ and then

$$P_j(z) = \pm\frac{2\sqrt{c}}{n+2}z^{(n+2)/2} + \cdots \tag{3.7.8}$$

The following result is obvious.

Theorem 3.7.4. *Let $w(z)$ be a meromorphic solution of (3.7.7). Then $T(r,w) \sim cr^{(n+2)/2}$ with $c > 0$ and $w(z)$ has exactly $n+2$ Stokes rays of order $\frac{n+2}{2}$ for some $a \in \widehat{\mathbb{C}}$ and equivalently, $w(z)$ has exactly $n+2$ Borel and T directions. If w_1 and w_2 are linearly independent meromorphic solutions of (3.7.7), setting $f = w_1/w_2$, then $T(r,f) \sim br^{(n+2)/2}$ with some $b > 0$ and $f(z)$ has exactly $n+2$ Borel and T directions.*

The final part of Theorem 3.7.4 follows from the equality $N(r, f = a) = N(r, w_1 - aw_2 = 0)$ and that $w_1 - aw_2$ has the same Stokes rays for all but at most two values of a. For a general case, we pose the following

Question 3.7.1. *Does a meromorphic solution $w(z)$ of the equation (3.7.1) with rational coefficients have at most $n\lambda(w)$ Borel and T directions?*

There is a close relation between the equation (3.7.7) and the Schwarzian derivative. The Schwarzian derivative of a meromorphic function $f(z)$ is

$$S_f = \frac{f'''}{f'} - \frac{3}{2}\left(\frac{f''}{f'}\right)^2.$$

We can consider the equality as an algebraic differential equation. The reader is referred to Chapter 6 of Laine [17] for the basic knowledge. For a given meromorphic function S_f, meromorphic solution, if exists, of the algebraic differential equation is unique up to a Möbius transformation. If S_f is a polynomial, then f must be a ratio of two linearly independent solutions of (3.7.7) with $A(z) = \frac{1}{2}S_f(z)$. Therefore, in terms of Theorem 3.7.4, a meromorphic function $f(z)$ with the polynomial Schwarzian derivative has exactly $2\lambda(f)$ Borel and T directions. Furthermore, in terms of Theorem 3.7.1, the complex plane $\mathbb{C}$ is divided by the $2\lambda(f)$ Stokes rays $\arg z = \phi_j$ into $2\lambda(f)$ equal angular domains:

$$D_j = \{z : \phi_{j-1} < \arg z < \phi_j\},\ 1 \leqslant j \leqslant 2\lambda(f),\ \phi_{2\lambda} = \phi_0,$$

$\phi_j - \phi_{j-1} = \pi/\lambda$ and for some $a_j \in \widehat{\mathbb{C}}$ we have

$$\log \frac{1}{|f(re^{i\theta}) - a_j|} = \pi C r^\lambda \sin \lambda(\theta - \phi_{j-1}) + o(r^\lambda),\ r \to +\infty, \tag{3.7.9}$$

uniformly with respect to θ in any angle being inside D_j (where $1/(f-a_j)$ is replaced with f for $a_j = \infty$) and $T(r,f) \sim Cr^\lambda$, where c is a positive number which can be found from the coefficient of the first term of $\frac{1}{2}S_f(z)$ in terms of (3.7.8). Actually, (3.7.9) follows from the following implication. Let $w_j, j = 1,2$, be the formal solutions (3.7.5) of (3.7.7). Thus

$$f = \frac{\alpha_1 w_1 + \alpha_2 w_2}{\beta_1 w_1 + \beta_2 w_2},\ \alpha_i, \beta_i \in \mathbb{C}$$

in D_j. Assume that $\mathrm{Re}P_1(z) > 0$ in D_j. If $\beta_1 \neq 0$, we take a_j such that $\alpha_1 - a_j\beta_1 = 0$ and thus

$$\frac{1}{f - a_j} = \frac{\beta_1}{\alpha_2 - a_j\beta_2}\frac{w_1}{w_2} + \frac{\beta_2}{\alpha_2 - a_j\beta_2}.$$

From the representation (3.7.5) of w_1 and w_2, and (3.7.8), we have (3.7.9). If $\beta_1 = 0$, then we have for a_j chosen to be ∞

$$f = \frac{\alpha_1}{\beta_2}\frac{w_1}{w_2} + \frac{\alpha_2}{\beta_2}.$$

This yields (3.7.9) with $a_j = \infty$. Finally, from the representation of f we always have

$$T(r,f) \sim m\left(r, \frac{w_1}{w_2}\right) \sim m\left(r, \frac{w_2}{w_1}\right) \sim cr^\lambda.$$

From (3.7.9), we have

$$\begin{aligned} m\left(r,\frac{1}{f-a}\right) &= \frac{1}{2\pi}\int_0^{2\pi}\log^+\frac{1}{|f(re^{i\theta})-a|}d\theta \\ &= \frac{1}{2}cr^\lambda \sum_{j:\ a_j=a}\int_{\phi_{j-1}}^{\phi_j}\sin\lambda(\theta-\phi_{j-1})d\theta+o(r^\lambda) \\ &= cr^\lambda\frac{p(a)}{\lambda}+o(r^\lambda) \end{aligned}$$

so that $\delta(a,f)=p(a)/\lambda$ where $p(a)$ is the number of a_j's equal to a and $\sum_a \delta(a,f)=2$. These results belong to F. Nevanlinna [21] and the reader is referred to Chapter 7 of this book for the discussion of general case, i.e., the Nevanlinna conjecture.

Theorem 3.7.5. *Let $f(z)$ be a transcendental meromorphic function with polynomial Schwarzian derivative. Then $T(r,f)\sim cr^\lambda$ for some $c>0$ with 2λ being a natural number, f has exactly 2λ Borel and T directions $\arg z=\phi_j(1\leqslant j\leqslant 2\lambda)$ with $\phi_{i+1}-\phi_i=\frac{\pi}{\lambda}$, $\phi_{2\lambda}=\phi_1+2\pi$ and asymptotic representation (3.7.9) in D_j and $\sum_a \delta(a,f)=2$.*

Below we shall mean Stokes rays of $w(z)-a$ by Stokes rays of $w(z)$ with respect to a in order to stress what value-points are considered. Note that if the linear representation of $w(z)$ in terms of the formal fundamental system contains at least two exponential polynomials with distinct first terms, then $w(z)$ has the same Stokes rays of order $\lambda(w)$ as those of $w(z)-a$ with the exception of at most one of a and therefore its Borel and T directions are exactly and completely determined by the argument distribution of almost its zeros.

Since the product $F_m=f_1f_2\cdots f_m$ of m meromorphic solutions of (3.7.1) with rational coefficients can be linearly represented in terms of items of the set produced by the products of elements of the formal fundamental system. Then the above discussion about one solution $w(z)$ including Theorem 3.7.1 and Theorem 3.7.2 applies to the $F_m(z)$ and so Theorem 3.7.3 is also true for the $F_m(z)$.

For a FS $\{f_1,f_2,\cdots,f_n\}$ of an equation (3.7.1), it is well known (see Proposition 1.4.8 of Laine [17]) that

$$W'+a_{n-1}W=0,$$

where $W=W(f_1,f_2,\cdots,f_n)$ is the Wronskian determinant of $f_1,f_2,\cdots,f_n$, and thus for $a_{n-1}(z)\equiv 0$, we have $W\equiv c\neq 0$. This yields

$$m\left(r,\frac{1}{E}\right)=m\left(r,\frac{W}{E}\right)+m\left(r,\frac{1}{W}\right)=O(\log r), \tag{3.7.10}$$

$E=f_1f_2\cdots f_n$, and then if $E(z)$ is transcendental, $\delta(0,E)=\Delta(0,E)=0$ and $E(z)$ has the exponent $\lambda(E)$ of convergence of zeros. Thus we can establish the following result.

Theorem 3.7.6. *Let $E=f_1f_2\cdots f_n$ be given as in above with $a_{n-1}(z)\equiv 0$. Assume that $E(z)$ is transcendental. Then*

(1) $E(z)$ has only finitely many and at least one Borel and T directions;

(2) *a ray is a Borel and T direction of E if and only if it is a Stokes ray of $E(z)$ with order $\lambda(E)$.*

Proof. In view of Theorem 3.7.3, we need only to prove (2) here. Let $\arg z = \theta_j (1 \leqslant j \leqslant m)$ be all Stokes rays of $E(z)$ with order $\lambda(E)$. It is obvious that every $\arg z = \theta_j$ is a Borel and T direction of E. In view of Theorem 3.7.2 and (3.7.4), there exists a $\varepsilon_0 > 0$ such that for arbitrary $0 < \varepsilon < \varepsilon_0$, we have $V_j(\varepsilon) = \{z : |\arg z - \theta_j| < \varepsilon\}$ are disjoint and

$$N(r, \mathbb{C} \setminus \cup_{j=1}^{m} V_j(\varepsilon), E = 0) = O(r^{\lambda - \eta})$$

for some $0 < \eta < \lambda$ and

$$N(r, V_j(\varepsilon), E = 0) \sim N(r, V_j(\varepsilon), E = a)$$

as $r \to \infty$ for all but at most one value of a. From (3.7.10) it follows that

$$\begin{aligned} T(r,E) &= N\left(r, \frac{1}{E}\right) + O(\log r) \\ &= \sum_{j=1}^{m} N(r, V_j(\varepsilon), E = 0) + O(r^{\lambda - \eta}) \\ &\sim \sum_{j=1}^{m} N(r, V_j(\varepsilon), E = a) + O(r^{\lambda - \eta}) \\ &\leqslant N\left(r, \frac{1}{E - a}\right) + O(r^{\lambda - \eta}) \\ &\leqslant T(r,E) + O(r^{\lambda - \eta}), \end{aligned}$$

so that

$$N(r, \mathbb{C} \setminus \cup_{j=1}^{m} V_j(\varepsilon), E = a) = O(r^{\lambda - \eta}).$$

This implies that any ray $\arg z = \theta$ with $\theta \neq \theta_j (1 \leqslant j \leqslant m)$ cannot be a Borel direction of E, from which Theorem 3.7.6 follows. □

Therefore, by using the same argument as in the proof of Theorem 3.7.6, we can prove that if $\delta(0, w) = \Delta(0, w) = 0$ for a meromorphic solution w of (3.7.1), then a ray is a Borel direction of $w(z)$ if and only if it is a Stokes ray of order $\lambda(w)$ with respect to zeros, and equivalently (3.7.4) holds for any sector containing this ray. This result also holds for $F_m(z)$.

Consider the equation of (3.7.1) with $a_{n-1} \equiv 0$ and polynomial coefficients $a_j(z) (1 \leqslant j \leqslant n-2)$. For a FS of meromorphic solutions $\{f_1, f_2, \cdots, f_n\}$, in view of Theorem 2 of [28], $E = f_1 f_2 \cdots f_n$ is transcendental if and only if

$$T(r,E) \neq o(\max_{1 \leqslant j \leqslant n} \{T(r, f_j)\}),$$

that is to say, in view of (3.7.10),

$$N\left(r,\frac{1}{E}\right)\neq o(\max_{1\leqslant j\leqslant n}\{T(r,f_j)\}).$$

By using Theorem 3.7.6 and (3.7.4) it is equivalent to that

$$T(r,E)=c(1+o(1))r^{\lambda},\ \lambda=\max_{1\leqslant j\leqslant n}\{\lambda(f_j)\}$$

for some $c>0$. And $E=f_1f_2\cdots f_n$ is transcendental if and only if $E(z)$ has the exponent λ of convergence of zeros. Theorem 2 of [28] asserts that if at least one of $a_j(z)(1\leqslant j\leqslant n-1)$ is not a constant, then $E=f_1f_2\cdots f_n$ is transcendental for any FS $\{f_1,f_2,\cdots,f_n\}$.

We remark that actually, Theorem 3.7.6 for the case when $n=2$ with $a_0(z)$ being a non-constant polynomial is essentially Theorem 1 proved in Wu [40], while our proof is completely different from Wu's.

In what follows we discuss singular directions of solutions of (3.7.1) with meromorphic function coefficients at least one of which is transcendental. It is well-known that an admissible meromorphic solution of such an equation (3.7.1) is of infinite order. Here a meromorphic solution of (3.7.1) is admissible if all coefficients are its small functions. In view of Theorem 2.7.3, however it is easily seen that a ray is a Borel direction of infinite order if the convergent exponent of a-points for some fixed $a\in\hat{\mathbb{C}}$ is infinite for any angular domain containing this ray. In fact, for the case of infinite order, we have further result.

Theorem 3.7.7. *Let $f(z)$ be a transcendental meromorphic function with $\lambda(f)=\infty$ and let $\lambda(r)$ be a Hiong's infinite order of $f(z)$. Then a ray will be a Borel direction of $\lambda(r)$ order of $f(z)$ if it is of $\lambda(r)$ order for a-points for some fixed $a\in\widehat{\mathbb{C}}$, that is to say, $n(r,\Omega,f=a)$ for every Ω which contains this ray has the infinite order $\lambda(r)$.*

Proof. Consider any angular domain $\Omega(\alpha,\beta)$. In view of the Nevanlinna second fundamental inequality (2.2.6) for Ω and for $a,a_j\in\widehat{\mathbb{C}}$ $(j=1,2,3)$ we easily get

$$C_{\alpha,\beta}\left(r,\frac{1}{f-a}\right)\leqslant\sum_{j=1}^{3}C_{\alpha,\beta}\left(r,\frac{1}{f-a_j}\right)+O(\log rT(r,f)),\ r\notin E(f).$$

From Lemma 2.2.2 for small $\varepsilon>0$ it follows that

$$\begin{aligned}N(r,\Omega_\varepsilon,f=a)\leqslant&\ 2(\sin(\omega\varepsilon))^{-1}N(r,\Omega,f=a_1,a_2,a_3)\\&+\omega(\sin(\omega\varepsilon))^{-1}r^{\omega}\int_1^r\frac{N(t,\Omega,f=a_1,a_2,a_3)}{t^{\omega+1}}dt\\&+O(r^{\omega}\log rT(r,f)),r\notin E(f)\end{aligned}\qquad(3.7.11)$$

with $\omega=\frac{\pi}{\beta-\alpha}$. Thus

$$\begin{aligned}\log N(r,\Omega_\varepsilon,f=a)\leqslant&\ \log N(r,\Omega,f=a_1,a_2,a_3)+2\omega\log r\\&+\log\log r+\log\log T(r,f)+O(1),\ r\notin E(f)\end{aligned}$$

so that letting $\lambda(r)$ be a Hiong's infinite order of $f(z)$, we have

$$\limsup_{r\to\infty}\frac{\log N(r,\Omega_\varepsilon,f=a)}{\lambda(r)\log r}\leqslant\limsup_{r\to\infty}\frac{\log N(r,\Omega,f=a_1,a_2,a_3)}{\lambda(r)\log r}. \tag{3.7.12}$$

This yields our desired result. □

The above result is used to admissible meromorphic solutions of (3.7.1) with meromorphic function coefficients at least one of which is transcendental to get that a ray will be a Borel direction of $\lambda(r)$ order of such a meromorphic solution if it is of $\lambda(r)$ order for zeros. But the inverse of the result is not always true by observing the example $f(z)=\mathrm{e}^{\mathrm{e}^z}$ which satisfies

$$f''-(\mathrm{e}^z+\mathrm{e}^{2z})f=0.$$

This leads us to pose a question:

Question 3.7.2. *Is $\rho_\Omega(0)$ infinite for any angle Ω containing a Borel direction of infinite order if the convergent exponent of zeros of the meromorphic solution is infinite for the complex plane?*

Let $f(z)$ be an admissible meromorphic solution of an equation (3.7.1) with meromorphic function coefficients at least one of which is transcendental. It is easy to see that

$$N\left(r,\frac{f'}{f}\right)=\overline{N}\left(r,\frac{1}{f}\right)+\overline{N}(r,f)\leqslant\overline{N}\left(r,\frac{1}{f}\right)+N(r),$$

where $N(r)=\max\{N(r,a_k):0\leqslant k\leqslant n-1\}$. In view of Lemma 7.6 of Laine [17], we have

$$m\left(r,\frac{f'}{f}\right)=O\left(\log^+T\left(r,\frac{f'}{f}\right)+\Phi(r)+\log r\right),\ r\notin E(f)$$

and hence

$$\begin{aligned}m\left(r,\frac{f'}{f}\right)&=O\left(\log^+N\left(r,\frac{f'}{f}\right)+\Phi(r)+\log r\right)\\&=O\left(\log^+\overline{N}\left(r,\frac{1}{f}\right)+\log^+N(r)+\Phi(r)+\log r\right),\ r\notin E(f),\end{aligned}$$

where $\Phi(r)=\max\{m(r,a_k):0\leqslant k\leqslant n-1\}$. As in Lemma 7.6 of Laine [17] and in view of Theorem 2.5.1, for an angular domain $\Omega(\alpha,\beta)$ we can get

$$
\begin{aligned}
(A_{\alpha,\beta}+B_{\alpha,\beta})\left(r,\frac{f^{(j)}}{f}\right) &= O\left(\log^+ S_{\alpha-\delta,\beta+\delta}\left(r,\frac{f'}{f}\right)+\Phi_{\alpha,\beta}(r)+\log r\right)\\
&= O\left(\log^+ T\left(r,\frac{f'}{f}\right)+\Phi_{\alpha,\beta}(r)+\log r\right)\\
&= O\left(\log^+ \overline{N}\left(r,\frac{1}{f}\right)+\log^+ T(r)\right.\\
&\quad \left.+\,\Phi_{\alpha,\beta}(r)+\log r\right), r\notin E(f), \qquad (3.7.13)
\end{aligned}
$$

where $T(r)=\max\{T(r,a_k):0\leqslant k\leqslant n-1\}$ and $\Phi_{\alpha,\beta}(r)=\max\{(A_{\alpha,\beta}+B_{\alpha,\beta})(r,a_k):0\leqslant k\leqslant n-1\}$.

For a meromorphic FS $\{f_1,f_2,\cdots,f_n\}$ of such an equation (3.7.1) with $a_{n-1}\equiv 0$, letting $E=f_1f_2\cdots f_n$, as in (3.7.10) and in view of (3.7.13) we easily get

$$
\begin{aligned}
S_{\alpha,\beta}(r,E) &= C_{\alpha,\beta}\left(r,\frac{1}{E}\right)+O\left(\log^+ S_{\alpha-\delta,\beta+\delta}(r)+\Phi_{\alpha,\beta}(r)+\log r\right)\\
&\leqslant C_{\alpha,\beta}\left(r,\frac{1}{E}\right)+O\left(\sum_{j=1}^{n}\log^+ \overline{N}\left(r,\frac{1}{f_j}\right)+\log^+ T(r)+\Phi_{\alpha,\beta}(r)+\log r\right)\\
&\leqslant C_{\alpha,\beta}\left(r,\frac{1}{E}\right)+O(\log^+ T(r,E)+\log^+ T(r)+\Phi_{\alpha,\beta}(r)+\log r)
\end{aligned}
$$

where we have used the inequality $\overline{N}(r,1/f_j)\leqslant N(r,1/E)+N(r)$ and $S_{\alpha-\delta,\beta+\delta}(r)=\max\{S_{\alpha-\delta,\beta+\delta}(r,f_k'/f_k):1\leqslant k\leqslant n\}$, so that for $a\in\widehat{\mathbb{C}}$

$$
C_{\alpha,\beta}\left(r,\frac{1}{E-a}\right)\leqslant C_{\alpha,\beta}\left(r,\frac{1}{E}\right)+O(\log^+ T(r,E)+\log^+ T(r)+\Phi_{\alpha,\beta}(r)+\log r)
$$

(When $a=\infty$, $C_{\alpha,\beta}\left(r,\frac{1}{E-a}\right)$ is replaced by $C_{\alpha,\beta}(r,E)$).

In view of Theorem 2.4.7, for a meromorphic function on the complex plane we have

$$
\log^+ S_{\alpha,\beta}(r,f)\leqslant \log^+ T(r,f)+O(\log r)
$$

so that

$$
\log^+ \Phi_{\alpha,\beta}(r)\leqslant \log^+ T(r)+O(\log r).
$$

By the same argument as above to produce (3.7.12) and the above inequalities, we easily show

$$
\limsup_{r\to\infty}\frac{\log N(r,\Omega_\varepsilon,E=a)}{\lambda(r)\log r}\leqslant \limsup_{r\to\infty}\frac{\log N(r,\Omega,E=0)}{\lambda(r)\log r}
$$

if $\log T(r)=o(\lambda(r)\log r)$ and further, a ray is of $\lambda(r)$ order for zeros of $E(z)$ if it is a Borel direction of $\lambda(r)$ order. This together with Theorem 3.7.7 establishes the following theorem.

Theorem 3.7.8. *Consider equation (3.7.1) with $a_{n-1} \equiv 0$. Assume that $\lambda(E) = \infty$ and $\lambda(r)$ is a Hiong's infinite order of $E(z)$ such that $\log T(r) = o(\lambda(r)\log r)$. Then a ray $\arg z = \theta$ is a Borel direction of $\lambda(r)$ order of $E(z)$ if and only if it is of $\lambda(r)$ order for zeros of $E(z)$.*

Obviously, for a function $a(z)$ of finite order, $\log T(r,a) = o(\lambda(r)\log r)$. Therefore we directly get a consequence of Theorem 3.7.8.

Corollary 3.7.1. *The result of Theorem 3.7.8 holds, if each a_k is of finite order.*

This result with entire coefficients of finite order instead was proved in [42] for the nth order equation and [40] for the second order equation. In fact, it is easy to see that if each a_k is of finite order, then for sufficiently small ε, $\Phi_{\theta-\varepsilon,\theta+\varepsilon}(r) = O(1)$ and $S_{\theta-\varepsilon,\theta+\varepsilon}(r) = O(1)$. For this case of that each a_k is entire of finite order, Yi [50] got by the Wiman-Valiron theory for sufficiently small $\varepsilon > 0$

$$S_{\theta-\varepsilon,\theta+\varepsilon}(r,E) = C_{\theta-\varepsilon,\theta+\varepsilon}\left(r,\frac{1}{E}\right) + O(1).$$

Using this equation is easy to deduce Corollary 3.7.1.

We can directly use the lemma of logarithmic derivative (see Lemma 2.5.3) to get the following inequality

$$\begin{aligned} S_{\alpha,\beta}(r,E) &= C_{\alpha,\beta}\left(r,\frac{1}{E}\right) + (A_{\alpha,\beta}+B_{\alpha,\beta})\left(r,\frac{1}{E}\right) + O(1) \\ &= C_{\alpha,\beta}\left(r,\frac{1}{E}\right) + (A_{\alpha,\beta}+B_{\alpha,\beta})\left(r,\frac{W}{E}\right) + O(1) \\ &= C_{\alpha,\beta}\left(r,\frac{1}{E}\right) + O(\log^+ \sum_{j=1}^{n} T(r,f_j) + \log r), \end{aligned}$$

possibly outside a set of r with finite linear measure. If $\sum_{j=1}^{n} T(r,f_j) = O(T(r,E))$, then

$$C_{\alpha,\beta}\left(r,\frac{1}{E-a}\right) \leqslant C_{\alpha,\beta}\left(r,\frac{1}{E}\right) + O(\log^+ T(r,E) + \log r)$$

so that a ray $\arg z = \theta$ is a Borel direction of $\lambda(r)$ order of $E(z)$ if and only if it is of $\lambda(r)$ order for zeros of $E(z)$. We stress that we do not explicitly impose any conditions on the coefficients $a_k(z)$. The readers are suggested to study the condition $\sum_{j=1}^{n} T(r,f_j) = O(T(r,E))$.

Here we suggest to investigate the T directions of admissible solutions of (3.7.1) with meromorphic function coefficients at least one of which is transcendental. We also ask whether a ray is a T direction of an admissible meromorphic solution if it is a Borel direction of Hiong's infinite order.

By means of the above discussion, we can establish the following result, whose proof is left to the reader.

Theorem 3.7.9. *Consider equation (3.7.1) with $a_{n-1} \equiv 0$. Assume that $\lambda(E) = \infty$ and $T(r) = o(T(r,E))$. Let $\{r_n\}$ be a sequence of positive numbers determined in*

Lemma 1.1.3 with $s_n \to \infty$ as $n \to \infty$. Then

$$\limsup_{r\to\infty} \frac{N(r, Z_\varepsilon(\theta), E=0)}{T(r,E)} > 0$$

if the ray $\arg z = \theta$ *is a T direction of $E(z)$ for $\{r_n\}$.*

The study of singular directions of meromorphic solutions of a difference equation is an important and interesting topic, either. Let us mention the Schröder equation as an example. Consider a rational function $R(z)$ of degree $d \geqslant 2$ and a complex number s. The following is the Schröder equation

$$f(sz) = R(f(z)). \tag{3.7.14}$$

Suppose that $f(z)$ is a non-constant meromorphic solution. Then $f(z)$ is transcendental and

$$T(|s|r, f) = (d + o(1))T(r, f), \ r \to \infty.$$

This implies that $|s| > 1$ and $\lambda(f) = \mu(f) = \log d / \log |s|$.

Assume that there exists a point ζ such that $R(\zeta) = \zeta$ and $s = R'(\zeta)$. If $|s| > 1$, in view of Poincaré's Theorem (see Chapter VII of Valiron [37]), there exists a unique meromorphic solution $f(z)$ with $f(0) = \zeta$ and $f'(0) = 1$. The solution f is called the Poincaré function. In the complex dynamics, it has been revealed that there exist a close connection between the value distribution of the Poincaré function f and the distribution of roots of the equation $R^n(z) = a$, where R^n is the nth iterates of R.

Ishizaki and Yanagihara in [16] proved that every Julia direction of a meromorphic solution of the Schröder equation must be a Borel direction and T direction as well. They in [15] discussed singular directions of meromorphic solutions of non-autonomous Schröder equation

$$f(sz) = R(z, f(z)),$$

where $R(z,w)$ is a fixed rational function in z and w with $\deg_w[R(z,w)] \geqslant 2$ and proved that each direction of its transcendental solution for $|s| > 1$ and $\arg s/(2\pi) \notin Q$ must be a Borel direction and a T direction as well. They obtained their result by using the fact that $2n\pi \arg s \bmod (2\pi)$ is dense in $[0,1]$. Therefore, we ask if a transcendental solution has only finitely many singular directions for $|s| > 1$ and $\arg s/(2\pi) \in Q$. And is the previous result about the Schröder equation correct for the non-autonomous Schröder equation?

3.8 Value Distribution of Algebroid Functions

We consider the ν-valued algebroid functions which are determined by the equation

$$\psi(z,w) = A_\nu(z)w^\nu + A_{\nu-1}(z)w^{\nu-1} + \cdots + A_0(z) = 0, \tag{3.8.1}$$

where A_j's are entire functions without common zeros. For the basic knowledge of value distribution of algebroid functions, the reader is referred to He and Xiao's book [13]. Following Valiron [34] and [35], define the characteristic of ν-valued algebroid function w by

$$T(r,w)=\frac{1}{2\pi}\int_0^{2\pi}\log A(re^{\mathrm{i}\theta})\mathrm{d}\theta,$$

where $A(z)=\max_{0\leqslant j\leqslant \nu}|A_j(z)|$, and the integrated counting function for $a\in\mathbb{C}$ by

$$N(r,w=a)=N\left(r,\frac{1}{\psi(z,a)}\right)$$

and

$$N(r,w=\infty)=N\left(r,\frac{1}{A_\nu}\right).$$

We define $\overline{N}(r,w=a)$ and $\overline{N}(r,w=\infty)$ by the above equalities with N replaced by $\overline{N}$ (Note: the definition of N and $\overline{N}$ here is different from that in some of other literatures). When $\nu=1$, $w=-A_0/A_1$ is a meromorphic function and its Nevanlinna characteristic and integrated counting function agree with those defined above.

If $w(z)$ is non-constant, then $T(r,w)$ is increasing and convex in $\log r$ so that $T(r,w)\to\infty$ as $r\to\infty$. This is because $\log A$ is subharmonic on $\mathbb{C}$ (see Chapter 7). $T(r,w)=O(\log r)$ if and only if $w(z)$ is an algebraic function, that is, every A_j is a polynomial. The order and lower order of w is defined by those of $T(r,w)$.

Obviously, $T(r,w)=T(r,1/w)$. For $a\in\mathbb{C}$, set $f=w-a$ and substituting $w=f+a$ into (3.8.1) yields an algebraic equation

$$B_\nu f^\nu+B_{\nu-1}f^{\nu-1}+\cdots+B_1f+B_0=0$$

with $B_\nu=A_\nu$ and $B_j=\sum_{k=1}^{\nu}c_k^{(j)}A_k$. Then for some constant $c>0$, we have

$$B(z)=\max_{0\leqslant j\leqslant\nu}|B_j(z)|\leqslant cA(z)$$

so that

$$T(r,w-a)\leqslant T(r,w)+O(1).$$

The above inequality also yields

$$T(r,w)=T(r,w-a+a)\leqslant T(r,w-a)+O(1)$$

and thus we get the first fundamental inequality for an algebroid function

$$T\left(r,\frac{1}{w-a}\right)=T(r,w)+O(1).$$

The second fundamental theorem for an algebroid function is stated as follows.

Theorem 3.8.1. *Let $w(z)$ be a v-valued algebroid function defined by (3.8.1). Then for q distinct values $a_j \in \widehat{\mathbb{C}}$ $(j = 1, 2, \cdots, q)$, we have*

$$(q-2v)T(r,w) \leqslant \sum_{j=1}^{q} \overline{N}(r, w = a_j) + S(r,w), \tag{3.8.2}$$

where $S(r,w) = O(\log r)$ if w is of finite order and $S(r,w) = O(\log rT(r,w))$ for all but a set of r with finite linear measure.

Theorem 3.8.1 was established by Valiron [33] without the bar over N in (3.8.2) and by Yu [52] who improved the Valiron result.

For uniqueness of algebroid functions, Valiron [33] claimed the $4v+1$ CM shared value unique theorem for v-valued algebroid functions, which was improved by He [11] to establish the following

Theorem 3.8.2. *Let $w(z)$ and $\widehat{w}(z)$ be v-valued and s-valued algebroid functions respectively and $s \leqslant v$. If $w(z)$ and $\widehat{w}(z)$ have $4v+1$ distinct IM shared values, then $w(z) \equiv \widehat{w}(z)$.*

Actually, he only proved the above result for shared-value points counted once for each circle and the complete proof of Theorem 3.8.2 was given by Yu in [51] with help of Theorem 3.8.1 and by He and Li in [12] who gave a different proof from Yu's. For further information, the reader is referred to He [11] and Yu [51].

For singular directions of algebroid functions, Toda [31] discussed Borel directions of an algebroid function and his result on the existence of Borel directions was improved by Lü and Gu [18]. The following is the result of Lü and Gu.

Theorem 3.8.3. *Let $w(z)$ be a v-valued algebroid function with the order $0 < \lambda(w) < \infty$. Then there exists at least one Borel direction $\arg z = \theta$ of $w(z)$, that is, for any $\varepsilon > 0$,*

$$\limsup_{r\to\infty} \frac{\log n(r, Z_\varepsilon(\theta), w = a)}{\log r} = \lambda,$$

possibly except at most $2v$ values of a.

Here $n(r, Z_\varepsilon(\theta), w = a) = n(r, Z_\varepsilon(\theta), \psi(z,a) = 0)$. We naturally consider the T direction for an algebroid function. We first of all give out the definition of T directions (see [57]).

Definition 3.8.1. *Let $w(z)$ be a v-valued algebroid function. A ray $\arg z = \theta$ is called a T direction of w, if for arbitrarily small $\varepsilon > 0$, we have*

$$\limsup_{r\to\infty} \frac{N(r, Z_\varepsilon(\theta), w = a)}{T(r,w)} > 0$$

for all $a \in \widehat{C}$, possibly except at most $2v$ values of a.

Here $N(r, Z_\varepsilon(\theta), w = a) = N(r, Z_\varepsilon(\theta), \psi(z,a) = 0)$. The first question we should solve is the existence of T directions of algebroid functions. We conjectured in [57] that an algebroid function with $T(r,w)/(\log r)^2 \to \infty$ as $r \to \infty$ would have at least one T direction. This conjecture was recently solved by Wu[41], Xuan [43] and Wang and Gao [38].

Finally, we mention that it is interesting to discuss corresponding aspects for an algebroid function to those for a meromorphic function in this book. For example, establish the characteristic of algebroid functions for an angular domain and the basic results; discuss the growth of algebroid function when some restriction is imposed on arguments of certain a-points of it; study relation between singular directions and Nevanlinna deficient values of algebroid functions.

References

1. Brüggemann, F., On solutions of linear differential equations with real zeros; Proof of a conjecture of Hellerstein and Rossi, Proc. Amer. Math. Soc., (2)**113**(1991), 371-379.
2. Chen, H. H., Singular directions related to the Hayman inequality, Progress in Math., (1)**16**(1987), 73-79.
3. Chen, H. H., Singular directions of meromorphic functions with zero order corresponding to the Hayman inequality, Acta Math. Sinica, (2)**30**(1987), 234-237.
4. Chen, H. H. and Guo, H., On singular directions concerning the proximate order, J. Nanjing Normal Univ., (3)**15** (1992), 3-16 (in Chinese).
5. Chen T. Y., Common Borel directions of a meromorphic function with zero order and its derivative, Proc. Amer. Math. Soc., (4)**132**(2003), 1171-1175.
6. Chuang, C. T., Un théoréme relatif aux directions de Borel des fonctions méromorphe d'ordre fini, C. R. Acad. Sci., **204**(1937), 951-952.
7. Chuang, C. T., Sur la comparaison de la croissance d'une fonction méromorphe et de celle de sa dérivée, Bull. Sci. Math., **75**(1951), 171-190.
8. Dietrich, V., Newton-Puiseux-Diagramm für system linearer differentialgleichungen, Complex Variables Theory Appl., **7**(1987), 256-296.
9. Drasin, D. and Weitsman, A., On the Julia directions and Borel directions of entire functions, Proc. London Math. Soc., **32**(1976), 199-212.
10. Guo, H., Zheng, J. H. and Ng, T., On a new singular direction of meromorphic functions, Bull. Austral. Math. Soc., **69**(2004), 277-287.
11. He, Y. Z., Sur un probléme d'unicité relitif aux fonctions algebroides, Sci. Sinica, **14**(1965), 174-180.
12. He, Y. Z. and Li, Y. Z., Some results on algebroid functions, Complex Variables, **43**(2001), 299-313.
13. He, Y. Z. and Xiao, X. Z., Algebroid functions and ordinary differential equations, Science Press, Beijing, 1988 (in Chinese).
14. Hiong King-Lai, Sue les fonctions entiéres et les fonctions méromorphes d'order infini, J. de Math., **14** (1935), 233-308.
15. Ishizaki, K. and Yanagihara, N., Singular directions of meromorphic solutions of some non-autonomous Schröder euqations, Potential theory in Matsue, Adv. Stud. Pure Math., **44**, Math. Soc. Japan, Tokyo, 2006, 155-166.
16. Ishizaki, K. and Yanagihara, N., Borel and Julia directions of meromorphic Shröder functions II, Arch. Math. (Basel), (2)**87**(2006), 172-178.
17. Laine, I., Nevanlinna theory and complex differential equations, de Gruyter Studies in Math., 15, Walter de Gruyter, Berlin-New York, 1993.
18. Lü Y. N. and Gu Y. X., On the existence of Borel direction for algebroid function, Sci. Bull, **28**(1983), 264-266.
19. Lü, Y. N. and Zhang, G. H., On the Nevanlinna directions of meromorphic functions, Science in China, Series A., **3**(1983), 215-224 (in Chinese).

20. Milloux, H., Sur les directions de Borel des fonctions entiéres, de leurs dérivées et de leurs integrales, J. Analyse Math., **1**(1951), 244-330.
21. Nevanlinna, R., Analytic functions, Springer, Berlin, Heidelberg, New York, 1970.
22. Ostrovskii, A., Ueber Folgen analytischer Funktionen und einige Verscharfungen des Picardschen Satzes, Math. Zeit., **24**(1926), 215-258.
23. Rauch, A., Cas oú une direction de Borel d'une fonction entiére $f(z)$ d'ordre fini est aussi direction de Borel pour $f'(z)$,C. R. Acad. Sci., **199**(1934), 1014-1016.
24. Rossi, J., A sharp result concerning cercles de remplissage, Ann. Acad. Sci. Fenn. Ser. A.I., **20**(1995), 179-185.
25. Sauer, A., Julia directions of meromorphic functions and their derivatives, Arch. Math., **79**(2002), 182-187.
26. Steinmetz, N., Eine Verallgemeimerung des zweiten Nevanlinnaschen Hauptsatzes, J. Reine Angew Math., **386**(1986), 134-141.
27. Steinmetz, N., Exceptional values of solutions of linear differential equations, Math. Z., **201**(1989), 317-326.
28. Steinmetz, N., Linear differential equations with exceptional fundamental sets. Analysis, **11**(1991), 119-128.
29. Steinmetz, N., Linear differential equations with exceptional fundamental sets. II, Proc. Amer. Math. Soc., (2)**117**(1993), 355-358.
30. Sternberg, W., Über die asymptotische Integration von Differentialgleichungen. Math. Ann., **81**(1920), 119-186.
31. Toda, N., Sur les directions de Julia et de Borel des fonctions algebroides, Nagoya Math. J., **34**(1969), 1-23.
32. Tsuji, M., On Borel's directions of meromorphic functions of finite order, Tohoku Math. J., **2**(1950), 97-112.
33. Valiron, G., Recherches sur le théoréme de M. Borel dans la théorie des fonctions méromorphes, Acta Math., **52**(1928), 67-92.
34. Valiron, G., Sur quelques propriétés des fonctions algébroides, C. R. Acad. Sc. Paris, **189**(1929), 729.
35. Valiron, G., Sur la dérivée des fonctions algébroides, Bull. Soc. Math. France, **95**(1929), 17-39.
36. Valiron, G., Directions de Borel des fonctions méromorphes, Mémor.Sci. Math., fasc. **89**, Paris, 1938.
37. Valiron, G., Fonctions Analytiques, Presses Univ. de France, Pairs, 1954.
38. Wang, S. G. and Gao, Z. S., On a new singular direction of algebroid functions, Bull. Austral. Math. Soc., **75**(2007), 459-468.
39. Wasow, W., Asymptotic expansions for ordinary differential equations, New York: Wiley 1965.
40. Wu S. J., Argument distribution in the complex oscilation, Science in China, Series A, Math., (5)**34**(2004), 567-573.
41. Wu Z. J., On T direction of algebroidal function, J. Math. Kyoto Univ., (4)**47**(2007), 767-779.
42. Wu Z. J. and Sun D. C., On argument distribution in complex oscillation, preprint.
43. Xuan Z. X., On the existence of T direction of algebroid functions: a problem of J. H. Zheng, J. Math. Anal. Appl., (1)**341**(2008), 540-547.
44. Yang, L., Borel directions of meromorphic functions in an angular domain, Science in China, Math. Series (I) (1979), 149-163.
45. Yang, L., Common Borel directions of a meromorphic function and its derivative, Sci. Sinica, Special Issue (II), 1979, 91-104.
46. Yang, L., New researches on value distribution, Science Press of China, 1982.
47. Yang, L., Meromorphic functions and their derivatives, J. London Math. Soc., (2)**25**(1982), 288-296.
48. Yang, L. and Zhang, G. H., Construction of a meromorphic function with the given singular directions, Science in China, **19**(1976), 445-459.
49. Yang, L. and Zhang, Q. D., New singular directions of meromorphic functions, Science in China, Series A, (4)**26**(1984), 352-366.

50. Yi, C. F., Argument distribution of solutions of higher order differential equation, Acta Math. Sinica, (1)**48**(2005), 133-140.
51. Yu, J. Y., On the second fundamental inequality of algebroid functions, Ann. Academia Sci. Fennica, Series A. Dessertationes, Helsinki, 1994.
52. Yu, J. Y., On the problem of uniqueness and determination of algebraic functions, J. Graduate School, Academia Sinica, **8**(1991), 10-22.
53. Zhang, G. H., Theory of entire and meromorphic functions (in Chinese) , Science Press Sinica, 1986.
54. Zhang, G. H., On researches of common Borel directions of meromorphic functions and their derivatives of any order or integrals, Acta Math. Sinica, **20**(1977), (I) 73-98; (II) 157-177; (III) 237-247.
55. Zhang, Q. D., On the Nevanlinna and Borel directions of meromorphic functions, Acta Math. Sinica, (4)**29**(1986), 550-554 (in Chinese).
56. Zhang, Q. D., T directions and Borel directions of meromorphic functions with finite positive order, Acta Math. Sinica, (2)**50**(2007), 413-420.
57. Zheng, J. H., On value distribution of meromorphic functions with respect to argument I, Complex Variables and Elliptic Equations, 2009.
58. Zheng, J. H., On transcendental meromorphic functions with radially distributed values, Science in China Ser. A. Math., (3)**47**(2004), 401-416.
59. Zheng, J. H., Linear differential equations with rational coefficients, J. Math. Soc. Japan, (3)**48**(1996), 501-510.
60. Zheng, J. H. and Wu, N., Hayman T directions of meromorphi functions, to appear in Taiwanese Journal of Mathematics.

Chapter 4
Argument Distribution and Deficient Values

Jianhua Zheng
Department of Mathematical Sciences, Tsinghua University, Beijing 100084, P. R. China
jzheng@math.tsinghua.edu.cn

Abstract: We investigate above bound of total number of deficient values of a transcendental meromorphic function and its derivatives of every order if most of its zeros and poles distribute along finitely many rays starting from the origin and prove that the bound is the number of the rays under some assumption, for example, the function considered is of finite lower order. Next we discuss relations between the numbers of deficient values and common T directions of the functions and their every order derivatives, and demonstrate that total number of deficient values of a meromorphic function and its every order primitive is not greater than the number of common T directions of the functions and its every order derivatives. We then go to the case of entire functions. For example, we verify that a bound of total number of deficient values of the function and its derivatives and primitives of all orders is two times its lower order for an entire function with the finite positive lower order. Finally, we make a simple survey of some of the celebrated related works of Edrei and Fuchs' and Yang and Zhang's and others.

Key words: Deficient value, T directions, Derivatives, Harmonic measures

Many results obtained in 1950's drop a hint on non-existence of the Nevanlinna deficient values under some restriction imposed on the distribution of arguments of points of some value a (for example, compare the results in next chapter). Noting that the deficiency is an important object in the study of the module distribution of a meromorphic function, this actually hints vaguely some relations between the module distribution and the argument distribution of a meromorphic function. It is the celebrated results obtained in a series of papers by Edrei and Fuchs [1] [2] in 1962 that made the relations clear and distinct. As we have known, the singular directions are the study object of argument distribution of a meromorphic function and then it is worth to investigate relation between deficient values and singular directions. Since 1975, L. Yang and G. H. Zhang attained a series of results about connection of deficient values and Borel or Julia directions. We will retrospect those works in Section 4.2, while in Section 4.1 we will discuss relations between deficient values and T directions.

4.1 Deficient Values and T Directions

The purpose of this section devotes to discussion of relations between the numbers of deficient values and T directions of a meromorphic function, which is motivated by works of Edrei and Fuchs', Yang and Zhang's.

First of all, we establish the following important lemma, some of whose ideas come from those of Edrei and Fuchs [1] and the section 3.5 of Zhang [11]. Let $f(z)$ be a transcendental meromorphic function. For four positive numbers R, H,ε and η with $4\mathrm{e}H<\eta$ and two complex numbers $a\neq 0$ and ξ with $|\xi|=R$, let

$$\begin{aligned}\mathfrak{X}_\varepsilon(f,a)&=\mathfrak{X}_\varepsilon(R,\eta,\xi;f,a)\\&=\left\{z: M_R\leqslant \varepsilon C\log^+\frac{1}{|f(z)-a|}\text{ and }|z-\xi|<\eta\right\},\end{aligned}\tag{4.1.1}$$

where

$$C=\min\left\{\frac{1}{18},\frac{1}{100}Q\right\},\quad M_R=\log RT(R+6\eta,f)+N\log^+\frac{6\eta}{H}+1$$

and

$$Q=\left(\log\frac{4\eta}{H}\right)^{-1},\quad N=n(5\eta,\xi,f=0)+n(5\eta,\xi,f=\infty).$$

Lemma 4.1.1. *Let $f(z)$ be a transcendental meromorphic function. Given positive numbers R, H and η with $4\mathrm{e}H<\eta\leqslant R$ and $H\geqslant T^{-1}(R,f)$, consider the disk $\{z:|z-\xi|<5\eta\}$, $\xi=R\mathrm{e}^{\mathrm{i}\theta}$.*

We denote by (γ) the set of the Boutroux-Cartan exceptional disks for the zeros and poles of $f(z)$, by (γ') one for the zeros of $f'(z)$ in $|z-\xi|<5\eta$ and H and by $(\gamma)_a$ the set of

$$n=n(R+6\eta,f=0)+n(R+6\eta,f=\infty)+n(R+6\eta,f=a)$$

disks centered at zeros, poles and $a(a\neq 0)$-points of f in $\{z:|z|<R+6\eta\}$ with radius $\frac{H}{n}$.

Then there exists a positive constant κ such that if for $0<\varepsilon<\kappa$, $\mathfrak{X}_\varepsilon(f,a)\setminus((\gamma)\cup(\gamma')\cup(\gamma)_a)\neq\varnothing$, then for $z_0\in\mathfrak{X}_\varepsilon(f,a)\setminus((\gamma)\cup(\gamma')\cup(\gamma)_a)$, we have

$$\frac{C}{2}\log^+\frac{1}{|f(z_0)-a|}\leqslant\log^+\frac{1}{|f(z)-a|}+2\log^+2+\log^+|a|+8\pi\eta\tag{4.1.2}$$

and

$$\log|f'(z)|\leqslant-\frac{C}{2}\log^+\frac{1}{|f(z_0)-a|}+\log^+|f(z_0)|+4\pi\eta\tag{4.1.3}$$

for all z in $\{z:|z-\xi|<\eta\}$ outside $(\gamma)\cup(\gamma')\cup(\gamma)_a$.

Proof. First of all we estimate $\log\left|\frac{f'(z)}{f(z)}\right|$ in $\{z:|z-\xi|<\eta\}$. In view of Lemma 2.5.2, we have

$$\log\left|\frac{f'(z)}{f(z)}\right| \leqslant \frac{u+\eta}{u-\eta}m\left(u,\xi,\frac{f'}{f}\right)-\frac{u-\eta}{u+\eta}m\left(u,\xi,\frac{f}{f'}\right)$$
$$+(n(u,\xi,f=0)+\overline{n}(u,\xi,f=\infty))\log\frac{u+\eta}{H}$$
$$-\frac{(u-t)^2}{u^2+t^2}n(t,\xi,f'=0) \tag{4.1.4}$$

inside $|z-\xi|\leqslant\eta<t<u<5\eta$ outside (γ).

A simple application of Lemma 2.5.1 implies that

$$\log^+\left|\frac{f'(z)}{f(z)}\right|+\log^+\left|\frac{f'(z)}{f(z)-a}\right|\leqslant K_1\log RT(R+6\eta,f), \tag{4.1.5}$$

for z in $\{z:|z|<R+5\eta\}$ except $(\gamma)_a$ the total sum of whose diameters equals to $2H$, by noting that $\delta(z)\geqslant\frac{H}{n}$, where we have used the inequality $n(R+5\eta,*)\leqslant\frac{R+6\eta}{\eta}N(R+6\eta,*)$ and $T^{-1}(R,f)\leqslant H<\eta\leqslant R$. Since $4eH<\eta$, there exist $2\eta<u_0<3\eta$ and $4\eta<u_1<5\eta$ such that $\{z:|z-\xi|=u_i\}\cap(\gamma)_a=\varnothing$ $(i=0,1)$. Thus integrating the first item of the above inequality along $|z-\xi|=u_i$ $(i=0,1)$ shows

$$m\left(u_i,\xi,\frac{f'}{f}\right)\leqslant K_1\log RT(R+6\eta,f)\ (i=0,1), \tag{4.1.6}$$

here and below we denote by $K_1,K_2,\cdots$ positive constants depending only on a.

Now we estimate $m\left(u,\xi,\frac{f}{f'}\right)$. An application of the Poisson-Jensen Formula (2.1.28) to $f(z)/f'(z)$ deduces that

$$\log^+\left|\frac{f(z)}{f'(z)}\right|\leqslant\frac{u+\eta}{u-\eta}m\left(u,\xi,\frac{f}{f'}\right)+n(u,\xi,f'=0)\log\frac{u+\eta}{H}, \tag{4.1.7}$$

for $2\eta<u<5\eta$ and $|z-\xi|\leqslant\eta$ outside (γ'). From the identical equality

$$\frac{1}{f(z)-a}=\frac{1}{a}\frac{f(z)}{f'(z)}\left(\frac{f'(z)}{f(z)-a}-\frac{f'(z)}{f(z)}\right),$$

it follows, in virtue of (4.1.5) and (4.1.7), that

$$\log^+\left|\frac{1}{f(z)-a}\right|\leqslant\log^+\left|\frac{f(z)}{f'(z)}\right|+\log^+\frac{1}{|a|}+\log^+\left|\frac{f'(z)}{f(z)}\right|$$
$$+\log^+\left|\frac{f'(z)}{f(z)-a}\right|+\log 2$$
$$\leqslant K_2\log RT(R+6\eta,f)+3m\left(u,\xi,\frac{f}{f'}\right)$$
$$+n(u,\xi,f'=0)\log\frac{u+\eta}{H}, \tag{4.1.8}$$

for $2\eta < u < 5\eta$ and $|z-\xi| \leqslant \eta$ outside $(\gamma') \cup (\gamma)_a$. Substituting (4.1.8) for $z = z_0 \in \mathfrak{X}_\varepsilon(f,a) \setminus \{(\gamma) \cup (\gamma') \cup (\gamma)_a\}$ and $u = u_0$ into (4.1.4) for $u = u_0$ implies that

$$\begin{aligned}\log\left|\frac{f'(z)}{f(z)}\right| &\leqslant 3K_1 \log RT(R+6\eta,f) - \frac{1}{3}m\left(u_0,\xi,\frac{f}{f'}\right) + N\log\frac{u_0+\eta}{H}\\ &\leqslant K_3 \log RT(R+6\eta,f) + \frac{1}{9}n(u_0,\xi,f'=0)\log\frac{u_0+\eta}{H}\\ &\quad -\frac{1}{9}\log^+\frac{1}{|f(z_0)-a|} + N\log\frac{u_0+\eta}{H} \qquad (4.1.9)\end{aligned}$$

for $|z-\xi| \leqslant \eta$ outside $(\gamma) \cup (\gamma') \cup (\gamma)_a$.

Below we need to treat two cases.

(I) If $n(u_0,\xi,f'=0)\log\frac{u_0+\eta}{H} > \frac{1}{2}\log^+\frac{1}{|f(z_0)-a|}$, then from (4.1.4) for $u = u_1$ and $t = u_0$, we have

$$\begin{aligned}\log\left|\frac{f'(z)}{f(z)}\right| &\leqslant K_4 \log RT(R+6\eta,f) + N\log\frac{u_1+\eta}{H}\\ &\quad -\frac{1}{68}\left(\log\frac{4\eta}{H}\right)^{-1}\log^+\frac{1}{|f(z_0)-a|} \qquad (4.1.10)\end{aligned}$$

for $|z-\xi| \leqslant \eta$ outside (γ).

(II) If $n(u_0,\xi,f'=0)\log\frac{u_0+\eta}{H} \leqslant \frac{1}{2}\log^+\frac{1}{|f(z_0)-a|}$, then from (4.1.9), we have (4.1.10) for K_3 in place of K_4 and with the coefficient $\frac{1}{18}$ in front of $\log^+\frac{1}{|f(z_0)-a|}$ instead of $\frac{1}{68}Q$ and for $|z-\xi| \leqslant \eta$ outside $(\gamma) \cup (\gamma') \cup (\gamma)_a$. Thus we always have

$$\log\left|\frac{f'(z)}{f(z)}\right| \leqslant K_4 \log RT(R+6\eta,f) + N\log\frac{6\eta}{H} - C\log^+\frac{1}{|f(z_0)-a|}$$

for $|z-\xi| \leqslant \eta$ outside $(\gamma) \cup (\gamma') \cup (\gamma)_a$, where $C = \min\{\frac{1}{18}, \frac{1}{68}Q\}$. In view of the definition (4.1.1) of $\mathfrak{X}_\varepsilon(f,a)$, then for $0 < \varepsilon < 1/(2K_4)$ we have

$$\log\left|\frac{f'(z)}{f(z)}\right| \leqslant -\frac{C}{2}\log^+\frac{1}{|f(z_0)-a|} \qquad (4.1.11)$$

and hence

$$\left|\frac{f'(z)}{f(z)}\right| < 1$$

for $|z-\xi| \leqslant \eta$ outside $(\gamma) \cup (\gamma') \cup (\gamma)_a$.

Since

$$|\log|f(z)| - \log|f(z_0)|| \leqslant \int_{L(z,z_0)}\left|\frac{f'(z)}{f(z)}\right||dz| \leqslant 4\pi\eta,$$

where $L(z,z_0)$ is a curve in $|z-\xi| < \eta$ connecting z and z_0 which does not intersect $(\gamma) \cup (\gamma') \cup (\gamma)_a$ with the length not greater than $4\pi\eta$, we have in virtue of (4.1.11)

$$\begin{aligned}|f'(z)| &= |f(z)|\left|\frac{f'(z)}{f(z)}\right| \\ &\leqslant |f(z_0)|\exp\left(4\pi\eta+\log\left|\frac{f'(z)}{f(z)}\right|\right) \\ &\leqslant e^{4\pi\eta}|f(z_0)|\exp\left(-\frac{C}{2}\log^+\frac{1}{|f(z_0)-a|}\right),\end{aligned} \tag{4.1.12}$$

which produces (4.1.3).

Furthermore, in virtue of (4.1.12), we have

$$\begin{aligned}|f(z)-a| &\leqslant |f(z)-f(z_0)|+|f(z_0)-a| \\ &\leqslant 4\pi\eta e^{4\pi\eta}|f(z_0)|\exp\left(-\frac{C}{2}\log^+\frac{1}{|f(z_0)-a|}\right)+|f(z_0)-a| \\ &\leqslant |f(z_0)-a|\left[1+4\pi\eta\left(1+\frac{|a|}{|f(z_0)-a|}\right)\exp\left(4\pi\eta\right.\right. \\ &\quad \left.\left. -\frac{C}{2}\log^+\frac{1}{|f(z_0)-a|}\right)\right]\end{aligned}$$

for $|z-\xi|\leqslant\eta$ outside $(\gamma)\cup(\gamma')\cup(\gamma)_a$. Thus

$$\begin{aligned}\log^+\frac{1}{|f(z_0)-a|} &\leqslant \log^+\frac{1}{|f(z)-a|}+2\log^+2+\log^+4\pi\eta+\log^+|a| \\ &\quad +4\pi\eta-\left(\frac{C}{2}-1\right)\log^+\frac{1}{|f(z_0)-a|}.\end{aligned}$$

This yields (4.1.2).

Lemma 4.1.1 follows. □

By noting the following equality

$$\log|f^{(k)}(z)| = \log\left|\frac{f^{(k)}(z)}{f^{(k-1)}(z)}\right|+\cdots+\log\left|\frac{f'(z)}{f(z)-a}\right|+\log|f(z)-a|, \tag{4.1.13}$$

it is easy to see that if $\log|f(z)-a|$ is very small, then $\log|f^{(k)}(z)|$ is also to do so. Thus (4.1.3) essentially also follows from (4.1.2). To deal with deficient values of derivatives of a meromorphic function we need the following lemma.

Lemma 4.1.2. *Let $f(z)$ be a transcendental meromorphic function. Consider the disk $\{z:|z-\xi|<5\eta\}$ with $\xi=Re^{i\theta}$ and $H\geqslant T^{-1}(R,f)$ and $4eH<\eta\leqslant R$. Then*

$$\begin{aligned}n(\eta,\xi,f^{(m)}=0) &\leqslant 5(m\bar{n}(4\eta,\xi,f=\infty)+n(4\eta,\xi,f=0))\log\frac{4\eta}{H} \\ &\quad +K\log RT(R+5\eta,f)-5\log\left|\frac{f^{(m)}(z_0)}{f(z_0)}\right|,\end{aligned} \tag{4.1.14}$$

for arbitrary $z_0 \in \{z : |z-\xi| < \eta\}$ *outside* (γ) *where* (γ) *is the set of Boutroux-Cartan exceptional disks for the zeros and poles of* $f(z)$ *and* H.

Proof. We can find a $2\eta < R_0 < 3\eta$ such that $\{z : |z-z_0| = R_0\}$ has the distance at least $\frac{\eta}{4n}$, n is the number of zeros and poles of $f(z)$ in $\{z : |z-\xi| < 4\eta\}$, from the zeros and poles of $f(z)$. It is easily seen that $\{z : |z-z_0| < R_0\} \subset \{z : |z-\xi| < 4\eta\}$. In view of Lemma 2.5.1, we have

$$m\left(R_0, z_0, \frac{f^{(m)}}{f}\right) \leqslant K \log RT(R+5\eta, f).$$

For $z_0 \notin (\gamma)$, using Lemma 2.5.2 to $\{z : |z-z_0| < R_0\}$ with $R = R_0, r = t = \eta$, we have

$$\log\left|\frac{f^{(m)}(z_0)}{f(z_0)}\right| \leqslant 3K \log RT(R+5\eta, f) + (m\bar{n}(4\eta, \xi, f=\infty)$$
$$+ n(4\eta, \xi, f=0)) \log\frac{4\eta}{H} - \frac{1}{5} n(\eta, \xi, f^{(m)} = 0).$$

This yields immediately (4.1.14). □

Now we come to establish our first main result of this section.

Theorem 4.1.1. *Let* $f(z)$ *be a transcendental meromorphic function. Consider a sequence of annuli* A_n *defined by*

$$\alpha_n r_n \leqslant |z| \leqslant \kappa r_n, \quad n = 1, 2, 3, \cdots$$

with $r_n \to \infty$, $0 < \alpha_n \leqslant 1$ *and* $\kappa > 1$ *and* $T(\kappa r_n, f) \leqslant KT(r_n, f)$ *for some positive constant* K.

With each A_n *it is possible to associate* $s(\geqslant 1)$ *arguments*

$$0 \leqslant \alpha_{n1} < \alpha_{n2} < \cdots < \alpha_{ns} < \alpha_{n1} + 2\pi$$

such that there are at most $o(T(r_n, f))$ *zeros and poles of* $f(z)$ *in the portion of* A_n *and outside the* s *sectors*

$$|\arg z - \alpha_{nj}| \leqslant \varepsilon, \quad j = 1, 2, \cdots, s,$$

for arbitrary fixed $\varepsilon > 0$. *Then* $p_0 \leqslant s$.

If, in addition, $\alpha_n \to 0$ *as* $n \to \infty$, $T(\eta_n r_n, f) = o(T(r_n, f))$ *for any sequence* $\{\eta_n\}$ *such that* $\eta_n \to 0$ *and* $\kappa > 12$, *then*

$$\sum_{j=0}^{\infty} p_j \leqslant s, \tag{4.1.15}$$

where p_j *is the number of finite non-zero deficient values of* $f^{(j)}(z)$.

Proof. Here we only prove (4.1.15). Let c be a real number greater than 1 and sufficiently close to 1. For any non-negative integer m and sufficiently large n, in view of (2.6.1) and (2.6.2), we have

$$\begin{aligned} T(c^2r_n, f^{(m)}) &\leqslant 2(m+1)T(2r_n, f) \\ &\leqslant 2K(m+1)T(r_n, f) \\ &\leqslant K(m+1)K_{m,c}T(cr_n, f^{(m)}). \end{aligned} \tag{4.1.16}$$

Set

$$A_n^{(j)} = \{z : r_n/2 \leqslant |z| \leqslant 2r_n, \alpha_{nj} < \arg z < \alpha_{n(j+1)}\}, \quad j = 1, 2, \cdots, s.$$

Let a_l be a non-zero and finite Nevanlinna deficient value of $f^{(l)}(z)$. In virtue of Lemma 2.8.1 it follows that for all sufficiently large n, $\mathrm{mes}E_n(a_l, f^{(l)}) \geqslant t(a_l) > 0$ for some $R_n \in (cr_n, c^2r_n)$. Take a $\varepsilon > 0$ such that $40s\varepsilon < t(a_l)$ and $\kappa > 2d$, $d = 6 \times 8^{1/\omega}(\sin(\omega\varepsilon))^{-1/\omega}$, $\omega = \frac{\pi}{10\varepsilon}$, and the d is that in Lemma 3.5.1. Then there exists at least one j_0 such that

$$\mathrm{mes}\{E_n(a_l, f^{(l)}) \cap (\alpha_{nj_0} + 20\varepsilon, \alpha_{n(j_0+1)} - 20\varepsilon)\} > 0$$

for all sufficiently large n, which we can assume without any loss of generalities. Now let us restrict our discussion to this angle $A_n^{(j_0)}$. Set $\theta_i = \alpha_{nj_0} + (i+19)\varepsilon$ $(i = 1, \cdots, q)$, where $q = \left[\frac{\alpha_{n(j_0+1)} - \alpha_{nj_0}}{\varepsilon}\right] - 39$. There exists a i_0 such that $\mathrm{mes}\{E_n(a_l, f^{(l)}) \cap (\theta_{i_0} - \varepsilon, \theta_{i_0} + \varepsilon)\} > 0$.

Choose $\delta_n \geqslant \alpha_n$ with $\delta_n \to 0$ as $n \to \infty$ such that letting $M_n = 1/\sqrt{\delta_n}$, we have

$$M_n^{3\omega} \log T(2dR_n, f) + \log(2dR_n) + N(R_n)M_n^{2\omega} \log M_n = o(T(r_n, f))$$

where $N(R_n) = N(2dR_n, (A_n^{(j_0)})_\varepsilon, f = 0) + N(2dR_n, (A_n^{(j_0)})_\varepsilon, f = \infty)$. Set $P_n = 2\sqrt{\delta_n}R_n$ and so $M_nP_n = 2R_n$ and $M_n^{-1}P_n = 2\delta_nR_n$.

We consider the sector $Z_{5\varepsilon}(\theta_{i_0})$. We need to treat two cases below.

(I) Assume that

$$n(Z_{5\varepsilon}(\theta_{i_0})[M_n, P_n], f^{(l)} = 0) \neq o(T(r_n, f)).$$

In view of Lemma 3.5.1 and using the same argument as in the proof of Theorem 3.5.5, we have

$$\log|f(z)| = o(T(r_n, f))$$

for all $z \in Z_{5\varepsilon}(\theta_{i_0})[M_n, P_n]$ outside the union (γ) of disks the total sum of whose radius is not greater than $\frac{\varepsilon}{10}M_n^{-1}P_n = \frac{\varepsilon}{5}\delta_nR_n$. Since for a point z_0 with $\arg z_0 \in E_n(a_l, f^{(l)}) \cap (\theta_{i_0} - \varepsilon, \theta_{i_0} + \varepsilon)$ we have

$$\log^+ \frac{1}{|f^{(l)}(z_0) - a_l|} \geqslant \frac{\delta}{4}T(R_n, f^{(l)}) \geqslant K_0T(r_n, f)$$

with $\delta = \delta(a_l, f^{(l)})$ and furthermore, $|f^{(l)}(z_0) - a_l| < \frac{|a_l|}{2}$ and so $\frac{|a_l|}{2} \leqslant |f^{(l)}(z_0)|$ and $-\log|f^{(l)}(z_0)| < -\log\frac{|a_l|}{2}$. In view of Lemma 4.1.2, we have

$$n(B_n, f^{(l)} = 0) = o(T(r_n, f)),\ B_n = B(R_n e^{i\theta_{i_0}}, 5\varepsilon R_n),$$

and hence using Lemma 4.1.1 yields

$$\log|f^{(l)}(z) - a_l| \leqslant -K_1 T(r_n, f) \tag{4.1.17}$$

for z in the arc $\{z : |z| = R_n\} \cap Z_\varepsilon(\theta_{i_0})$ possibly except the set of the measure not greater than $\frac{\varepsilon}{16} R_n$.

(II) Assume that

$$n(Z_{5\varepsilon}(\theta_{i_0})[M_n, P_n], f^{(l)} = 0) = o(T(r_n, f)).$$

The same argument as in above deduces (4.1.17). Thus we always have (4.1.17).

Notice that the arc $\{z : |z| = R_n\} \cap (A_n^{(j_0)})_{20\varepsilon}$ can be covered by a finite number of disks whose cardinality is independent of n. Thus we have (4.1.17) for z in the arc $\{z : |z| = R_n\} \cap (A_n^{(j_0)})_{20\varepsilon}$ outside a set of disks the total sum of whose diameters does not exceed $L/16$ where L is the length of the arc.

Let b_k be another non-zero and finite Nevanlinna deficient value of $f^{(k)}(z)$. Then we also have the similar inequality to (4.1.17) for b_k and $f^{(k)}$ and arc $\{z : |z| = R_n\} \cap (A_n^{(i)})_{20\varepsilon}$ for some i (if necessary, we shall shrink ε). If $k \neq l$, assume that $k > l$ and then in view of (4.1.13), in the arc $\{z : |z| = R_n\} \cap (A_n^{(j_0)})_{20\varepsilon}$ associated to a_l,

$$\log|f^{(k)}(z)| \leqslant -K_2 T(r_n, f).$$

Since $b_k \neq 0$, the two domains $A_n^{(i)}$ and $A_n^{(j_0)}$ associated to a_l and b_k respectively do not coincide; If $k = l$ but $a_l \neq b_k$, then $A_n^{(i)}$ and $A_n^{(j_0)}$ are also distinct. Thus it is obvious that Theorem 4.1.1 follows. □

Theorem 4.1.1 still holds even if the ring A_n is divided by s B-regular curves, which is easily attained when the derivatives are not considered, while in general, we need to modify Lemma 3.5.1 to be available for this case.

We remark that under the assumption in Theorem 4.1.1, $f(z)$ is usually of the finite lower order and that a sequence $\{r_n\}$ of relaxed Pólya peak of positive order will satisfy the requirement of Theorem 4.1.1.

In what follows, we come to estimate the number of deficient values in term of the number of T directions. In order to discuss the case of deficient values of the $|l|$th primitive of $f(z)$ for a negative integer l, which is denoted by $f^{(l)}(z)$, we need the following lemma, some of whose idea comes essentially from Lemma 4 of Yang [7]. Recall that in Lemma 4.1.1, we require $a \neq 0$, while no restriction is imposed on a in Theorem 2.1.7. Set

$$\mathfrak{Y}_\varepsilon(f, a) = \left\{z :\ \hat{M}_R < \varepsilon \log^+ \frac{1}{|f(z) - a|},\ |z - \xi| < \eta\right\},$$

$$\hat{M}_R = \log RT(R+6\eta, f) + N\log^+ \frac{6\eta}{H} + 1,$$

$\xi = Re^{i\theta}$ and $4eH < \eta$.

Lemma 4.1.3. *Under the assumption of Lemma 4.1.1, assume l is a negative integer, $a \in \widehat{\mathbb{C}}$ and*

$$N = \sum_{j=1}^{3} n(5\eta, \xi, f = a_j)$$

instead where $a_j (j = 1, 2, 3)$ are three distinct complex numbers or ∞. Then there exists a positive number κ such that for $0 < \varepsilon < \kappa$ and $z_0 \in \mathfrak{Y}_\varepsilon(f^{(l)}, a) \setminus ((\gamma) \cup (\gamma') \cup (\gamma)_a) \neq \varnothing$, we have

$$K \log^+ \frac{1}{|f^{(l)}(z_0) - a|} \leqslant \log^+ \frac{1}{|f^{(l)}(z) - a|} \tag{4.1.18}$$

for all z in $\{z : |z - \xi| < \eta\}$ outside $(\gamma) \cup (\gamma') \cup (\gamma)_a$, where K is a positive constant depending only on a and l.

Proof. Below we always denote by $K_1, K_2, \cdots$ positive constants only depending on a and l. As in (4.1.5), for $z_0 \in \{z : |z - \xi| < \eta\} \setminus ((\gamma) \cup (\gamma') \cup (\gamma)_a)$, we can get

$$\begin{aligned} \log |f^{(l+j)}(z_0)| &\leqslant -\log \frac{1}{|f^{(l)}(z_0) - a|} + K_1 \log RT(R+6\eta, f^{(l)}) \\ &= -\log^+ \frac{1}{|f^{(l)}(z_0) - a|} + K_1 \log RT(R+6\eta, f^{(l)}), \end{aligned}$$

$j = 1, 2, \cdots, -l$. Specially, we have

$$\log \frac{1}{|f(z_0)|} \geqslant \log^+ \frac{1}{|f^{(l)}(z_0) - a|} - K_1 \log RT(R+6\eta, f^{(l)}).$$

Thus from the proof of Theorem 2.1.7 it follows that

$$\begin{aligned} \log^+ \frac{1}{|f(z)|} &\geqslant K_2 \log^+ \frac{1}{|f(z_0)|} - N\log^+ \frac{6\eta}{H} - 1 \\ &\geqslant K_2 \log^+ \frac{1}{|f^{(l)}(z_0) - a|} - K_2 K_1 \log RT(R+6\eta, f^{(l)}) - N\log^+ \frac{6\eta}{H} - 1 \end{aligned}$$

and then there exists a positive number $\kappa < \frac{1}{2K_1}$ such that for $0 < \varepsilon < \kappa$ and for $z_0 \in \mathfrak{Y}_\varepsilon(f^{(l)}, a) \setminus ((\gamma) \cup (\gamma') \cup (\gamma)_a) \neq \varnothing$, we have

$$\log^+ \frac{1}{|f(z)|} \geqslant \frac{K_2}{2} \log^+ \frac{1}{|f^{(l)}(z_0) - a|}$$

that is,

$$\log |f(z)| \leqslant -\frac{K_2}{2} \log^+ \frac{1}{|f^{(l)}(z_0) - a|},$$

with $0 < K_2 < 1$, for all z in $\{z : |z-\xi| < \eta\}$ outside $(\gamma)\cup(\gamma')\cup(\gamma)_a$. Notice the following equality

$$f^{(l)}(z)-a = \frac{1}{(|l|-1)!}\int_{z_0}^{z}(z-\zeta)^{-l-1}f(\zeta)\mathrm{d}\zeta + \sum_{j=1}^{|l|-1}\frac{f^{(l+j)}(z_0)}{j!}(z-z_0)^j + (f^{(l)}(z_0)-a),$$

where the path of the integral is from z_0 to z along Γ which is constructed from $\overline{z_0z}$ by replacing the part of $\overline{z_0z}$ in $(\gamma)\cup(\gamma')\cup(\gamma)_a$ with the minimum arcs. We estimate every term in the right side of above equality. We have

$$\left|\int_{z_0}^{z}(z-\zeta)^{-l-1}f(\zeta)\mathrm{d}\zeta\right| \leqslant (2\eta)^{-l-1}L(\Gamma)\exp\left(-\frac{K_2}{2}\log^+\frac{1}{|f^{(l)}(z_0)-a|}\right) \leqslant \exp\left(-\frac{K_2}{3}\log^+\frac{1}{|f^{(l)}(z_0)-a|}\right)$$

and

$$\begin{aligned}&\sum_{j=1}^{|l|-1}\frac{|f^{(l+j)}(z_0)|}{j!}|z-z_0|^j + |f^{(l)}(z_0)-a| \\ &\leqslant \mathrm{e}\max\{|z-z_0|^j|f^{(l+j)}(z_0)| : 1\leqslant j\leqslant |l|-1\} + |f^{(l)}(z_0)-a| \\ &\leqslant \exp\left(-\frac{1}{2}\log^+\frac{1}{|f^{(l)}(z_0)-a|}\right) + \exp\left(-\log^+\frac{1}{|f^{(l)}(z_0)-a|}\right) \\ &\leqslant \exp\left(-\frac{1}{3}\log^+\frac{1}{|f^{(l)}(z_0)-a|}\right).\end{aligned}$$

The above inequalities can hold if we suitably shrink κ.

Therefore

$$\log|f^{(l)}(z)-a| \leqslant -\frac{K_2}{4}\log^+\frac{1}{|f^{(l)}(z_0)-a|},$$

so that (4.1.18) follows. □

Now we are in position to establish the second main result of this section.

Theorem 4.1.2. *Let $f(z)$ be a transcendental meromorphic function with $\mu(f) < \infty$ and $\lambda(f) > 0$. We denote by q the number of common T directions of $f(z)$ and its every order derivative. Then*

$$\sum_{j=-\infty}^{0} p_j \leqslant q;$$

If, in addition, $\delta(\infty, f) = 1$, we have

$$\sum_{j=-\infty}^{\infty} p_j \leqslant q.$$

Here when j is negative, $f^{(j)}(z)$ stands for the $|j|$th primitive of $f(z)$.

Proof. Since $f(z)$ has the finite lower order and non-zero order, we can find a sequence $\{r_n\}$ of common relaxed Pólya peak of $f^{(j)}(z)$ for finitely many integers j with positive order such that for all sufficiently large n, $\text{mes}E_n(a_l, f^{(l)}) \geqslant t(a_l) > 0$ for r_n for finitely many fixed Nevanlinna deficient values a_l of $f^{(l)}(z)$ (see Theorem 2.6.3 and Lemma 2.8.1).

Let $\arg z = \theta_j (1 \leqslant j \leqslant q)$ be all common T directions of $f(z)$ and its every order derivative. We can choose a $\varepsilon > 0$ such that for some j_0, $\text{mes}(E_n(a_l, f^{(l)}) \cap (\theta_{j_0} + 20\varepsilon, \theta_{j_0+1} - 20\varepsilon)) > 0$. We can choose finitely many ϕ_i such that $(\phi_i - \tau, \phi_i + \tau)$ is a open covering of $[\theta_{j_0} + 20\varepsilon, \theta_{j_0+1} - 20\varepsilon]$ with $\tau \leqslant \varepsilon$ and for each i, there exist three distinct complex numbers $b_k (k = 1, 2, 3)$ such that

$$\sum_{k=1}^{3} N(r, Z_{5\tau}(\phi_i), f^{(m_i)} = b_k) = o(T(r, f^{(m_i)})),$$

as $r \to \infty$. Then for some i_0, $\text{mes}(E_n(a_l, f^{(l)}) \cap (\phi_{i_0} - \tau, \phi_{i_0} + \tau)) > 0$. If $l \leqslant m_{i_0}$, then using Lemma 4.1.3 yields (4.1.17) on $\{z : |z| = r_n, \phi_{i_0} - \tau \leqslant \arg z \leqslant \phi_{i_0} + \tau\}$ possibly except a set with measure not greater than $\frac{\tau}{16} r_n$; If $l > m_{i_0}$, then using the method in the proof of Theorem 4.1.1 under the additional assumption $\delta(\infty, f) = 1$ implies the above result. Furthermore, we therefore have (4.1.17) on the arc $\{z : |z| = r_n, \theta_{j_0} + 20\varepsilon \leqslant \arg z \leqslant \theta_{j_0+1} - 20\varepsilon\}$ possibly except a set with measure not greater than $\frac{L}{16}$, where L is the length of the arc.

Thus using the same method as in the proof of Theorem 4.1.1, we can deduce the result of Theorem 4.1.2. □

We guess that the result in Theorem 4.1.2 would not be true without the assumption about the growth of $f(z)$, but here we do not know how one get it.

The following is a result on the majorant of the harmonic measure, which follows from Theorem III.67, Tsuji[5] and will be used in the below discussion.

Lemma 4.1.4. *Let D be a domain in $\{z : r < |z| < R\}$ $(0 < r < R < \infty)$ such that both $\Gamma = \overline{D} \cap \{z : |z| = R\}$ and $\gamma = \overline{D} \cap \{z : |z| = r\}$ contain segment arcs and $\omega(z, \Gamma, D)$ and $\omega(z, \gamma, D)$ be the harmonic measures of, respectively, Γ and γ with respect to D at $z \in D$. Then for $z \in D$ we have*

$$\omega(z, \Gamma, D) \leqslant \frac{3}{\sqrt{1-\eta}} \frac{c+1}{c-1} \exp\left(-\pi \int_{c|z|}^{\eta R} \frac{dt}{t\Theta(t)}\right) \tag{4.1.19}$$

and

$$\omega(z, \gamma, D) \leqslant \frac{3}{\sqrt{1-\eta}} \frac{c+1}{c-1} \exp\left(-\pi \int_{r/\eta}^{|z|/c} \frac{dt}{t\Theta(t)}\right), \tag{4.1.20}$$

$0 < \eta < 1$ and $1 < c$, where $\Theta(t)$ is defined in this way: when $\{z : |z| = t\}$ is wholly in D, $\Theta(t) = \infty$; otherwise, $\Theta(t)$ is the quantity such that $t\Theta(t)$ is the arc length of the part of the circle $\{z : |z| = t\}$ in D.

Proof. In fact, (4.1.19) follows from the proof of Corollary of Theorem III.67, Tsuji[5] and however we deduce (4.1.20) in view of (4.1.19).

Define transformation $w = T(z) = Rr/z$. Then using the result on (4.1.19) yields that

$$\begin{aligned}\omega(z,\gamma,D) &= \omega(T(z),T(\gamma),T(D))\\ &\leqslant \frac{3}{\sqrt{1-\eta}}\frac{c+1}{c-1}\exp\left(-\pi\int_{c|T(z)|}^{\eta R}\frac{\mathrm{d}t}{t\Theta^*(t)}\right),\end{aligned}$$

where $\Theta^*(t) = \Theta(Rr/t)$. By means of the formula for integration by transformation $x = Rr/t$, we have

$$\int_{c|T(z)|}^{\eta R}\frac{\mathrm{d}t}{t\Theta^*(t)} = -\int_{|z|/c}^{r/\eta}\frac{\mathrm{d}x}{x\Theta(x)} = \int_{r/\eta}^{|z|/c}\frac{\mathrm{d}t}{t\Theta(t)}.$$

Then (4.1.20) follows. □

It is easy to see that the harmonic measure in Lemma 4.1.4 will become small provided that $\Theta(t)$ is small and/or $\frac{kr}{2|z|}$ is large by noting $\Theta(t) \leqslant 2\pi$. For the domain $D = \{z : r < |z| < R, \alpha < \arg z < \beta\}$, elementary estimates of harmonic measures in Lemma 4.1.4 are given in Lemma 7.4 of Yang [6]. However, Lemma 4.1.4 is able to be used in discussion of the annuli in question divided by B-regular curves in below theorems and in problem on asymptotic values and direct singularities in Chapter 6.

The method used to prove the below Theorems 4.1.3 and 4.1.4 is that to estimate $\log|f(z) - a|$ in terms of its values on the boundary by applying the two constant theorem of harmonic measure, whose idea is essentially due to Edrei and Fuchs [1].

Theorem 4.1.3. *Under the same assumption as in Theorem 4.1.1, assume in addition that $f(z)$ is analytic in each A_n with sufficiently large κ and $\{r_n\}$ is a sequence of relaxed Pólya peak with order $\sigma > 0$. Then we have*

$$\sum_{j=0}^{\infty} p_j \leqslant \min\{s/2,\ 2\sigma\}. \tag{4.1.21}$$

Proof. We continue to use the notations in the proof of Theorem 4.1.1. We have known that a_l and b_k are associated to, respectively, curvilinear quadrilaterals $(A_n^{(j)})_{20\varepsilon}$ and $(A_n^{(i)})_{20\varepsilon}$ in which but a small set (4.1.17) and its alternation for b_k hold for arbitrarily fixed sufficiently small $0 < \varepsilon < \frac{\pi}{200\sigma}$. Since $f(z)$ is analytic, the above-mentioned inequalities hold in the whole domains $(A_n^{(j)})_{20\varepsilon}$ and $(A_n^{(i)})_{20\varepsilon}$ respectively, that is,

$$\log|f^{(l)}(z) - a_l| \leqslant -K_0 T(r_n, f),\ z \in (A_n^{(j)})_{20\varepsilon} \tag{4.1.22}$$

and

$$\log|f^{(k)}(z) - b_k| \leqslant -K_0 T(r_n, f),\ z \in (A_n^{(i)})_{20\varepsilon}, \tag{4.1.23}$$

where K_0 is a positive constant only depending on ε. It is obvious that the associated $A_n^{(j)}$ and $A_n^{(i)}$ to a_l and b_k are distinct. We want to prove that they are not next to each other. Suppose they are next to each other, that is to say $A_n^{(i)} = A_n^{(j+1)}$ or $A_n^{(j)} = A_n^{(i+1)}$. Without any loss of generalities we assume $A_n^{(i)} = A_n^{(j+1)}$. Then $(\alpha_{nj}, \alpha_{n(j+1)})$ and $(\alpha_{n(j+1)}, \alpha_{n(j+2)})$ are associated to a_l and b_k respectively. We assume that $k \geqslant l$ and consider the domain

$$U_n = \{z : \alpha_{n(j+1)} - 20\varepsilon < \arg z < \alpha_{n(j+1)} + 20\varepsilon, c^{-3}r_n < |z| < c^3 r_n\},$$

where $c = \sqrt[4]{K}$. We denote by Γ_1 the part on the circles $\{z : |z| = c^3 r_n\}$ and $\{z : |z| = c^{-3}r_n\}$ of the boundary of U_n and set $\Gamma_2 = \partial U_n \setminus \Gamma_1$. In virtue of Lemma 4.1.4, on the arc $\{z : |z| = r_n\} \cap U_n$ we have

$$\begin{aligned}\omega(z, \Gamma_1, U_n) \leqslant & \frac{3}{\sqrt{1-1/c}} \frac{c+1}{c-1} \left(\exp\left(-\pi \int_{cr_n}^{c^3 r_n/c} \frac{dt}{40\varepsilon t} \right) \right. \\ & \left. + \exp\left(-\pi \int_{c^{-3}r_n/c^{-1}}^{r_n/c} \frac{dt}{40\varepsilon t} \right) \right) \\ = & \frac{6\sqrt{c}}{\sqrt{c-1}} \frac{c+1}{c-1} \exp\left(-\frac{\pi}{40\varepsilon} \log c \right).\end{aligned}$$

Let Γ_1' be the boundary of $U_n' = \{z : \alpha_{n(j+1)} - 20\varepsilon < \arg z < \alpha_{n(j+1)} + 20\varepsilon, r_n/2 < |z| < 2r_n\}$ on the circles $\{z : |z| = 2r_n\}$ and $\{z : |z| = r_n/2\}$. Then as in above estimate with $c^3 = 2$ we have

$$\omega(z, \Gamma_1', U_n') \leqslant C \exp\left(-\frac{\pi}{120\varepsilon} \log 2 \right)$$

on the arc $\{z : |z| = r_n\} \cap U_n$, where C is an absolute constant. According to the basic properties of harmonic functions, for small $\varepsilon > 0$ we have

$$\omega(z, \Gamma_1, U_n) + \omega(z, \Gamma_2 \setminus \Gamma_2', U_n) \leqslant \omega(z, \Gamma_1', U_n') < \frac{1}{2},$$

where $\Gamma_2' = \Gamma_2 \cap \{z : r_n/2 \leqslant |z| \leqslant 2r_n\}$ and hence $\omega(z, \Gamma_2', U_n) > \frac{1}{2}$. Using the method in the proof of Theorem 4.1.1 we have that (4.1.22) and (4.1.23) hold respectively on the two segments of Γ_2 with the coefficients K_0 depending on c and ε. Since $k \geqslant l$, in view of (4.1.13), (4.1.22) and (4.1.23), we have

$$\log |f^{(k+1)}(z)| \leqslant 0, \ z \in \Gamma_2$$

and

$$\log |f^{(k+1)}(z)| \leqslant -K_1 T(r_n, f), \ z \in \Gamma_2',$$

where K_1 is a positive constant depending only on ε. On $\{z : c^{-3}r_n \leqslant |z| \leqslant c^3 r_n\}$, in virtue of Lemma 2.1.3, we have

$$\log|f^{(k+1)}(z)| \leqslant K_2 T(c^3\sqrt{c}r_n, f^{(k+1)}) \leqslant K_2 K_3 \kappa^\sigma (k+2) T(r_n, f),$$

where K_2 and K_3 do not depend on c when $c \geqslant 2$. From the two constant theorem of harmonic measure, by noting that $0 < \varepsilon < \frac{\pi}{200\sigma}$ we have

$$\begin{aligned}\log|f^{(k+1)}(z)| &\leqslant \omega(z,\Gamma_1,U_n)K_2K_3\kappa^\sigma(k+2)T(r_n,f) - \omega(z,\Gamma_2',U_n)K_1T(r_n,f),\\ &\leqslant K_2K_3(k+2)\frac{6\sqrt{c}}{\sqrt{c}-1}\frac{c+1}{c-1}\exp(-\sigma\log c)T(r_n,f) - \frac{K_1}{2}T(r_n,f),\\ &< -\frac{K_1}{3}T(r_n,f), \text{ on } |z| = r_n \text{ and } z \in U_n\end{aligned}$$

for the small fixed ε and sufficiently large κ.

If $k > l$, from (4.1.22) and (4.1.13) we have

$$\log|f^{(k)}(z)| \leqslant -\frac{K_0}{2}T(r_n,f),\ z \in (A_n^{(j)})_{20\varepsilon}.$$

Then

$$\begin{aligned}|b_k| &\leqslant |f^{(k)}(r_n\exp(\mathrm{i}(\alpha_{n(j+1)} - 20\varepsilon))) - f^{(k)}(r_n\exp(\mathrm{i}(\alpha_{n(j+1)} + 20\varepsilon)))|\\ &\quad + |f^{(k)}(r_n\exp(\mathrm{i}(\alpha_{n(j+1)} - 20\varepsilon)))| + |f^{(k)}(r_n\exp(\mathrm{i}(\alpha_{n(j+1)} + 20\varepsilon))) - b_k|\\ &\leqslant 80\varepsilon r_n \exp\left(-\frac{K_1}{3}T(r_n,f)\right) + \exp(-\frac{K_0}{2}T(r_n,f)) + \exp(-K_0T(r_n,f))\\ &\to 0, \text{ as } n \to \infty,\end{aligned}$$

that is, $b_k = 0$, a contradiction is derived. Thus $k = l$. In this case, in view of the same method as above, we can deduce $a_l = b_k$, a contradiction is also derived. Thus $A_n^{(j)}$ and $A_n^{(i)}$ associated to a_l and b_k are not next to each other. It is obvious that

$$\sum_{j=0}^{\infty} p_j \leqslant \frac{s}{2}.$$

The inequality $\sum_{j=0}^{\infty} p_j \leqslant 2\sigma$ is able to follow from the proof of below Theorem 4.1.4. □

A ray $\arg z = \theta$ is called a T cluster line of $f(z)$ for a-points if for arbitrary $\varepsilon > 0$

$$\limsup_{r\to\infty} \frac{N(r, Z_\varepsilon(\theta), f = a)}{T(r,f)} > 0.$$

Then we have the following consequence of Theorem 4.1.3.

Corollary 4.1.1. *Let $f(z)$ be a transcendental entire function with the finite positive lower order μ. Suppose that s is the number of T cluster lines for zeros and poles of $f(z)$ and $s < \infty$. Then*

$$\sum_{j=0}^{\infty} p_j \leqslant \min\left\{\frac{s}{2}, 2\mu\right\}.$$

If we consider T directions, then we have the following

Theorem 4.1.4. *Let $f(z)$ be a transcendental entire function with the finite positive lower order μ. Set*

$$\Omega = \{\theta \in [0, 2\pi) : \arg z = \theta \text{ is a common } T \text{ direction of } f(z) \text{ and every } f^{(j)}\}.$$

If Ω has measure zero, then we have

$$\sum_{j=-\infty}^{\infty} p_j \leqslant 2\mu. \tag{4.1.24}$$

Proof. Suppose that (4.1.24) does not hold. Then we choose p non-zero values a_{kj} $(k = 1, 2, \cdots, p; p > 2\mu)$ which are Nevanlinna deficient values of $f^{(j)}(z)$ $(-\infty < j < \infty)$ and $a_{sj} \neq a_{tj}$ for $s \neq t$ (However, a_{kj} is allowed to be equal for distinct j). Write

$$\delta = \min_{k,j}\{\delta(a_{kj}, f^{(j)})\} > 0 \text{ and } q = \max\{j : a_{kj} \text{ is a deficient value of } f^{(j)}\}.$$

Since the lower order μ of $f(z)$ is finite, in virtue of Theorem 2.6.3 and Lemma 2.8.1, there exist a sequence of common relaxed Pólya peaks $\{r_n\}$ of order μ for each $f^{(j)}$ which is decided by a_{kj} such that

$$\min_{k,j}\{\text{mes}(E_n(a_{kj}, f^{(j)}))\} \geqslant B > 0,$$

where B is a constant independent of n. Since Ω is a compact set, $\Phi = [0, 2\pi) \setminus \Omega$ consists of at most countably infinite number of maximum open intervals and there exist such s maximum open intervals I_i $(1 \leqslant i \leqslant s)$ such that $\text{mes}(\Phi \setminus \bigcup_{i=1}^{s} I_i) < p^{-1}B$ and hence from $\text{mes}\Omega = 0$, $\text{mes}([0, 2\pi) \setminus \bigcup_{i=1}^{s} I_i) < p^{-1}B$. We denote by $\Omega_i = \Omega(\alpha_i, \beta_i)$ the angular domain corresponded by I_i and then no common T directions are contained in Ω_i $(1 \leqslant i \leqslant s)$. As in the proof of Theorem 4.1.1 and Theorem 4.1.3, each a_{kj} is associated to at least one $I_{i(kj)}$ such that

$$\log|f^{(j)}(z) - a_{kj}| < -K(M)T(r_n, f), \tag{4.1.25}$$

$$e^{-M}r_n \leqslant |z| \leqslant e^{M}r_n \text{ and } \arg z \in \Omega(\alpha_{i(kj)} + 20\varepsilon, \beta_{i(kj)} - 20\varepsilon),$$

where $K(M)$ is a positive constant depending on M (Please see the proof of Theorems 4.1.1 and 4.1.3). As in the proof of Theorem 4.1.1, $I_{i(kj)}$ does not intersect each other. According to the increasing order of $\alpha_{i(kj)} + 20\varepsilon$ and $\beta_{i(kj)} - 20\varepsilon$, we write them in γ_i and θ_i in turn. We consider the angle $S_i = \Omega(\theta_i, \gamma_{i+1})$ bounded by the rays $\arg z = \theta_i$ and $\arg z = \gamma_{i+1}$. Set

$$\Theta_i(t) = \gamma_{i+1}(t) - \theta_i(t)$$

where $\theta_i(t)$ is the argument of intersecting point of ray $\arg z = \theta_i$ and circle $|z| = t$ and the same significance is given to $\gamma_{i+1}(t)$ (Here in fact, $\theta_i(t) \equiv \theta_i$, while the statement is required in consideration of B regular curves). Since applying the Schwarz's inequality and the obvious fact $\sum \Theta_i(t) \leqslant 2\pi$ yields

$$p^2 = \left\{ \sum_{i=1}^{p} (\Theta_i(t))^{1/2} (\Theta_i(t))^{-1/2} \right\}^2 \leqslant 2\pi \sum_{i=1}^{p} (\Theta_i(t))^{-1},$$

we therefore have

$$\frac{1}{2} p^2 (M - 2\log 2) = \frac{1}{2} p^2 \int_{2r_n}^{\frac{1}{2}e^M r_n} \frac{dt}{t} \leqslant \sum_{i=1}^{p} \pi \int_{2r_n}^{\frac{1}{2}e^M r_n} \frac{dt}{t\Theta_i(t)}$$

and there exists at least one angle S_{i_0} such that

$$\pi \int_{2r_n}^{\frac{1}{2}e^M r_n} \frac{dt}{t\Theta_{i_0}(t)} \geqslant \frac{1}{2} p(M - 2\log 2). \tag{4.1.26}$$

In what follows, we confine our discussion to the angle $S = S_{i_0}$. Denote by Γ_1 the part of the boundary $\arg z = \theta = \theta_{i_0}$ and $\arg z = \gamma = \gamma_{i_0+1}$ of S in $e^{-M} r_n \leqslant |z| \leqslant e^M r_n$, by Γ_2 the arc of $|z| = e^M r_n$ in S and by Γ_3 the arc of $|z| = e^{-M} r_n$ in S. Let D be the domain bounded by Γ_1, Γ_2 and Γ_3 and for a $1 < Q \leqslant e^M$, set

$$U = \{z : \theta < \arg z < \gamma,\ r_n/Q < |z| < Qr_n\},$$

$\Gamma_1' = \Gamma_1 \cap \{z : r_n/Q \leqslant |z| \leqslant Qr_n\}$ and $\Gamma = \partial U \setminus \Gamma_1'$. From Lemma 4.1.4, it follows, by noting $\Theta_{i_0}(t) \leqslant 2\pi$, that

$$\begin{aligned}\omega(z, \Gamma, U) &\leqslant 9\sqrt{2} \exp\left(-\frac{1}{2} \int_{2|z|}^{\frac{1}{2}Qr_n} \frac{dt}{t} \right) + 9\sqrt{2} \exp\left(-\frac{1}{2} \int_{2r_n/Q}^{|z|/2} \frac{dt}{t} \right) \\ &= 36\sqrt{\frac{2}{Q}}, \quad \text{on } |z| = r_n,\end{aligned}$$

and therefore choosing a sufficiently large Q, we have

$$\omega(z, \Gamma_1', D) \geqslant 1 - \omega(z, \Gamma, U) > \frac{1}{2}, \quad \text{on } |z| = r_n.$$

It follows from (4.1.26) that

$$\begin{aligned}\omega(z,\Gamma_2,D) &\leqslant 9\sqrt{2}\exp\left(-\pi\int_{2|z|}^{\frac{1}{2}e^M r_n}\frac{dt}{t\Theta_{i_0}(t)}\right)\\ &\leqslant 9\sqrt{2}\exp\left(-\frac{1}{2}p(M-2\log 2)\right)\\ &= 9\sqrt{2}\times 2^p\exp(-\frac{p}{2}M),\ \text{on } |z|=r_n,\end{aligned}$$

From discussion of the first paragraph and (4.1.25), for simple writing we have

$$\log|f^{(l)}(z)-a| < -K(Q)T(r_n,f),\ \text{on } \Gamma_1'\cap\{z:\arg z=\theta\}$$

$$\log|f^{(k)}(z)-b| < -K(Q)T(r_n,f),\ \text{on } \Gamma_1'\cap\{z:\arg z=\gamma\}$$

and $\log|f^{(l)}(z)-a|<0$, on $\Gamma_1\cap\{z:\arg z=\theta\}$; $\log|f^{(k)}(z)-b|<0$, on $\Gamma_1\cap\{z:\arg z=\gamma\}$ for two non-zero values a and b and two integers l and k. If $k>l$, then in view of the equality (4.1.13), we have

$$\log|f^{(k)}(z)| < -K_1(Q)T(r_n,f),\ \text{on } \Gamma_1'\cap\{z:\arg z=\theta\}.$$

Therefore, we can assume that $k=l$ and $a\neq b$, because the below argument is also available in treating the above case.

Now we estimate $\log|f^{(k+1)}(z)|$ on the part of $|z|=r_n$ in D. In this time, we have

$$\log|f^{(k+1)}(z)| < -K_1(Q)T(r_n,f),\ \text{on } \Gamma_1'.$$

From Lemma 2.1.3, (2.6.1) and 3) of Definition 1.1.1 for relaxed Pólya peak, it follows that

$$\log M(e^M r_n, f^{(k+1)}) \leqslant 3m(2e^M r_n, f^{(k+1)}) \leqslant K_2 e^{\mu M}T(r_n,f)$$

and

$$\log M(e^{-M} r_n, f^{(k+1)}) \leqslant 3m(2e^{-M} r_n, f^{(k+1)}) \leqslant K_2 e^{-\mu M}T(r_n,f),$$

where K_2 is independent of n. Application of the two constant theorem of harmonic measure yields that

$$\begin{aligned}\log|f^{(k+1)}(z)| &\leqslant -\omega(z,\Gamma_1',D)K_1T(r_n,f)\\ &\quad+\omega(z,\Gamma_2,D)K_2e^{\mu M}T(r_n,f)+\omega(z,\Gamma_3,D)K_2e^{-\mu M}T(r_n,f)\\ &\leqslant -\frac{K_1}{2}T(r_n,f)+9\sqrt{2}\times 2^pK_2\exp((\mu-\frac{p}{2})M)T(r_n,f)\\ &\quad+K_2\exp(-\mu M)T(r_n,f)\\ &\leqslant -\frac{K_1}{4}T(r_n,f),\ \text{on } \{|z|=r_n\}\cap D\end{aligned}$$

holds for sufficiently large M. Thus for any two points $z_1,z_2\in\{z:|z|=r_n\}\cap D$,

$$|f^{(k)}(z_1) - f^{(k)}(z_2)| \leqslant 2\pi r_n \exp\left(-\frac{K_1}{4}T(r_n,f)\right) \to 0, \text{ as } n \to \infty,$$

this deduces definitely that $a = b$, a contradiction is derived. Theorem 4.1.4 follows. □

4.2 Retrospection

In 1962, A. Edrei and W. H. J. Fuchs published two papers [1] [2] in which they estimated the number of deficient values in terms of the distribution of argument of zeros and poles of a meromorphic function. Here we state some of their celebrated results.

Theorem 4.2.1. *(Edrei and Fuchs, 1962) Let $f(z)$ be a transcendental meromorphic function. Assume that there exist a number δ $(0 \leqslant \delta < 1)$ and a positive, increasing, unbounded sequence $\{\rho_k\}$ such that the annuli A_k defined by*

$$\frac{\rho_k}{\sigma_k} \leqslant |z| \leqslant \sigma_k\rho_k, \quad k = 1,2,3,\cdots$$

with $\sigma_k = 1 + \{\log T(\rho_k, f)\}^{-\delta}$ have the following property.

With each A_k it is possible to associate $s(\geqslant 1)$ arguments

$$0 \leqslant \alpha_{k1} < \alpha_{k2} < \cdots < \alpha_{ks} < \alpha_{k1} + 2\pi$$

such that there are at most $O(T^d(\rho_k, f))$ $(1 > d = \text{constant})$ zeros and poles of $f(z)$ in the portion of A_k and outside the s sectors

$$|\arg z - \alpha_{kj}| \leqslant \{\log T(\sigma_k\rho_k, f)\}^{-1-\eta}, \quad j = 1,2,\cdots,s;\ \delta < \eta = \text{constant}.$$

Then $f(z)$ has at most $s+1$ deficient values. Moreover, if $s+1$ values are exactly deficient, then 0 and ∞ are among them.

Theorem 4.2.2. *(Edrei and Fuchs, 1962) Let $f(z)$ be an entire function of the finite order λ and let $L_1, L_2, \cdots, L_s$ $(L_j : z = z(t) = te^{i\alpha_j(t)})$ be the s B- regular paths. Given a fixed $\delta > 0$, $\overline{n}_\delta(r)$ denotes the number of distinct zeros of f in $r_0 \leqslant |z| \leqslant r$ but outside the s sectors:*

$$\alpha_j(t) - \delta \leqslant \arg z \leqslant \alpha_j(t) + \delta, \ |z| = t.$$

Assume that for every fixed δ, we have

$$\lim_{r\to\infty} \frac{\overline{n}_\delta(r)}{T(r,f)} = 0.$$

If p is the number of finite non-zero deficient values of $f(z)$, then $p \leqslant \min\{2\lambda, s\}$.

Theorem 4.2.3. *(Edrei and Fuchs, 1962) Let $f(z)$ be an entire function of the finite order λ. Set*

$$\Omega = \{\theta \in [0, 2\pi) : z = |z|e^{i\theta} \text{ is a zero of } f\}.$$

If Ω is of measure zero, then f has at most 2λ deficient values other than 0 and ∞.

In 1954, A. A. Gol'dberg proved that for any λ and an at most countable set $E \subset \mathbb{C}$, there exits a meromorphic function with order λ and exactly with each element of E as its deficient value. N. U. Arakelyan in 1966 and A. E. Eremenko in 1987 obtained the same result for the case of an entire function. These describes significance of Theorems of Edrei and Fuchs in some extent.

Oum K. in [3] proved that an entire function of order $0 < \lambda < +\infty$ has at most 2λ finite deficient values if it is of completely regular growth. In 1975, Yang and Zhang [8] considered and revealed a relation between the number of deficient values and the number of the Borel directions.

Theorem 4.2.4. *(Yang L. and Zhang G. H., 1975) Let $f(z)$ be a transcendental meromorphic function with the finite positive order λ. If we denote by q the number of Borel directions and by p the number of deficient values, then $p \leqslant q$.*

In [9], for the case of entire function, they obtained more precise result than above Theorem 4.2.4.

Theorem 4.2.5. *(Yang L. and Zhang G. H., 1975) Let $f(z)$ be an entire function of order $\lambda (0 < \lambda < +\infty)$. If the number of Borel directions, denote by q, is finite, then*

$$\sum_{l=0}^{\infty} p_l \leqslant \min\left\{2\lambda, \frac{q}{2}\right\}.$$

Let $f(z)$ be a transcendental meromorphic function with the finite positive order λ. A ray $\arg z = \theta$ is called a cluster line of order λ of f for a-points, if for arbitrary $\varepsilon > 0$, (3.3.4) holds for $\rho = \lambda$.

Theorem 4.2.6. *(Yang L. and Zhang G. H. [10], 1982) Let $f(z)$ be a transcendental meromorphic function with the finite positive order λ. Suppose that q is the number of cluster lines of order λ of $f(z)$ for zeros and poles. Then*

$$\sum_{j=0}^{\infty} p_j \leqslant q.$$

Moreover, if f is entire and $q < \infty$, then

$$\sum_{j=0}^{\infty} p_j \leqslant \min\left\{\frac{q}{2}, 2\lambda\right\}. \tag{4.2.1}$$

Theorem 4.2.5 follows immediately from Theorem 4.2.6 and Valiron Theorem 2.7.5.

In 1985, Zhang Q. D. and Pang X. C. [14] and Pang X. C. and Ru M. [4] took into account this subject on small deficient functions together with (common) Borel directions of a meromorphic function and its derivatives. Let $f(z)$ be a transcendental meromorphic function with finite positive order. Let $U(r)$ be a type function of $f(z)$. For a small function $a(z)$ with respect to $f(z)$, define

$$\delta^*(a,f) = 1 - \limsup_{r\to\infty} \frac{N(r,f=a)}{U(r)}, \quad a(z) \not\equiv \infty;$$

$$\delta^*(a,f) = 1 - \limsup_{r\to\infty} \frac{N(r,f)}{U(r)}, \quad a(z) \equiv \infty.$$

If $\delta^*(a,f) > 0$, $a(z)$ is called a precise deficient function of $f(z)$.

Theorem 4.2.7. *(Zhang and Pang, 1985) Let $f(z)$ be a transcendental meromorphic function with finite positive order. Let p^* be the number of precise deficient functions of $f(z)$ and q^* the number of common Borel directions of $f^{(j)}(z)(j = 0,1,2,\cdots)$. Then $p^* \leqslant q^*$ or $p^* \leqslant 1$.*

In 1988, Yang [7] considered the case of the lower order.

Theorem 4.2.8. *(Yang L., 1988) Let $f(z)$ be a transcendental entire function with the finite positive lower order μ. If f has finitely many Borel directions of order μ, then*

$$\sum_{j=0}^{-\infty} p_j \leqslant 2\mu.$$

Since a T direction must be a Borel direction of the lower order μ, Theorem 4.1.4 is a generalization of Yang Lo's Theorem 4.2.8. In fact, under the assumption of Theorem 4.2.8, $f(z)$ has only finitely many T directions.

In 1978 and 1983, Zhang [13] and [12] discovered a relation among the numbers of deficient values, asymptotical values and the Julia directions of an entire function.

Theorem 4.2.9. *(Zhang G. H., 1978, 1983) Let $f(z)$ be a transcendental entire function with the finite lower order μ. Then (1) $2p+l \leqslant J$ and (2) $p+l \leqslant 2\mu$ if $J < +\infty$, where p is the number of finite deficient values, l the number of finite distinct asymptotic values which are not deficient values and J the number of the Julia directions.*

Therefore, an entire function of the finite lower order has finitely many asymptotic values if it has finitely many Julia directions. In 1986, Zhang (see Theorem 5.14, [11]) generalized in fact the above-mentioned Theorem 4.2.9 and obtained the second inequality in Theorem 4.2.9 with "$J < \infty$" replaced by J directions $\arg z = \theta_k$, $k = 1,2,\cdots,J$ such that

$$\limsup_{r\to\infty} \frac{\log^+ n(r, \cup_{k=1}^{J} \Omega(\theta_k+\varepsilon, \theta_{k+1}-\varepsilon), f=0)}{\log r} = 0.$$

In 1978, Zhang at the same paper [13] considered the case of meromorphic functions and established the inequality $p+l \leqslant J$ with f being meromorphic and with asymptotic values replaced by direct singularities of the inverse of f (see Theorem 6.2, [11]) and in 1983, the inequality (2) under some additional condition (see Theorem 6.1, [11]).

Finally we conclude this section with the following questions.

Is Theorem 4.2.9 still true, provided that "the Julia directions" is replaced by "the Borel directions"?

And could we consider the Nevanlinna deficient values of the derivatives and the primitive in Theorem 4.2.9?

References

1. Edrei, A. and Fuchs, W. H. J., Bounds for the number of deficient values of certain classes of meromorphic functions, Proc. London Math. Soc., **12**(1962), 315-344.
2. Edrei, A. and Fuchs, W. H. J., On meromorphic functions with regions free of poles and zeros, Acta Math., **108**(1962), 113-145.
3. Oum Ki-Chou, Bounds for the number of deficient values of entire functions whose zeros have angular densities, Pacific J. Math., **29**(1969), 187-202.
4. Pang, X. C. and Ru, M., On total sum of deficient entire functions of entire functions, Chin. Ann. Math., (4)**6A**(1985), 411-424.
5. Tsuji, M., Potential theory in modern function theory, Maruzen Co. LTD. Tokyo, 1959.
6. Yang, L., New researches on value distribution, Science Press of China, 1982.
7. Yang, L., Deficient values and angular distribution of entire functions, Trans. Amer. Math. Soc., (2)**308**(1988), 583-601.
8. Yang, L. and Zhang, G. H., Recherches sur le nombre des valeurs déficientes et le nombre des directions de Borel des fonctions méromorphes, Sci. Sinica, **18**(1975), 23-37.
9. Yang, L. and Zhang, G. H., On total sum of deficiencies of entire functions, Acta Math. Sinica, **18**(1975), 35-53.
10. Yang, L. and Zhang, G. H., Distribution of zeros and poles of meromorphic functions and their filling disks, Scientia Sinica (Series A), (4)**XXV**(1982), 371-383.
11. Zhang, G. H., Theory of entire and meromorphic functions (in Chinese) , Science Press Sinica, 1986.
12. Zhang, G. H., On entire functions with finitly many Julia directions, Science in China, **9**(1983), 775-786 (in Chinese).
13. Zhang, G. H., On the relations among deficient values, asymptotic values and Julia directions of entire and meromorphic functions, Science in China, Extesion 1, (1978), 1-80 (in Chinese).
14. Zhang, Q. D. and Pang, X. C., Relations of the numbers of precise deficient functions and common Borel directions of meromorphic functions, Chin. Ann. Math., (4)**6A**(1985), 455-466.

Chapter 5
Meromorphic Functions with Radially Distributed Values

Jianhua Zheng
Department of Mathematical Sciences, Tsinghua University, Beijing 100084, P. R. China
jzheng@math.tsinghua.edu.cn

Abstract: A value on the extended complex plane is a radially distributed value of a transcendental meromorphic function if most of points at which the value is assumed distribute closely along a finite number of rays from the origin. In this chapter, we study the growth order of a meromorphic function with two radially distributed values and a distinct deficient value (in other words, this hints a condition under which deficient values do not exist). We respectively treat two cases: one is without assumption about the growth of the function considered and the other is under assumption of the function being of the finite lower order. The Nevanlinna characteristic for an angle plays crucial role in the investigation of this subject. Actually, the idea to study this subject is the following: the Nevanlinna characteristic $T(r, f)$ for $\{|z| < r\}$ is controlled by the corresponding proximate function $m(r, *)$ to a deficient value; there exists a solid relation between $m(r, *)$ and the sum of $B(r, *)$ on the finitely many angular domains; and according to some fundamental theorems, we estimate $B(r, *)$ in terms of two $C(r, **)$, which describes the number of poles of the function $**$ in the angle. Thus $T(r, f)$ is controlled in terms of the number of value-points in the angles. This way also produces discussion of the growth order dealing with other type of radially distributed values. Finally, we simply survey the background and other main results of the subject.

Key words: Angle Nevanlinna characteristic, Growth order, Radially distributed value

This chapter is devoted to discussing how the growth of a meromorphic function could be affected by distribution of the arguments of its a-points (i.e., points at which the function assumes the value a). Radially distributed values mean such values that most of corresponding value-points distribute nearly along a finite number of rays from the origin. We shall determine an simple approach to make discussion of this subject, roughly speaking, certain radially distributed values will affect the growth of function provided that in an angle $B(r, *)$ can be controlled by $C(r, **)$ related to these values. Therefore, we shall proceed with the Nevanlinna characteristic on an angle. In the first part, we shall consider the functions with radially distributed values without any restriction on its growth and then in the second part, take those functions of finite lower order into account.

5.1 Growth of Such Meromorphic Functions

In this section, our basic idea is to use $B(r,*)$ on angles to control the Nevanlinna characteristic on disks and then use $C(r,**)$ and further the number of value-points on angles to estimate $B(r,*)$ so that the Nevanlinna characteristic on disks can be controlled by the number of value-points on angles. Thus we attain the purpose that the order of the function with suitable radially distributed values is bounded from above in term of the arguments of rays in question.

Given

$$-\pi \leqslant \alpha_1 < \alpha_2 < \cdots < \alpha_q < \pi, \ \alpha_{q+1} = \alpha_1 + 2\pi, \tag{5.1.1}$$

we set

$$D(\alpha_1, \alpha_2, \cdots, \alpha_q) = \cup_{j=1}^{q} \{z : \arg z = \alpha_j\}$$

and consider the following quantity

$$W(r, D, f = a) = \max\left\{ r^{\omega_j} B_{\alpha_j, \alpha_{j+1}}\left(r, \frac{1}{f-a}\right) : 1 \leqslant j \leqslant q \right\},$$

where $\omega_j = \frac{\pi}{\alpha_{j+1}-\alpha_j}$ and $\omega = \max\{\omega_j : 1 \leqslant j \leqslant q\}$. Our first result is to control the characteristic in terms of W, i.e., B.

Theorem 5.1.1. *Let $f(z)$ be a transcendental meromorphic function. Assume that a is a Nevanlinna deficient value of $f^{(p)}$ for an integer p. Then for a fixed $\tau > 0$ and for all r possibly outside a set of finite logarithmic measure, we have*

$$T(r, f) \leqslant K_1 (\log T(r, f))^{2+\tau} W(r, D, f^{(p)} = a) \tag{5.1.2}$$

and if, in addition, $W(r, D, f^{(p)} = a)$ is of finite order, then given $\varepsilon > 0$, we have

$$T(r, f) \leqslant K_2 W(r, D, f^{(p)} = a), \ r \notin E$$

(for negative p, $f^{(p)}$ stands for the $|p|$th primitive of f, if it exists), where K_1 and K_2 are positive constants and K_2 depends on ε and $\overline{\log \mathrm{dens}}(E) < \varepsilon$.

Proof. Here we assume $p \geqslant 0$, while we leave the proof of the case $p < 0$ to the reader. For each r let $\varepsilon(r)$ be a positive number which will be determined in the sequel. From the definition of $B_{\alpha,\beta}(r,*)$, we have

$$\begin{aligned} \int_{\alpha_j+\varepsilon(r)}^{\alpha_{j+1}-\varepsilon(r)} \log^+ \frac{1}{|f^{(p)}(re^{i\theta}) - a|} d\theta &\leqslant \frac{\pi}{2\omega_j \sin(\varepsilon(r)\omega_j)} r^{\omega_j} B_{\alpha_j, \alpha_{j+1}}\left(r, \frac{1}{f^{(p)} - a}\right) \\ &\leqslant \frac{\pi}{2\omega_j \sin(\varepsilon(r)\omega_j)} W(r, D, f^{(p)} = a). \end{aligned} \tag{5.1.3}$$

In view of Lemma 2.1.5 for $R = r(1+(\log T(r,f^{(p)}))^{-1-\tau})$, it follows that

$$\begin{aligned}\int_{\alpha_j-\varepsilon(r)}^{\alpha_j+\varepsilon(r)} \log^+ \frac{1}{|f^{(p)}(re^{i\theta})-a|}d\theta &\leqslant 14[1+(\log T(r,f^{(p)}))^{1+\tau}]T(R,f^{(p)})\\ &\quad\times 2\varepsilon(r)\left(1+\log^+\frac{1}{2\varepsilon(r)}\right)\\ &\leqslant 29e(\log T(r,f^{(p)}))^{1+\tau}T(r,f^{(p)})\\ &\quad\times\varepsilon(r)\left(1+\log^+\frac{1}{\varepsilon(r)}\right), \end{aligned} \tag{5.1.4}$$

where the second inequality follows from Corollary 1.1.1 for r outside a set of finite logarithmic measure. Now choose

$$\varepsilon(r) = \frac{\delta}{4}\frac{1}{29eq}(\log T(r,f^{(p)}))^{-1-2\tau},\ \delta = \delta(a,f^{(p)})$$

and hence

$$\varepsilon(r)\left(1+\log^+\frac{1}{\varepsilon(r)}\right) \leqslant \frac{\delta}{4}\frac{1}{29eq}(\log T(r,f^{(p)}))^{-(1+\tau)}$$

for sufficiently large r. Thus combining (5.1.3) and (5.1.4), by noting that a is a deficient value of $f^{(p)}$ we have

$$\begin{aligned}\frac{\delta}{2}T(r,f^{(p)}) &\leqslant m\left(r,\frac{1}{f^{(p)}-a}\right)\\ &\leqslant \frac{\delta}{4}T(r,f^{(p)})+K(\log T(r,f^{(p)}))^{1+2\tau}W(r,D,f^{(p)}=a),\end{aligned}$$

for a positive constant K and therefore

$$T(r,f^{(p)}) \leqslant \frac{4K}{\delta}(\log T(r,f^{(p)}))^{1+2\tau}W(r,D,f^{(p)}=a).$$

In virtue of Theorem 2.6.2, we have

$$T(r,f) \leqslant T(r,f^{(p)})(\log T(r,f^{(p)}))^{1+\tau}$$

for all r outside a set of finite logarithmic measure. Thus noting $T(r,f^{(p)}) < (p+2)T(r,f)$, we have

$$\begin{aligned}T(r,f) &\leqslant \frac{4K}{\delta}(\log T(r,f^{(p)}))^{2+3\tau}W(r,D,f^{(p)}=a)\\ &\leqslant 2\frac{4K}{\delta}(\log T(r,f))^{2+3\tau}W(r,D,f^{(p)}=a)\end{aligned}$$

for all r outside a set of finite logarithmic measure. This yields (5.1.2) by replacing 3τ with τ.

Now assume that $W(r,D,f^{(p)}=a)$ is of finite order. Employing (5.1.2) deduces that $T(r,f)$ and hence $T(r,f^{(p)})$ is of finite order $\lambda(f)$. Take a number $C>1$ with $\lambda(f)\frac{\log 2}{\log C}<\varepsilon$ and set

$$E=\{r:\ T(2r,f^{(p)})\geqslant CT(r,f^{(p)})\}.$$

In virtue of Lemma 1.1.8 we know that $\overline{\log\operatorname{dens}}E<\varepsilon$.

In the above discussion we fix the $\varepsilon(r)\equiv\eta$, a positive constant, such that

$$56C\eta\left(1+\log^+\frac{1}{\eta}\right)=\frac{\delta}{4}.$$

Then as in above we have

$$\begin{aligned}\frac{\delta}{2}T(r,f^{(p)})&\leqslant m\left(r,\frac{1}{f^{(p)}-a}\right)\\&\leqslant 14\times 2T(2r,f^{(p)})2\eta\left(1+\log^+\frac{1}{2\eta}\right)+KW(r,D,f^{(p)}=a)\\&\leqslant\frac{\delta}{4}T(r,f^{(p)})+KW(r,D,f^{(p)}=a),\ r\notin E,\end{aligned}$$

so that, in view of Chuang's inequality (see Theorem 2.6.1),

$$T(r,f)\leqslant C_2T(2r,f^{(p)})\leqslant C_2CT(r,f^{(p)})\leqslant\frac{4C_2CK}{\delta}W(r,D,f^{(p)}=a),r\notin E.$$

Thus Theorem 5.1.1 follows. □

From Theorem 5.1.1 and a modified version of the Milloux inequality (2.2.8), we come to estimate the growth of meromorphic functions in terms of distribution of arguments of points of two values. This is one of our main purposes of this chapter. This process immediately deduces some remarkable results of Ostrovskii and actually we can get more. However, the proof of Theorem 5.1.1 is simpler and more elementary than that of Ostrovskii's Theorem 5.3.4 (I). The reader is referred to Section Retrospection for further review about that.

Let us begin with the following lemma which follows from the Milloux inequality (2.2.8).

Lemma 5.1.1. *Let $f(z)$ be a meromorphic function in $\mathbb{C}$ and consider an angular domain $\Omega(\alpha,\beta)$. Then*

(1) *for two integers $p\geqslant k\geqslant 0$ and $a\in\mathbb{C}\setminus\{0\}$, we have*

$$B_{\alpha,\beta}\left(r,\frac{1}{f^{(p)}-a}\right)\leqslant(p-k+1)\overline{C}_{\alpha,\beta}(r,f)+(2p+1-2k)\overline{C}_{\alpha,\beta}\left(r,\frac{1}{f^{(k)}}\right)+R_{\alpha,\beta}(r,f);\tag{5.1.5}$$

(2) *for $k > p$ and $a \in \mathbb{C}$, we have*

$$B_{\alpha,\beta}\left(r, \frac{1}{f^{(p)}-a}\right) \leqslant \overline{C}_{\alpha,\beta}(r,f) + \overline{C}_{\alpha,\beta}\left(r, \frac{1}{f^{(k)}-1}\right) + R_{\alpha,\beta}(r,f). \tag{5.1.6}$$

Proof. We first prove the case (1). From (2.2.8), we have

$$\begin{aligned}
B\left(r, \frac{1}{f^{(p)}-a}\right) &\leqslant S(r, f^{(p)}) - C\left(r, \frac{1}{f^{(p)}-a}\right) + O(1) \\
&\leqslant S\left(r, \frac{f^{(p)}}{f^{(k)}}\right) + S(r, f^{(k)}) - C\left(r, \frac{1}{f^{(p)}-a}\right) + O(1) \\
&\leqslant (p-k)\overline{C}(r,f) + (p-k)\overline{C}\left(r, \frac{1}{f^{(k)}}\right) + \overline{C}(r, f^{(k)}) \\
&\quad + C\left(r, \frac{1}{f^{(k)}}\right) - C\left(r, \frac{1}{f^{(p+1)}}\right) + R(r,f) \\
&\leqslant (p+1-k)\overline{C}(r,f) + (2p+1-2k)\overline{C}\left(r, \frac{1}{f^{(k)}}\right) + R(r,f).
\end{aligned}$$

This is (5.1.5). To prove the case (2), by (2.2.8), we have

$$\begin{aligned}
B\left(r, \frac{1}{f^{(p)}-a}\right) &= S(r, f^{(p)}-a) - C\left(r, \frac{1}{f^{(p)}-a}\right) + O(1) \\
&\leqslant \overline{C}(r, f^{(p)}-a) + C\left(r, \frac{1}{f^{(k)}-1}\right) - C\left(r, \frac{1}{f^{(k+1)}}\right) + R(r,f) \\
&\leqslant \overline{C}(r,f) + \overline{C}\left(r, \frac{1}{f^{(k)}-1}\right) + R(r,f).
\end{aligned}$$

This is (5.1.6) □

Now we consider the quantity concerning value points

$$V(r, D, f=a) = \max\{r^{\omega_j} C_{\alpha_j,\alpha_{j+1}}(r, f=a) : 1 \leqslant j \leqslant q\}$$

and $\overline{V}(r,D,f=a)$ for $\overline{C}_{\alpha_j,\alpha_{j+1}}(r,f=a)$. Combining Theorem 5.1.1 and Lemma 5.1.1 yields the following

Theorem 5.1.2. *Let $f(z)$ be a transcendental meromorphic function and $a \in \widehat{\mathbb{C}} \setminus \{0,\infty\}$. Assume that a is a Nevanlinna deficient value of $f^{(p)}(z)$ $(p \geqslant k \geqslant 0)$. Then for $\tau > 0$,*

$$\begin{aligned}
T(r,f) \leqslant K(\log T(r,f))^{2+\tau}(&\overline{V}(r,D,f^{(k)}=0) + \overline{V}(r,D,f=\infty) \\
&+ r^{\omega}\log r T(r,f)),
\end{aligned} \tag{5.1.7}$$

for all r outside a set of finite logarithmic measure.

If, in addition, $\overline{V}(r,D,f^{(k)}=0) + \overline{V}(r,D,f=\infty)$ is of finite order, then for $\varepsilon > 0$

$$T(r,f) \leqslant K(\overline{V}(r,D,f^{(k)}=0)+\overline{V}(r,D,f=\infty)+r^{\omega}),\ r\notin E,$$

where K is a positive constant and $\overline{\log\mathrm{dens}}E<\varepsilon$.

Proof. From (5.1.5) it is easy to see that

$$W(r,D,f^{(p)}=a)\leqslant(2p+1)(\overline{V}(r,D,f^{(k)}=0)+\overline{V}(r,D,f=\infty))+r^{\omega}R(r,f),$$

where $R(r,f)=\max\{R_{\alpha_j,\alpha_{j+1}}(r,f):1\leqslant j\leqslant q\}$. In view of Lemma 2.5.3 we have $R(r,f)=O(\log rT(r,f))$ and furthermore if $f(z)$ is of finite order, then $R(r,f)=O(1)$. Thus application of Theorem 5.1.1 yields our desired results. □

Obviously, the same argument as in above yields a result corresponding to the case (2) of Lemma 5.1.1, that is to say, we also have the inequalities in Theorem 5.1.2 for $k>p$ and $a\in\mathbb{C}$ with $\overline{V}(r,D,f^{(k)}=1)+\overline{V}(r,D,f=\infty)$ in the place of $\overline{V}(r,D,f^{(k)}=0)+\overline{V}(r,D,f=\infty)$.

In what follows, we connect the growth of meromorphic functions with the number of value points in the angular domains. To the end we need

Lemma 5.1.2. *Let $f(z)$ be a meromorphic function in an angular domain $\overline{\Omega}(\alpha,\beta)$. If for $\rho>0$,*

$$\limsup_{r\to\infty}\frac{\log\overline{n}(r,\Omega,f=a)}{\log r}\leqslant\rho, \tag{5.1.8}$$

then for arbitrary small $\varepsilon>0$ with $\rho+\varepsilon\neq\omega(=\frac{\pi}{\beta-\alpha})$ and all sufficiently large r, we have

$$\overline{C}_{\alpha,\beta}(r,f=a)\leqslant\left(4\omega+\frac{2\omega^2}{|\rho+\varepsilon-\omega|}\right)r^{\rho-\omega+\varepsilon}+O(1).$$

Proof. In view of Lemma 2.2.2 it suffices to estimate the integrate in (2.2.14). This is because that it follows from (5.1.8) that

$$\limsup_{r\to\infty}\frac{\log\overline{N}(r,\Omega,f=a)}{\log r}\leqslant\rho$$

and thus for all $r\geqslant r_0\geqslant 1$, $\overline{N}(r,\Omega,f=a)<r^{\rho+\varepsilon}$. This also implies

$$\int_{r_0}^{r}\frac{\overline{N}(t,\Omega,f=a)}{t^{\omega+1}}\mathrm{d}t<\int_1^r t^{\rho+\varepsilon-\omega-1}\mathrm{d}t<\frac{1}{|\rho+\varepsilon-\omega|}(r^{\rho-\omega+\varepsilon}+1).$$

Thus Lemma 5.1.2 follows. □

Lemma 5.1.2 still holds for n and C in the places of $\bar{n}$ and $\bar{C}$.

Theorem 5.1.3. *Let $f(z)$ be a transcendental meromorphic function such that for some $a\in\widehat{\mathbb{C}}\setminus\{0,\infty\}$ and an integer $p\geqslant 0$, $\delta=\delta(a,f^{(p)})>0$. Given q radii $\arg z=\alpha_j$ $(1\leqslant j\leqslant q)$ satisfying (5.1.1), set $Y=\mathbb{C}\setminus\bigcup_{j=1}^{q}\{z:\ \arg z=\alpha_j\}$. Let k be an integer with $0\leqslant k\leqslant p$.*

(1) *If for some* $\rho \geqslant 0$,

$$\limsup_{r\to\infty} \frac{\log(\overline{n}(r,Y,f^{(k)}=0)+\overline{n}(r,Y,f=\infty))}{\log r} \leqslant \rho, \tag{5.1.9}$$

then $\lambda(f) \leqslant \max\{\omega,\rho\}$;

(2) *If for some* $\tau > 0$

$$\overline{N}(r,Y,f^{(k)}=0)+\overline{N}(r,Y,f=\infty) = o(T(r,f)(\log T(r,f))^{-2-\tau}),\ r \notin F, \tag{5.1.10}$$

where $\mathrm{dens}F < 1$, *then* $\lambda(f) \leqslant \omega$.

Here $\omega = \max\{\omega_j : 1 \leqslant j \leqslant q\}$.

Proof. (1) Under (5.1.9), in view of Lemma 5.1.2 we have for arbitrary $\varepsilon > 0$ and all sufficiently large $r > 0$

$$\overline{V}(r,D,f^{(k)}=0)+\overline{V}(r,D,f=\infty) \leqslant r^{\rho+\varepsilon}+O(r^{\omega}).$$

This with the help of Theorem 5.1.2 implies that $\lambda(f) \leqslant \max\{\omega,\rho\}$.

(2) Set

$$N(r) = \overline{N}(r,Y,f^{(k)}=0)+\overline{N}(r,Y,f=\infty) = o(T(r,f)(\log T(r,f))^{-2-\tau}).$$

We first of all want to show that $N(r)$ is of order not greater than ω. Suppose that it fails. Then from Lemma 1.1.3 and Theorem 1.1.3 for some $\sigma > \omega$, there exists an unbounded sequence $\{r_n\}$ of positive numbers such that

$$N(t) \leqslant \mathrm{e}\left(\frac{t}{r_n}\right)^{\sigma} N(r_n),\ \text{for } 1 \leqslant t \leqslant r_n.$$

Let F_1 be the except set outside which $R_{\alpha,\beta}(r,f) = O(\log rT(r,f))$ for the error term in (5.1.5) holds and then $\mathrm{dens}(F\cup F_1) = \mathrm{dens}F < 1$. For d with $d > (1-\mathrm{dens}F)^{-1}$ we can find $r'_n \in [r_n, dr_n]\setminus(F\cup F_1)$. Thus

$$\begin{aligned}\int_1^{r'_n}\frac{N(t)}{t^{\omega+1}}\mathrm{d}t &= \int_1^{r_n}\frac{N(t)}{t^{\omega+1}}\mathrm{d}t + \int_{r_n}^{r'_n}\frac{N(t)}{t^{\omega+1}}\mathrm{d}t\\ &< \frac{\mathrm{e}}{\sigma-\omega}\frac{N(r_n)}{r_n^{\omega}} + \frac{1}{\omega}\frac{N(r'_n)}{r_n^{\omega}}\\ &\leqslant \left(\frac{\mathrm{e}}{\sigma-\omega}+\frac{1}{\omega}\right)d^{\omega}\frac{N(r'_n)}{r'^{\omega}_n}\end{aligned}$$

so that in view of (2.2.14) and then by using (5.1.10) we have

$$\overline{V}(r'_n,D,f^{(k)}=0)+\overline{V}(r'_n,D,f=\infty) = o(T(r'_n,f)(\log T(r'_n,f))^{-2-\tau}).$$

It follows from (5.1.7) that

$$T(r'_n,f)=o(T(r'_n,f))+O(r_n'^{\omega}(\log T(r'_n,f))^{2+\tau}\log r'_nT(r'_n,f))$$

and noting that the order of $T(r'_n,f)$ is at least σ, we have $T(r'_n,f)=o(T(r'_n,f))$, this is impossible.

Thus we have proved that $N(r)$ is of order at most ω and employing (1) yields that $\lambda(f)\leqslant\omega$.

Theorem 5.1.3 follows. □

We remark we can obtain the corresponding results to above Theorems if that $a(\neq 0,\infty)$ is a Nevanlinna deficient value is replaced by that $a(\neq 0,\infty)$ is a Borel exceptional value.

Recall that $\overline{n}(r,Y,f=a)$ is the number of distinct roots of $f(z)=a$ in $\{z:|z|<r\}\cap Y$, that is, ignoring the roots lying on the rays $\arg z=\alpha_j (j=1,2,\cdots,q)$. Actually from the proof of Theorem 5.1.1 we can redefine $W(r,D,*)$ and $V(r,D,*)$ with α_j and α_{j+1} replaced respectively by $\alpha_j+\varepsilon(r)$ and $\alpha_{j+1}-\varepsilon(r)$, where $\varepsilon(r)$ is chosen suitably such that $\varepsilon(r)=O(\log T(r,f))^{-1-\tau}$. Therefore, we can replace Y in Theorem 5.1.3 with

$$Z=\cup_{j=1}^{q}\{z=re^{\mathrm{i}\theta}:\alpha_j+\varepsilon(r)<\theta<\alpha_{j+1}-\varepsilon(r)\}.$$

Finally, according to a result of Ostrovskii (see Theorem 5.3.4(2) in below Section 5.3) we get

Theorem 5.1.4. *Let $f(z)$ be a transcendental meromorphic function such that for some $a\in\widehat{\mathbb{C}}\setminus\{0,\infty\}$ and an integer $p\geqslant 0$, $\delta=\delta(a,f^{(p)})>0$. Assume that*

$$N(r,\Omega_j,f=0)+N(r,\Omega_j,f=\infty)+N(r,\Omega_j,f=a)=O\left(\frac{r^{\omega}}{(\log r)^{\tau}}\right), \tag{5.1.11}$$

where $\tau>1$, $\Omega_j=\Omega(\alpha_j,\alpha_{j+1})$ and $\omega=\max\{\omega_j:1\leqslant j\leqslant q\}$. Then

$$\log|f(re^{\mathrm{i}\theta})|=r^{\omega_j}c_j\sin(\omega_j(\theta-\alpha_j))+o(r^{\omega_j})$$

uniformly relative to θ, $\alpha_j\leqslant\theta\leqslant\alpha_{j+1}$, with $c_j\in\mathbb{R}$ as $r\to\infty$ perhaps passing outside a set of finite logarithmic measure.

Proof. In view of (2.2.14) and (5.1.11), we have

$$\begin{aligned}
r^{\omega_j-\omega}C_{\alpha_j,\alpha_{j+1}}(r,f=X)&\leqslant 4\omega_jN(r,\Omega_j,f=X)r^{-\omega}\\
&\quad+2\omega_j^2r^{\omega_j-\omega}\int_1^r\frac{N(t,\Omega_j,f=X)}{t^{\omega_j+1}}\mathrm{d}t\\
&=O((\log r)^{-\tau})+O\left(r^{\omega_j-\omega}\int_1^r\frac{t^{\omega-\omega_j}}{t(\log t)^{\tau}}\mathrm{d}t\right)\\
&=O((\log r)^{-\tau})+O\left(\int_1^{\infty}\frac{\mathrm{d}t}{t(\log t)^{\tau}}\right)\\
&=O(1),
\end{aligned}$$

$X = 0, \infty, a$, and therefore the condition in (2) of Theorem 5.3.4 in below Section 5.3 is satisfied. Theorem 5.1.4 follows. □

5.2 Growth of Such Meromorphic Functions with Finite Lower Order

It is noted that in Theorems 5.1.1-5.1.4, no restriction is imposed on the growth of the meromorphic function considered. In what follows, we always assume that transcendental meromorphic function $f(z)$ has the finite lower order and thus we can impair the requirement on the argument distribution of value points.

Let

$$-\pi \leqslant \alpha_1 < \beta_1 \leqslant \alpha_2 < \beta_2 \leqslant \cdots < \alpha_q < \beta_q \leqslant \pi,\ \alpha_{q+1} = \alpha_1 + 2\pi, \tag{5.2.1}$$

and by D we denote the corresponding ray system $D(\alpha_1, \beta_1, \cdots, \alpha_q, \beta_q)$. Define for D

$$\omega' = \omega'(D) = \max\{\pi/(\beta_j - \alpha_j) : 1 \leqslant j \leqslant q\}$$

and

$$\omega'' = \omega''(D) = \min\{\pi/(\alpha_{j+1} - \beta_j) : 1 \leqslant j \leqslant q\}.$$

In [14], we studied this topic and established two fundamental theorems which are formulated in the following form, which make the discussion of this subject very simple and elementary.

Theorem 5.2.1. *Let $f(z)$ be a transcendental meromorphic function with the finite lower order $\mu(f)$ and for N distinct values $a_i \in \widehat{\mathbb{C}}$ $(1 \leqslant i \leqslant N)$ and an integer $p \geqslant 0$, $\delta_i = \delta(a_i, f^{(p)}) > 0$. For q pair of real numbers $\{\alpha_j, \beta_j\}$ satisfying (5.2.1) and arbitrary sequence of Pólya peak $\{r_n\}$ of $f^{(p)}(z)$ of any order σ outside $E(f)$ such that $\mu(f) \leqslant \sigma \leqslant \lambda(f)$ and $\sigma > \omega_j = \frac{\pi}{\beta_j - \alpha_j}$, $1 \leqslant j \leqslant q$ (if exists), for each i we have*

$$B_{\alpha_j,\beta_j}\left(r_n, \frac{1}{f^{(p)} - a_i}\right) = o\left(\frac{T(r_n, f)}{r_n^{\omega_j}}\right) + K_j \log(r_n T(r_n, f)),\ 1 \leqslant j \leqslant q, \tag{5.2.2}$$

(we replace $B_{\alpha_j,\beta_j}(r, 1/(f^{(p)} - a))$ by $B_{\alpha_j,\beta_j}(r, f^{(p)})$ if $a = \infty$.), where K_j is a positive constant only depending on j and σ. If

$$\sum_{j=1}^{q} (\alpha_{j+1} - \beta_j) < \frac{4}{\gamma} \sum_{i=1}^{N} \arcsin\sqrt{\frac{\delta(a_i, f^{(p)})}{2}}, \tag{5.2.3}$$

$\gamma = \max\{\omega'(D), \mu\}$, then

$$\lambda(f) \leqslant \omega'(D).$$

Proof. We assume that $a \in \mathbb{C}$. By the same argument we can show Theorem 5.2.1 for the case when $a = \infty$. Suppose conversely that $\lambda(f) > \omega'$. We need to treat two cases.

(I) $\lambda(f) > \mu$. Then $\lambda(f^{(p)}) = \lambda(f) > \gamma \geqslant \mu(f) = \mu(f^{(p)})$. And by the inequality (5.2.3), we can take a real number $\varepsilon > 0$ such that

$$\sum_{j=1}^{q}(\alpha_{j+1} - \beta_j + 2\varepsilon) + 2\varepsilon < \frac{4}{\gamma + 2\varepsilon}\sum_{i=1}^{N}\arcsin\sqrt{\frac{\delta_i}{2}}, \tag{5.2.4}$$

where $\alpha_{q+1} = 2\pi + \alpha_1$, and

$$\lambda(f^{(p)}) > \gamma + 2\varepsilon > \mu.$$

Applying Theorem 1.1.3 to $f^{(p)}(z)$ determines the existence of a sequence $\{r_n\}$ of the Pólya peaks of order $\gamma + 2\varepsilon$ of $f^{(p)}$ outside $E(f)$. Set $\Lambda(r) = \Gamma^{1/2}(r)$ and

$$\Gamma(r) = \max\left\{ r_n^{\omega_j} \frac{B_{\alpha_j,\beta_j}(r_n, 1/(f^{(p)} - a_i))}{T(r_n, f^{(p)})} : 1 \leqslant j \leqslant q, 1 \leqslant i \leqslant N \right\}, \tag{5.2.5}$$

$$r_{n-1} < r \leqslant r_n.$$

From the Chuang's inequality and (3) in Definition 1.1.1 it follows that

$$T(r_n, f) = O(T(2r_n, f^{(p)})) = O(T(r_n, f^{(p)})). \tag{5.2.6}$$

Thus from (2) in Definition 1.1.1 and by noting $\gamma + \varepsilon > \omega' \geqslant \omega_j$, we have

$$r_n^{\omega_j} \log(r_n T(r_n, f)) = o(T(r_n, f^{(p)})), \quad n \to \infty.$$

From this, using (5.2.2) to the sequence of Pólya peak $\{r_n\}$ of $f^{(p)}$ of order $\sigma = \gamma + 2\varepsilon$, we can deduce that as $r \to +\infty$, $\Gamma(r) \to 0$ and $\Lambda(r) \to 0$. Then from Theorem 2.8.1 for sufficiently large n we have

$$\mathrm{mes}\bigcup_{i=1}^{N} D_\Lambda(r_n, a_i) = \sum_{i=1}^{N}\mathrm{mes}D_\Lambda(r_n, a_i) > \frac{4}{\gamma + 2\varepsilon}\sum_{i=1}^{N}\arcsin\sqrt{\frac{\delta_i}{2}} - \varepsilon, \tag{5.2.7}$$

since $\gamma + 2\varepsilon > 1/2$. We can assume for all the n (5.2.7) holds. Set

$$K_n = \mathrm{mes}\left(\bigcup_{i=1}^{N} D_\Lambda(r_n, a_i) \cap \bigcup_{j=1}^{q}(\alpha_j + \varepsilon, \beta_j - \varepsilon)\right).$$

Then from (5.2.4) and (5.2.7) it follows that

$$\begin{aligned}K_n &\geqslant \operatorname{mes}\bigcup_{i=1}^{N} D_\Lambda(r_n,a_i) - \operatorname{mes}\left([-\pi,\pi)\setminus\bigcup_{j=1}^{q}(\alpha_j+\varepsilon,\beta_j-\varepsilon)\right)\\ &= \operatorname{mes}\bigcup_{i=1}^{N} D_\Lambda(r_n,a_i) - \operatorname{mes}\left(\bigcup_{j=1}^{q}(\beta_j-\varepsilon,\alpha_{j+1}+\varepsilon)\right)\\ &= \operatorname{mes}\bigcup_{i=1}^{N} D_\Lambda(r_n,a_i) - \sum_{j=1}^{q}(\alpha_{j+1}-\beta_j+2\varepsilon) > \varepsilon > 0.\end{aligned}$$

It is easy to see that there exists a j_0 and a i_0 such that for infinitely many n, we have

$$\operatorname{mes}(D_\Lambda(r_n,a_{i_0})\cap(\alpha_{j_0}+\varepsilon,\beta_{j_0}-\varepsilon)) \geqslant \frac{K_n}{qN} > \frac{\varepsilon}{qN}. \tag{5.2.8}$$

We can assume for all the n (5.2.8) holds. Set $E_n = D_\Lambda(r_n,a)\cap(\alpha_{j_0}+\varepsilon,\beta_{j_0}-\varepsilon)$ with $a=a_{i_0}$. Thus from the definition of $D_\Lambda(r,a)$ it follows that

$$\begin{aligned}\int_{\alpha_{j_0}+\varepsilon}^{\beta_{j_0}-\varepsilon}\log^+\frac{1}{|f^{(p)}(r_n\mathrm{e}^{\mathrm{i}\theta})-a|}\mathrm{d}\theta &\geqslant \int_{E_n}\log^+\frac{1}{|f^{(p)}(r_n\mathrm{e}^{\mathrm{i}\theta})-a|}\mathrm{d}\theta\\ &\geqslant \operatorname{mes}(E_n)\Lambda(r_n)T(r_n,f^{(p)})\\ &\geqslant \frac{\varepsilon}{qN}\Lambda(r_n)T(r_n,f^{(p)}).\end{aligned} \tag{5.2.9}$$

On the other hand, by the definition of $B_{\alpha,\beta}(r,*)$ and (5.2.5), we have

$$\begin{aligned}\int_{\alpha_{j_0}+\varepsilon}^{\beta_{j_0}-\varepsilon}\log^+\frac{1}{|f^{(p)}(r_n\mathrm{e}^{\mathrm{i}\theta})-a|}\mathrm{d}\theta &\leqslant \frac{\pi}{2\omega_{j_0}\sin(\varepsilon\omega_{j_0})}r_n^{\omega_{j_0}}B_{\alpha_{j_0},\beta_{j_0}}\left(r_n,\frac{1}{f^{(p)}-a}\right)\\ &\leqslant \frac{\pi}{2\omega_{j_0}\sin(\varepsilon\omega_{j_0})}\Lambda^2(r_n)T(r_n,f^{(p)}).\end{aligned} \tag{5.2.10}$$

Combining (5.2.9) with (5.2.10) gives

$$0 < \frac{2\varepsilon\omega_{j_0}\sin(\varepsilon\omega_{j_0})}{\pi qN} \leqslant \Lambda(r_n) \to 0,\ n\to\infty.$$

This is impossible.

(II) $\lambda(f)=\mu$. Then $\gamma=\mu=\lambda(f)$. By the same argument as in (I) with all the $\gamma+2\varepsilon$ replaced by $\gamma=\mu$, we can derive a contradiction.

Theorem 5.2.1 follows. □

We remark on a condition of Theorem 5.2.1. If there exist no σ such that for each j, $\sigma>\omega_j$, then it is easy to see that $\lambda(f)\leqslant\omega'(D)$.

Theorem 5.2.2. *Let $f(z)$ and $a_i(i=1,2,\cdots,N)$ be given as in Theorem 5.2.1. For q pair of real numbers $\{\alpha_j,\beta_j\}$ satisfying (5.2.1) and arbitrary small $\varepsilon>0$, for each i we have*

$$B_{\alpha_j,\beta_j}\left(r,\frac{1}{f^{(p)}-a_i}\right) < K_j[r^{\rho-\omega_j+\varepsilon}+\log(rT(r,f))],\ r\notin E,\ 1\leqslant j\leqslant q, \quad (5.2.11)$$

(we replace $B_{\alpha_j,\beta_j}(r,1/(f^{(p)}-a))$ by $B_{\alpha_j,\beta_j}(r,f^{(p)})$ if $a=\infty$.), where $\omega_j=\frac{\pi}{\beta_j-\alpha_j}$, $1\leqslant j\leqslant q$, ρ is a positive number, K_j a positive constant only depending on j and ε. If (5.2.3) holds for $\gamma=\max\{\omega'(D),\rho,\mu\}$, then

$$\lambda(f)\leqslant\max\{\omega'(D),\rho\}.$$

Proof. Suppose conversely that $\lambda(f)>\max\{\omega',\rho\}$. We shall derive a contradiction by making a minor modification of the proof of Theorem 5.2.1, so below we put the same meanings on the same notations in the proof of Theorem 5.2.1.

(I) $\lambda(f)>\mu$. $\{r_n\}$ is a sequence of the Pólya peaks of order $\gamma+2\varepsilon$ of $f^{(p)}$ outside $E(f)$. Set $\Lambda(r)=[\log r]^{-1}$. Then we can deduce

$$\int_{\alpha_{j_0}+\varepsilon}^{\beta_{j_0}-\varepsilon}\log^+\frac{1}{|f^{(p)}(r_n\mathrm{e}^{\mathrm{i}\theta})-a|}\mathrm{d}\theta\geqslant\frac{\varepsilon}{qN}\frac{T(r_n,f^{(p)})}{\log r_n}. \quad (5.2.12)$$

On the other hand, from (5.2.11), we have for $r\notin E$

$$\int_{\alpha_{j_0}+\varepsilon}^{\beta_{j_0}-\varepsilon}\log^+\frac{1}{|f^{(p)}(r\mathrm{e}^{\mathrm{i}\theta})-a|}\mathrm{d}\theta\leqslant\frac{\pi}{2\omega_{j_0}\sin(\varepsilon\omega_{j_0})}r^{\omega_{j_0}}B_{\alpha_{j_0},\beta_{j_0}}\left(r,\frac{1}{f^{(p)}-a}\right)$$
$$<\tilde{K_{j_0}}[r^{\rho+\varepsilon}+r^{\omega_{j_0}}\log(rT(r,f))]. \quad (5.2.13)$$

Combining (5.2.12) with (5.2.13) gives

$$T(r_n,f^{(p)})\leqslant\frac{qN\tilde{K_{j_0}}}{\varepsilon}\log r_n[r_n^{\rho+\varepsilon}+r_n^{\omega_{j_0}}\log(r_nT(r_n,f))]$$

and then applying (5.2.6) gives that

$$\log T(r_n,f^{(p)})\leqslant 2\log\log r_n+\max\{\rho+\varepsilon,\omega_{j_0}\}\log r_n+\log\log T(r_n,f^{(p)})+O(1).$$

Thus from (2) in Definition 1.1.1 for $\gamma+2\varepsilon$, we have

$$\gamma+2\varepsilon\leqslant\limsup_{n\to\infty}\frac{\log T(r_n,f^{(p)})}{\log r_n}\leqslant\max\{\rho+\varepsilon,\omega_{j_0}\}\leqslant\gamma+\varepsilon.$$

This is impossible.

(II) $\lambda(f)=\mu$. Then $\gamma=\mu=\lambda(f)$. By the same argument as in (I) with all the $\gamma+2\varepsilon$ replaced by $\gamma=\mu$, we can derive

$$\mu=\gamma\leqslant\max\{\rho,\omega'\}+\varepsilon<\lambda(f).$$

This is impossible.

Theorem 5.2.2 follows. □

In what follows, we deal with the argument distribution of value points in term of Theorem 5.2.1 and Theorem 5.2.2. First of all let us establish connection between $C(r,*)$ and the number of corresponding value points in an angle.

Lemma 5.2.1. *Let $f(z)$ be a transcendental meromorphic function with the finite lower order μ and $0<\lambda=\lambda(f)\leqslant+\infty$. If for $d\geqslant 1$ and an integer $k\geqslant 0$,*

$$n(r,\Omega(\alpha,\beta),f^{(k)}=a)=o(T(dr,f)), \tag{5.2.14}$$

then for arbitrary sequence of the relaxed Pólya peaks $\{r_n\}$ of $f^{(p)}$ $(p\geqslant 0)$ of any order σ outside $E(f)$ such that $\mu\leqslant\sigma\leqslant\lambda$ and $\sigma>\omega=\frac{\pi}{\beta-\alpha}$, we have

$$C_{\alpha,\beta}\left(r_n,\frac{1}{f^{(k)}-a}\right)=o\left(\frac{T(r_n,f)}{r_n^{\omega}}\right). \tag{5.2.15}$$

(If in (5.2.14), $\overline{n}$ is in the place of n, then we have (5.2.15) for $\overline{C}$ in place of C).

Proof. From (3) in Definition 1.1.1 and the Chunag's inequality (2.6.2), we have

$$T(dr_n,f)\leqslant K_pT(2dr_n,f^{(p)})\leqslant K_pK_{d,\sigma}T(r_n,f^{(p)})$$

and since $r_n\notin E(f)$, in view of (2.6.1) and (2.5.1) we have

$$T(r_n,f^{(p)})\leqslant C_pT(r_n,f),$$

where K_p, $K_{d,\sigma}$ and C_p are constants depending on their subscripts.

Applying the Chunag's inequality (2.6.2) again and then (4) in Definition 1.1.1, we estimate the following integral

$$\begin{aligned}
\int_1^{r_n}\frac{T(\mathrm{d}t,f)}{t^{\omega+1}}\mathrm{d}t &\leqslant K_p\int_1^{r_n}\frac{T(2\mathrm{d}t,f^{(p)})}{t^{\omega+1}}\mathrm{d}t\\
&=K_p(2d)^{\omega}\int_{2d}^{2dr_n}\frac{T(t,f^{(p)})}{t^{\omega+1}}\mathrm{d}t\\
&\leqslant KK_p(2d)^{\omega}\int_{2d}^{2dr_n}\left(\frac{t}{r_n}\right)^{\sigma-\varepsilon_n}\frac{T(r_n,f^{(p)})}{t^{\omega+1}}\mathrm{d}t\\
&=KK_p(2d)^{\omega}\frac{T(r_n,f^{(p)})}{r_n^{\sigma-\varepsilon_n}}\int_{2d}^{2dr_n}t^{\sigma-\varepsilon_n-\omega-1}\mathrm{d}t\\
&=\frac{KK_p(2d)^{\sigma-\varepsilon_n}}{\sigma-\varepsilon_n-\omega}\frac{T(r_n,f^{(p)})}{r_n^{\sigma-\varepsilon_n}}(r_n^{\sigma-\varepsilon_n-\omega}-1)\\
&<\frac{KK_p(2d)^{\sigma-\varepsilon_n}}{\sigma-\varepsilon_n-\omega}\frac{T(r_n,f^{(p)})}{r_n^{\omega}}\\
&\leqslant\frac{KK_pC_p(2d)^{\sigma-\varepsilon_n}}{\sigma-\varepsilon_n-\omega}\frac{T(r_n,f)}{r_n^{\omega}}.
\end{aligned}$$

Now in view of (2.2.14) (also see the proof of Lemma 2.2.2) and (5.2.14), we obtain

$$C_{\alpha,\beta}\left(r_n, \frac{1}{f^{(k)}-a}\right) \leqslant 4\omega \int_1^{r_n} \frac{n(t,\Omega,f^{(k)}=a)}{t^{\omega+1}}\mathrm{d}t$$
$$= o\left(\int_1^{r_n} \frac{T(\mathrm{d}t,f)}{t^{\omega+1}}\mathrm{d}t\right) = o\left(\frac{T(r_n,f)}{r_n^{\omega}}\right),$$

this is (5.2.15). □

Combination of Lemmas 5.1.1 and 5.2.1 with Theorems 5.2.1 and 5.2.2 yields the following

Theorem 5.2.3. *Let $f(z)$ be given as in Theorem 5.2.1. Then the following two statements hold.*

(1) *If*

$$\sum_{j=1}^{q}(\alpha_{j+1}-\beta_j) < \frac{4}{\gamma}\sum_{a\neq 0,\infty} \arcsin\sqrt{\frac{\delta(a,f^{(p)})}{2}}, \tag{5.2.16}$$

$\gamma = \max\{\omega'(D),\mu\}$, *and for some $d \geqslant 1$*

$$\overline{n}(r,Y,f^{(k)}=0)+\overline{n}(r,Y,f=\infty) = o(T(dr,f)), \tag{5.2.17}$$

where $Y = \bigcup_{j=1}^{q}\{z:\ \alpha_j < \arg z < \beta_j\}$, *then* $\lambda(f) \leqslant \omega'(D)$.

(2) *If (5.2.16) holds for* $\gamma = \max\{\omega'(D),\rho,\mu\}$, *and*

$$\limsup_{r\to\infty} \frac{\log(\overline{n}(r,Y,f^{(k)}=0)+\overline{n}(r,Y,f=\infty))}{\log r} \leqslant \rho, \tag{5.2.18}$$

then $\lambda(f) \leqslant \max\{\omega'(D),\rho\}$.

Proof. Here we only provide the proof of (1) of Theorem 5.2.3. In fact, (5.2.3) follows from (5.2.16) for some N and hence it suffices to prove (5.2.2) under (5.2.17). Given arbitrarily a sequence of Pólya peak $\{r_n\}$ of $f^{(p)}(z)$ of any order σ outside $E(f)$ such that $\mu(f) \leqslant \sigma \leqslant \lambda(f)$ and $\sigma > \omega_j = \frac{\pi}{\beta_j-\alpha_j}$, $1 \leqslant j \leqslant q$ (if exists), in view of (5.2.17) and Lemma 5.2.1, we have

$$\overline{C}_{\alpha_j,\beta_j}\left(r_n, \frac{1}{f^{(k)}}\right) + \overline{C}_{\alpha_j,\beta_j}(r_n,f) = o\left(\frac{T(r_n,f)}{r_n^{\omega_j}}\right)$$

and then from Lemma 5.1.1 it follows that for any $a \in \widehat{\mathbb{C}}\setminus\{0,\infty\}$, we have

$$B_{\alpha_j,\beta_j}\left(r_n, \frac{1}{f^{(p)}-a}\right) = o\left(\frac{T(r_n,f)}{r_n^{\omega_j}}\right) + K_j\log(r_nT(r_n,f)),\ 1 \leqslant j \leqslant q.$$

This is (5.2.2) and hence Theorem 5.2.3 (1) follows. □

It is clear from the above proof that the result (2) in Theorem 5.2.3 still holds if the condition (5.2.18) is directly replaced by the inequality

$$\overline{C}_{\alpha_j,\beta_j}(r,f^{(k)}=0)+\overline{C}_{\alpha_j,\beta_j}(r,f=\infty)<K_j(r^{\rho-\omega_j+\varepsilon}+\log r),\ r\notin E.$$

Corollary 5.2.1. *Let $f(z)$ be given as in Theorem 5.2.1. Then in any angular domain $\Omega=\{z:\ \alpha<\arg z<\beta\}$ such that*

$$\beta-\alpha>\max\left\{\frac{\pi}{\mu},\ 2\pi-\frac{4}{\mu}\arcsin\sqrt{\frac{\delta}{2}}\right\},$$

$\delta=\delta(a,f)>0$, there exists a radial $\arg z=\theta$ such that for arbitrary small $\varepsilon>0$,

$$\limsup_{r\to\infty}\frac{\overline{n}(r,Z,f^{(k)}=0)+\overline{n}(r,Z,f=\infty)}{T(dr,f)}>0,$$

where $Z=\{z:\ \theta-\varepsilon<\arg z<\theta+\varepsilon\}$ and $d\geqslant 1$.

Let us discuss significance of Theorems 5.1.1, 5.2.1 and 5.2.2. Actually, they assert that as long as we can estimate $B(r,*)$ in terms of a few $C(r,**)$, we can establish the results on the growth order of a meromorphic function with suitable restriction imposed on distribution of arguments of value points expressed by the corresponding $C(r,**)$, and further by $n(r,\Omega(\alpha,\beta),**)$ by noticing the closed relation between $n(r,\Omega(\alpha,\beta),*)$ and $C_{\alpha,\beta}(r,*)$. The above statements prove important and thus it made those very simple and elementary the discussions on the growth of transcendental meromorphic functions dealing with some radially distributed values. Let us make it clear once more by establishing the following result.

Theorem 5.2.4. *Let $f(z)$ be given as in Theorem 5.2.1 with $\delta=\delta(\infty,f)>0$ instead of $\delta(a,f^{(p)})>0$ and*

$$\sum_{j=1}^{q}(\alpha_{j+1}-\beta_j)<\frac{4}{\gamma}\arcsin\sqrt{\frac{\delta}{2}},$$

$\gamma=\max\{\omega'(D),\mu\}$ or $\max\{\omega'(D),\rho,\mu\}$. Then the results of Theorem 5.2.3 hold, provided that (5.2.17) and (5.2.18) are respectively replaced by

$$\overline{n}(r,Y,f^2f'=1)=o(T(dr,f)),\ d\geqslant 1$$

and

$$\limsup_{r\to\infty}\frac{\log\overline{n}(r,Y,f^2f'=1)}{\log r}\leqslant\rho. \tag{5.2.19}$$

Proof. Analyzing the proof of Theorem 5.2.3, it suffices to show an inequality of that $B(r,f)$ is controlled by $C(r,f^2f'=1)$. Using the Milloux's fundamental inequality (2.2.8) on an angle to $f^3/3$, we have

$$\begin{aligned}3S(r,f) &= S\left(r,\frac{f^3}{3}\right)+O(1)\\ &\leqslant \overline{C}(r,f^3)+C(r,f^3=0)+C(r,(f^3/3)'=1)-C(r,(f^3/3)''=0)+R(r,f)\\ &\leqslant \overline{C}(r,f)+2\overline{C}(r,f=0)+\overline{C}(r,f^2f'=1)+R(r,f),\end{aligned}$$

and hence

$$B(r,f)=S(r,f)-C(r,f)\leqslant \overline{C}(r,f^2f'=1)+R(r,f).$$

Thus Theorem 5.2.4 follows. □

Yang and Yang [11] proved Theorem 5.2.4 for entire function $f(z)$ under (5.2.19) with "$f^2f'=1$" replaced by "$ff'=1$" and $Y=\cup_{j=1}^{q}\{z:\alpha_j+\varepsilon\leqslant \arg z\leqslant \alpha_{j+1}-\varepsilon\}$ for arbitrary small ε and $\beta_j=\alpha_{j+1}$.

5.3 Retrospection

The discussion on this subject considered in this chapter can go back to the result obtained by Bieberbach in 1919, which says that an entire function of finite order λ assumes infinitely often every finite value with the possible exception of one value in each angular domain $\Omega(\alpha,\beta)$ such that

$$\beta-\alpha>\max\left\{\frac{\pi}{\lambda},\ 2\pi-\frac{\pi}{\lambda}\right\}. \tag{5.3.1}$$

Therefore if an entire function has only finitely many zeros and 1-points in $\Omega(\alpha,\beta)$, then we have the inequality opposite to (5.3.1), which implies that

$$\lambda\leqslant\frac{\pi}{\beta-\alpha}\ \text{or}\ 2\pi+\alpha-\beta\geqslant\frac{\pi}{\lambda}.$$

We state the Bieberbach's result in the following way which is suitable to the point of view we consider in this paper.

Theorem 5.3.1. *(Bieberbach, 1919) Let $f(z)$ be a transcendental entire function with order $\lambda<+\infty$. If $f(z)$ has only finitely many zeros and 1-points in $\Omega(\alpha,\beta)$ and*

$$(2\pi+\alpha)-\beta<\frac{\pi}{\lambda}, \tag{5.3.2}$$

then

$$\lambda\leqslant\frac{\pi}{\beta-\alpha}.$$

The Bieberbach Theorem 5.3.1 reveals that the order of entire function can be estimated in term of distribution of points of its two values. There seems to be a little relationship between the Bieberbach Theorem 5.3.1 and Nevanlinna Theorem 2.7.7, the former asserts that distribution of its two value points in an angular domain determines the growth of an entire function not only in the angle considered but also in the complement of the angle whose opening is not too large and however, the latter impairs the restriction imposed on the number of value points in the angular domain considered. Actually, the Bieberbach Theorem 5.3.1 is able to follow from the Nevanlinna Theorem 2.7.7 and the Phragmén-Lindelöf Theorem demonstrated in 1908 (see Corollary 4.2, page 139 in [1] and Theorem 4.3.2, page 102, in [15]) by noting the fact that the condition (5.3.2) produces inequality $\log M(r,f) < r^{\pi/(2\pi+\alpha-\beta)-\varepsilon}$ for all sufficiently large r and a suitable small positive ε. Therefore, the Nevanlinna Theorem 2.7.7 is actually an extension of the Bieberbach Theorem 5.3.1.

The Bieberbach Theorem 5.3.1 was also extended by Valiron in 1932 and Cartwright in 1932 and 1935. Their results can be stated in the following format.

Theorem 5.3.2. *(Valiron, 1932 and Cartwright, 1932 and 1935) Let $f(z)$ be a transcendental entire function with order $\lambda < +\infty$. If $f(z)$ has no Borel direction of maximal kind in $\Omega(\alpha,\beta)$ and (5.3.2) holds, then $\lambda \leqslant \frac{\pi}{\beta-\alpha}$.*

Here we remark on that Valiron and Cartwright Theorem 5.3.2 without "maximal kind" follows immediately from Nevanlinna Theorem 2.7.7 and Valiron Theorem 2.7.5 and the Phragmén-Lindelöf Theorem as mentioned previously by noting the finite covering property of a compact set. Theorem 5.3.2 without "maximal kind" was extended by Yang Lo [10] to the case of meromorphic functions with some Nevanlinna deficient value (see Theorem 3.1.6 for the Borel directions) in terms of the spread relation proved by Baernstein II.

A. Edrei[2] in 1955 turned to this problem. He seems to be the first one who investigated this aspect dealing with meromorphic function, its derivative and Nevanlinna deficiency to extend in some extent the Bieberbach Theorem 5.3.1.

Theorem 5.3.3. *(Edrei, 1955) Let $f(z)$ be a transcendental meromorphic function and such that all but finitely many roots of the three equations*

$$f(z)=0,\ f(z)=\infty,\ f^{(n)}(z)=1$$

$(n \geqslant 0, f^{(0)}=f)$ lie on the radii $D(\alpha_1,\cdots,\alpha_q)$. If

$$\delta(0,f)+\delta(\infty,f)+\delta(1,f^{(n)})>0,$$

then

$$\lambda(f) \leqslant \max\left\{\frac{\pi}{\alpha_{j+1}-\alpha_j} : 1 \leqslant j \leqslant q\right\}.$$

After A. Edrei's work, many mathematicians revealed and discovered the new connection among the growth of a meromorphic function, distribution of arguments of a-points of it and / or of its derivative and the Nevanlinna deficiency of it and / or of its derivative in different approaches, and so essentially developed the Bieberbach's Theorem 5.3.1. From the Milloux inequality about a disk, when f has a Nevanlinna deficient value, the growth order of f can be controlled by the order of the number of points of other two values, while the role of a deficient value is not obvious in consideration of an angular domain. A. Ostrovskii in 1957-1961 and in

1970 was successful in generalizing Edrei's result (Theorem 5.3.3) in this direction. In his result, he used $C_{\alpha,\beta}(r,f=a)$ to characterize the distribution of argument of a-points, and actually, this quantity measures not only the distance of a-points from the sides of the angle, but also is related to the number of a-points in the angle.

In 1970, A. Ostrovskii first took the following quantity into account in the discussion of the growth of meromorphic functions with radially distributed values

$$U(r,D,f=a)=\max\left\{\sum_{j=1}^{q} t^{\omega_j-\omega}C_{\alpha_j,\alpha_{j+1}}(t,f=a):1\leqslant t\leqslant r\right\}$$

and $\overline{U}(r,D,f=a)$ for $\overline{C}_{\alpha_j,\alpha_{j+1}}(t,f=a)$ and proved the following

Theorem 5.3.4. *Let $f(z)$ be a transcendental meromorphic function and a,b and c three distinct points in $\widehat{\mathbb{C}}$. Assume that a is a Nevanlinna deficient value of $f^{(n)}(z)$ $(n\geqslant 0)$.*

(1) *Then for the fixed $\varepsilon>0$ and $d>1$,*

$$T(r,f)\leqslant Kr^{d\omega}(\overline{U}(r,D,f=b)+\overline{U}(r,D,f=c)+\log rT(r,f))^d,\ r\notin E,$$

where $\mathrm{dens}E<\varepsilon$.

If, in addition, $\overline{U}(r,D,f=X)$ is of finite order for $X=b$ and c, then for some $d>1$ and all sufficiently large r,

$$T(r,f)\leqslant Kr^{\omega}(\overline{U}(dr,D,f=b)+\overline{U}(dr,D,f=c)+1),$$

where K is a positive constant.

(2) *If*

$$U(r,D,f=a)+U(r,D,f=b)+U(r,D,f=c)=O(1),$$

then for $r\rightarrow\infty$ perhaps passing outside a set of finite logarithmic measure, we have

$$\log|f(re^{i\theta})|=r^{\omega_j}c_j\sin(\omega_j(\theta-\alpha_j))+o(r^{\omega_j})$$

uniformly relative to θ, $\alpha_j\leqslant\theta\leqslant\alpha_{j+1}$, with $c_j\in\mathbb{R}$.

We shall sketch the proof of result (1) of Ostrovskii's Theorem 5.3.4, for it is an excellent representation of the related results without any assumption imposed on the growth.

The sketch proof of the result (1) of Theorem 5.3.4. For the simple sake, assume $n=0$, $b=0$ and $c=\infty$. From the equality

$$\frac{1}{f(z)-a}=\frac{1}{a}\frac{f(z)}{f'(z)}\left(\frac{f'(z)}{f(z)-a}-\frac{f'(z)}{f(z)}\right)$$

and in view of the definition of the Nevanlinna deficiency, for all sufficiently large r we have

$$\frac{\delta}{2}T(r,f)\leqslant m\left(r,\frac{1}{f-a}\right)\leqslant m\left(r,\frac{f}{f'}\right)+S(r,f).$$

Below we estimate $m\left(r,\frac{f}{f'}\right)$ in terms of $\overline{U}(r,D,f=0,\infty)$. Applying Lemma 2.2.3 and the first Nevanlinna fundamental theorem for angular domains, we have

$$\begin{aligned} m\left(r,\frac{f}{f'}\right) &= \sum_{j=1}^{q} m_{\alpha_j,\alpha_{j+1}}\left(r,\frac{f}{f'}\right) \\ &\leqslant \sum_{j=1}^{q} K_j r^{\omega_j}\left(S_{\alpha_j,\alpha_{j+1}}\left(r,\frac{f}{f'}\right)+1\right)^d \\ &\leqslant \sum_{j=1}^{q} K_j r^{\omega_j}\left(S_{\alpha_j,\alpha_{j+1}}\left(r,\frac{f'}{f}\right)+O(1)\right)^d \\ &\leqslant K r^{d\omega}(\overline{U}(r,D,f=0)+\overline{U}(r,D,f=\infty)+\log rT(r,f))^d, \end{aligned}$$

$r\not\in E$. This completes the proof of Theorem 5.3.4. □

Obviously, we can also deduce the result (1) of Theorem 5.3.4 in terms of Theorem 5.1.2 by noting

$$\overline{V}(r,D,f=a)\leqslant r^{\omega}\overline{U}(r,D,f=a)$$

and for fixed $\varepsilon>0$ with $d=(1-\varepsilon)^{-1}$ and sufficiently large r, $(\log T(r,f))^3 < T(r,f)^{\varepsilon}$. Theorem 5.3.4 covers Theorem 5.3.3 since if all but finitely many a-points lie on the radii system D, then $U(r,D,f=a)=O(1)$.

A. Edrei and W. H. J. Fuchs [3] in 1962 considered the case of that a-points lie on a finite system of pairwise non-intersecting curves tending to ∞, that is, so-called B regular curves (for definition please see the paragraph after Definition 3.1.2) and proved the following.

Theorem 5.3.5. *(Edrei and Fuchs, 1962) Let $f(z)$ be a transcendental meromorphic function and D be a finite system of B-regular curves which divides $|z|\geqslant t_0$ into finitely many curvilinear sectors such that each sector has opening $\geqslant c>0$, that is, the intersection of every circle $|z|=r\geqslant t_0$ and the sector is an arc of length $\geqslant cr$.*

Assume that all but finitely many zeros and poles of f lie on the system D and $f^{(p)}$ $(p\geqslant 0)$ has a finite and non-zero Nevanlinna deficient value. Then

$$\lambda(f)\leqslant\frac{9\pi B^2}{c}.$$

There does not seem to be further results in improving Theorem 5.3.5 and developing the point of view of Edrei and Fuchs since 1962. It is interesting to consider weakening of restriction imposed on the number of zeros and poles of the function considered in the curvilinear sectors, for example, we take into account the Problem: under the assumption of that the number of zeros and poles in the intersection of the curvilinear sectors and the disk $\{z:|z|<r\}$ equals $o(T(r,f))$ instead, is Theorem 5.3.5 true? Up to now we do not know whether the upper bound obtained in Theorem 5.3.5 is precise. When the B-regular curve is a ray from the origin, the upper bound is not precise, while the precise upper bound is that divided by 9.

In what follows, we discuss the case of meromorphic functions with finite lower order. Let us consider the system of rays

$$\begin{aligned} D &= D(\alpha_1, \beta_1, \cdots, \alpha_q, \beta_q) \\ &= \cup_{j=1}^{q}(\{z : \arg z = \alpha_j\} \cup \{z : \arg z = \beta_j\}). \end{aligned}$$

Following Goldberg and Ostrovskii [5], we shall say that almost all a-points of $f(z)$ lie in small angles if

$$\min\{\beta_j - \alpha_j : 1 \leqslant j \leqslant q\} > \max\{\alpha_{j+1} - \beta_j : 1 \leqslant j \leqslant q\},$$

that is, $\omega''(D) > \omega'(D)$, and

$$\overline{C}_{\alpha_j,\beta_j}(r, f = a) < K_j r^{\omega' - \omega'_j},$$

where K_j is a positive constant. This implies that for at least one j, $\overline{C}_{\alpha_j,\beta_j}(r, f = a)$ is bounded.

Ostrovskii in 1960, Gol'dberg and Ostrovskii in 1970 and Glejzer in 1985 and 1990 investigated the growth of a meromorphic function most of whose a-points for two values of a lie in small angles. Most general results among them are ones obtained finally by Glejzer [4].

Theorem 5.3.6. *(Glejzer, 1985, 1990) Let $f(z)$ be a transcendental meromorphic function with finite lower order. Assume that almost all its zeros and poles lie in the small angles of the system D of rays. Then for $\mu \leqslant \rho \leqslant \lambda$ (in the case $\lambda = \infty$, naturally, $\mu \leqslant \rho < \lambda$), no one of the relations*

$$\omega'(D) < \rho < \frac{4}{\pi}\omega''(D) \arcsin\sqrt{\frac{\delta(a,f)}{2}}$$

and

$$\omega'(D) < \rho < \min\left\{2\omega''(D), \frac{4}{\sum\limits_{j=1}^{q}(\alpha_{j+1} - \beta_j)} \sum_{a \neq 0,\infty} \arcsin\sqrt{\frac{\delta(a,f)}{2}}\right\}$$

can hold.

We can restate Theorem 5.3.6 in the following equivalent form which is however convenient and natural for us in the point of view of this chapter.

Write $\eta = \frac{4}{\pi}\omega''(D) \arcsin\sqrt{\frac{\delta(a,f)}{2}}$. Assume that $\omega'(D) < \eta$. Theorem 5.3.6 asserts

$$(\mu, \lambda) \cap (\omega'(D), \eta) = \varnothing$$

and hence if $\mu < \eta$, we have $\lambda \leqslant \omega'(D)$. This yields the result that if

$$\max\{\alpha_{j+1}-\beta_j : 1\leqslant j\leqslant q\} < \frac{4}{\gamma}\arcsin\sqrt{\frac{\delta(a,f)}{2}}, \tag{5.3.3}$$

$\gamma=\max\{\omega'(D),\mu\}$, then $\lambda\leqslant\omega'(D)$.

Write

$$\tau=\min\left\{2\omega''(D), \frac{4}{\sum\limits_{j=1}^{q}(\alpha_{j+1}-\beta_j)}\sum_{a\neq 0,\infty}\arcsin\sqrt{\frac{\delta(a,f)}{2}}\right\}.$$

Assume $\omega'(D)<\tau$ and then in view of Theorem 5.3.6, $(\mu,\lambda)\cap(\omega'(D),\tau)=\varnothing$. Therefore, if $\mu<\tau$, we have $\lambda\leqslant\omega'(D)$. However, the inequalities $\omega'(D)<\tau$ and $\mu<\tau$ are equivalent to

$$\sum_{j=1}^{q}(\alpha_{j+1}-\beta_j) < \frac{4}{\gamma}\sum_{a\neq 0,\infty}\arcsin\sqrt{\frac{\delta(a,f)}{2}} \tag{5.3.4}$$

and $\mu<2\omega''(D)$, that is,

$$\max\{\alpha_{j+1}-\beta_j : 1\leqslant j\leqslant q\} < \frac{2\pi}{\mu}. \tag{5.3.5}$$

Thus we have the following

Theorem 5.3.7. *Under the assumption of Theorem 5.3.6, if (5.3.3) or (5.3.4) and (5.3.5) hold, then $\lambda\leqslant\omega'(D)$.*

Glejzer [4] considered the Pólya lower order and indeed, we can also take into account the Pólya lower order in the place of the lower order in Section 5.2. Theorem 5.2.3 improves the result of Theorem 5.3.7 under (5.3.4) and (5.3.5) because a stronger restriction is given in Theorem 5.3.7 to the number of zeros and 1-points in the angular domains. However, we do not know if we could get the results of Theorems 5.2.1, 5.2.2 and 5.2.3 under the assumption of (5.3.3) replacing (5.2.3) and (5.2.16). A few papers we do not mention here investigate this subject, some of which are listed in the Reference.

The well-known methods to treat this subject on the growth order of meromorphic functions with radially distributed values are mainly those by mapping conformally the angular domain onto the unit disk and then by using the Nevanlinna theory on the unit disk or by the Nevanlinna theory on the angular domains only and/or by that together with the Baernstein's spread theorem 2.8.1 or the Baernstein's *-function. It is obvious that Baernstein's Theorem 2.8.1 is available only to a transcendental meromorphic function with a sequence of Pólya peaks, so in this case, it should be assumed that the meromorphic function in question is of finite lower order or of finite Pólya lower order.

References

1. Conway, J. B., Functions of one complex variable, Springer-Verlag,1978.
2. Edrei, A., Meromorphic functions with three radially distributed values, Trans. Amer. Math. Soc., **78**(1955), 276-293.
3. Edrei, A. and Fuchs, W. H. J., Bounds for the number of deficient values of certain classes of meromorphic functions, Proc. London Math. Soc., **12**(1962), 315-344.
4. Gleizer, E. V., Meromorphic functions with zeros and poles in small angles, Sib. Mat. Zh., (4)**26**(1985), 22-37; II, Sib. Mat. Zh., (2)**31**(1990), 9-20.
5. Goldberg, A. A. and Ostrovskii, I. V., The distribution of values of meromorphic functions (in Russian), Izdat. Nauk. Moscow, 1970.
6. Muto, H., The value distribtion of meromorphic functions in an angular domain, J. London Math. Soc., (2)**13**(1976), 193-198.
7. Muto, H., The value distribtion of meromorphic functions in an angular domain II, J. London Math. Soc., (2)**18** (1981), 287-294.
8. Wu S. J., Arguemnt distribution and growth of meromorphic functions, Science in China, Series A, (6)**23**(1993), 565-575.
9. Wu S. J., Distribution of the $(0, \infty)$ accumulative lines of meromorphic functions, Chin. Ann. Math., **15B**: 4(1994), 453-462.
10. Yang, L., Borel directions of meromorphic functions in an angular domain, Science in China, Math. Series (I) (1979), 149-163.
11. Yang, L. and Yang, C. C., Agular distribution of values of ff', Science in China, Seires A, (3)**37**(1994), 284-294.
12. Zheng, J. H., On value distribution of meromorphic functions with respect to argument I, Complex Variables and Elliptic Equations, 2008.
13. Zheng, J. H., On the growth of meromorphic functions with two radially distributed values, J. Math. Anal. Appl., **206**(1997), 140-154.
14. Zheng, J. H., On transcendental meromorphic functions with radially distributed values, Science in China Ser. A. Math., (3)**47**(2004), 401-416.
15. Zheng, J. H., Complex Analysis, Tsinghua University Press, 2003 (in Chinese).

Chapter 6
Singular Values of Meromorphic Functions

Jianhua Zheng
Department of Mathematical Sciences, Tsinghua University, Beijing 100084, P. R. China
jzheng@math.tsinghua.edu.cn

Abstract: This chapter is devoted to discussing singular values of a transcendental meromorphic function. The singular value is that in any neighborhood of which the inverse of the function contains a multiple-valued branch. A value is a singular value if and only if it is an asymptotic value or a critical value. We show the construction of the parabolic simply connected Riemann surface associated with a fixed meromorphic function, and point out that every boundary point of the Riemann surface is an asymptotic value of the function. Next we consider dense properties of singularities of the inverse of a meromorphic function including relationships among singular values and between the number of direct singularities and the growth order. We then exhibit Eremenko's construction of a meromorphic function with every value on the extended complex plane as its asymptotic value. Finally, we discuss the existence of (repelling) fixed-points of a meromorphic function of finite type, that is, the set of its singular values is bounded, and consider the case when the singular values do not distribute along a sequence of annuli.

Key words: Riemann surface, Asymptotic values, Critical values, Fixed-points, Bounded type

Let $f(z)$ be a meromorphic function on $\mathbb{C}$. Associated with this meromorphic function there exists a Riemann surface. According to Nevanlinna, Ahlfors and Teichmüller, the center problem of meromorphic function theory is to investigate how we could determine properties of meromorphic functions in terms of geometric properties of their associated Riemann surfaces. The topological properties, such as the number of omitted values, asymptotic values and critical values, are parts of the geometric properties mentioned here. Indeed, corresponding to a boundary point of the associated Riemann surface is an asymptotic value of a meromorphic function. This chapter is devoted to discussing singular values, namely asymptotic values and critical values, of meromorphic functions composing of two aspects: one is the existence and properties of singular values; the other is to characterize meromorphic functions in terms of their singular values.

6.1 Riemann Surfaces and Singularities

Let us begin with the definition of Riemann surface.

Definition 6.1.1. *Let W be a connected Hausdorff space. The pair (W, Φ) is called a Riemann surface provided that W is equipped with a family of pairs $\Phi = \{(U_\alpha, \varphi_\alpha)\}$ satisfying the following items*

(1) *$\{U_\alpha\}$ is an open covering of W, that is, each U_α is open and $W = \bigcup_\alpha U_\alpha$;*

(2) *each φ_α is a homeomorphism of a domain of $\mathbb{C}$ from U_α;*

(3) *if $U_\alpha \cap U_\beta \neq \varnothing$, $\varphi_\beta \circ \varphi_\alpha^{-1} : \varphi_\alpha(U_\alpha \cap U_\beta) \to \varphi_\beta(U_\alpha \cap U_\beta)$ is analytic, which is called analytically compatible;*

(4) *Φ is maximal with respect to (2) and (3), that is, if a pair (U, φ) satisfies (2) and (3), then $(U, \varphi) \in \Phi$.*

A complex atlas means a family of pairs $\Phi = \{(U_\alpha, \varphi_\alpha)\}$ satisfying (1), (2) and (3) and an element in Φ is a complex chart. The maximal complex atlas Φ is called a complex structure on W. One usually writes briefly W instead of (W, Φ) whenever no confusion occurs in the context. Sometimes one also writes (W, Φ^*) where Φ^* is a complex atlas.

For a complex chart $(U_\alpha, \varphi_\alpha)$, U_α is called a local coordinate neighborhood of any point $p \in U_\alpha$ and φ_α a local coordinate and so for $p \in U_\alpha$, $z = \varphi_\alpha(p) \in \mathbb{C}$ is a coordinate of p. Obviously we can choose a special local coordinate φ_α which is required to map U_α onto the unit disk.

Definition 6.1.2. *Let (X, Φ) and (Y, Ψ) be two Riemann surfaces and let $f : Y \to X$ be a continuous mapping. If for arbitrary two pair of charts $(U_\alpha, \varphi_\alpha) \in \Psi$ and $(V_\beta, \psi_\beta) \in \Phi$ with $f(U_\alpha) \subset V_\beta$,*

$$\psi_\beta \circ f \circ \varphi_\alpha^{-1} : \varphi_\alpha(U_\alpha) \to \psi_\beta(V_\beta)$$

is analytic, that is, f, φ_α and ψ_β are analytically compatible, then we say that f is analytic, or holomorphic from Y to X.

Actually to show that the function $f(z)$ is analytic, it suffices to verify that f is analytically compatible with φ_α and ψ_β for corresponding complex atlas.

A mapping $f : Y \to X$ is called conformal if it is bijective and both $f : Y \to X$ and $f^{-1} : X \to Y$ are holomorphic and in this case we say Y and X are conformally equivalent.

By a meromorphic function on a Riemann surface X we mean that for an open subset X' of X with $X \setminus X'$ containing only isolated points, $f : X' \to \mathbb{C}$ is holomorphic and for each $p \in X \setminus X'$, $\lim_{x \to p} |f(x)| = \infty$, and p is called a pole of f. If $f : X \to \mathbb{C}$ is meromorphic, then defining $f(x) = \infty$ at all poles of f we have $f : X \to \widehat{\mathbb{C}}$ is holomorphic. Conversely, if $f : X \to \widehat{\mathbb{C}}$ is holomorphic, then f is either identically equal to ∞ or else $f^{-1}(\infty)$ consists of isolated points and $f : X \to \mathbb{C}$ is meromorphic.

Let X and Y be two Riemann surfaces. A mapping $f : Y \to X$ is called a covering map if for each point $x \in X$ there exists a neighborhood U of x in X such that

$$f^{-1}(U) = \bigcup_{j \in J} V_j, \tag{6.1.1}$$

where the V_j, $j \in J$, are disjoint open subsets of Y, and every restriction of mapping f in V_j is a homeomorphism of U from V_j. Therefore f is a local homeomorphism in Y. We say that Y is a covering space of X if there exists a covering mapping from Y onto X and if Y is simply connected, then covering mapping $f : Y \to X$ is called universal covering. In fact, the universal covering is determined by the following universal property. Let $f : Y \to X$ be a universal covering. For every covering $g : Z \to X$ and every pair of $y_0 \in Y$ and $z_0 \in Z$ with $f(y_0) = g(z_0)$, there exists unique continuous fiber-preserving mapping $h : Y \to Z$, namely $f = g \circ h$, such that $h(y_0) = z_0$. For any Riemann surface there must be its universal covering space. Any simply connected Riemann surface is conformally equivalent to the Riemann sphere $\mathbb{S}$ or the complex plane $\mathbb{C}$ or the unit disk Δ. Here "conformally equivalent" means that there exists a conformal analytic mapping between them. Therefore a Riemann surface is called in turn elliptic, parabolic or hyperbolic provided that the Riemann sphere $\mathbb{S}$, the complex plane $\mathbb{C}$ or the unit disk Δ is its universal covering space. In particular, $\widehat{\mathbb{C}} \setminus \{a\}$ and $\widehat{\mathbb{C}} \setminus \{a, b\}$ are parabolic Riemann surfaces and any domain X on $\widehat{\mathbb{C}}$ with $\widehat{\mathbb{C}} \setminus X$ containing at least three points is hyperbolic.

Now we come to discuss holomorphic map between two Riemann surfaces. At this time, we take branch points into account. Let X and Y be two Riemann surfaces and $f : Y \to X$ be a non-constant holomorphic map. A point $y \in Y$ is called a branch point or ramification point of f, if there is no neighborhood V of y such that f is injective on V. A holomorphic map then is unbranched if it has no branch points. What we mention is when we say a holomorphic map to be covering, it is allowed to have branch points, namely it may not be local homeomorphism on the whole surface, while it is local homeomorphism on remaining part from the Riemann surface punctured at branch points. If there exist no branch points at all, we shall specifically emphasize that the holomorphic map is unbranched.

Theorem 6.1.1. *Let X be a Riemann surface and $f : X \to \Delta^*$ is an unbranched holomorphic covering map where Δ^* is the punctured unit disk $\{z : 0 < |z| < 1\}$. Then one of the following statements holds:*

(1) *there exists a conformal mapping ψ of X onto the left half plane $H = \{z : \mathrm{Re}z < 0\}$ such that $f = \exp \circ \psi$;*

(2) *there exists a conformal mapping ψ of X onto Δ^* such that $f = (\psi)^n$ for some natural number n.*

Proof. It is clear that $\exp : H \to \Delta^*$ is the universal covering and since $f : X \to \Delta^*$ is a covering in the sense of (6.1.1), in view of the universal property we therefore have a holomorphic mapping $\phi : H \to X$ such that $\exp = f \circ \phi$. It is easy to show that $\phi : H \to X$ is also a universal covering. If ϕ is injective, then it is conformal and its inverse mapping ψ is the desired one stated in (1); If ϕ is not injective, then there exist two distinct point z_1 and z_2 in H such that $\phi(z_1) = \phi(z_2)$ and so $\exp(z_1) = \exp(z_2)$ and equivalently $z_1 - z_2 = 2m\pi i$ for some non-zero integer m. It follows from $\phi : H \to X$ being a universal covering that there exists unique holomorphic mapping h of H onto H such that $\phi(z) = \phi \circ h(z)$ and $h(z_2) = z_1$. Therefore it is easy to see that $h(z)$ is conformal and a Möbius transformation and from $\exp \circ h = f \circ \phi \circ h = f \circ \phi = \exp$, we have $h(z) = z + 2m\pi i$ and so $\phi(z) = \phi(z + 2m\pi i)$, $\forall z \in H$, namely $\phi(z)$ is periodic. Assume that n is its primitive period and hence there exists a bijective mapping $\psi : X \to \Delta^*$ with $\psi \circ \phi = \exp(z/n)$ and from $(\exp(z/n))^n = \exp z$ it follows that $(\psi \circ \phi)^n = f \circ \phi$, namely $f = (\psi)^n$, from which it is easy to show that ψ is holomorphic.

Thus Theorem 6.1.1 is proved. □

Given a fixed point $a \in \widehat{\mathbb{C}}$ and a function $f(z)$ meromorphic at a, we consider the pair (f, a) and introduce an equivalent relation in the family of all such pairs. Two pairs (f, a) and (g, b) are equivalent if $a = b$ and $f(z) \equiv g(z)$ at a neighborhood of a. It is clear that this is an equivalent relation. By notation $[f]_a$ we denote the equivalent class determined by (f, a), that is,

$$[f]_a = \{(g, D) : \ a \in D \text{ and } g(z) \equiv f(z) \text{ at a neighborhood of } a\},$$

where (g, D) is a meromorphic element, namely $g(z)$ is a meromorphic function in domain D. And $[f]_a$ is called the germ of f at a.

Let U be an open set on $\widehat{\mathbb{C}}$. Set

$$\mathscr{L}(U) = \{[f]_a : \ a \in U \text{ and } f \text{ is meromorphic at } a\},$$

namely, $\mathscr{L}(U)$ is the family of all germs at points of U and define a project of U from $\mathscr{L}(U)$:

$$\pi([f]_z) = z, \ \forall \, [f]_z \in \mathscr{L}(U).$$

Through π we can induce a topology to $\mathscr{L}(U)$ from the topology of $\widehat{\mathbb{C}}$ such that $\mathscr{L}(U)$ becomes a topological space and $\pi : \mathscr{L}(U) \to U$ is continuous. Indeed, a neighborhood of $[f]_a$ can be obtained in the following way. Since $f(z)$ is meromorphic in a domain D containing a,

$$\mathscr{N}(f, D) = \{[f]_z : \ \forall \, z \in D\}$$

is a neighborhood of $[f]_a$ and then π is a homeomorphism of D from $\mathscr{N}(f, D)$. Thus $\mathscr{L}(U)$ becomes a Hausdorff space.

In what follows, we consider $\mathscr{L}(\widehat{\mathbb{C}})$ and its connected components. Let $\mathscr{L}_0(\widehat{\mathbb{C}})$ be a connected component of $\mathscr{L}(\widehat{\mathbb{C}})$ and therefore $\mathscr{L}_0(\widehat{\mathbb{C}})$ is a connected Hausdorff space. Now to make $\mathscr{L}_0(\widehat{\mathbb{C}})$ be a Riemann surface we equip $\mathscr{L}_0(\widehat{\mathbb{C}})$ with the following complex structure: for each point $[f]_z \in \mathscr{L}_0(\widehat{\mathbb{C}})$, we have a neighborhood $\mathscr{N}(f, D)$ of $[f]_z$ and then $(\mathscr{N}(f, D), \pi|_{\mathscr{N}(f,D)})$ is a complex chart and we obtain a complex atlas

$$\Phi_0 = \{(\mathscr{N}(f, D), \pi|_{\mathscr{N}(f,D)}) : \ \forall \, [f]_z \in \mathscr{L}_0(\widehat{\mathbb{C}})\}$$

which decides a complex structure Φ. Importantly, associated with the Riemann surface $(\mathscr{L}_0(\widehat{\mathbb{C}}), \Phi)$ is a function $\mathscr{F} : \mathscr{L}_0(\widehat{\mathbb{C}}) \to \widehat{\mathbb{C}}$ which is defined as follows: for each $[f]_z \in \mathscr{L}_0(\widehat{\mathbb{C}})$,

$$\mathscr{F}([f]_z) = f(z).$$

Let us show that $\mathscr{F}$ is meromorphic. For each element $(\mathscr{N}(f,D), \pi|_{\mathscr{N}(f,D)})$ of Φ_0, it is clear that

$$\mathscr{F} \circ (\pi|_{\mathscr{N}(f,D)})^{-1}(z) = \mathscr{F}([f]_z) = f(z)$$

is a meromorphic function from D to $\widehat{\mathbb{C}}$ so that $\mathscr{F}$ is meromorphic over $\mathscr{L}_0(\widehat{\mathbb{C}})$. Since $\mathscr{L}_0(\widehat{\mathbb{C}})$ is a component of $\mathscr{L}(\widehat{\mathbb{C}})$, that is, maximal in the connected sense, we say that $\mathscr{F}$ is a complete meromorphic function determined by $\mathscr{L}_0(\widehat{\mathbb{C}})$ and $\mathscr{L}_0(\widehat{\mathbb{C}})$ is the Riemann surface associated with $\mathscr{F}$.

We can obtain the Riemann surface $\mathscr{L}_0(\widehat{\mathbb{C}})$ and the associated complete meromorphic function $\mathscr{F}$ by meromorphic continuation of a meromorphic function element (f,D) where D is a domain on $\widehat{\mathbb{C}}$ and $f(z)$ is a meromorphic function on D. In order to make this process clear let us recall the concepts and basic results of meromorphic continuation.

Let D_j $(j = 1,2,\cdots,n)$ be n domains on $\widehat{\mathbb{C}}$ and $\{D_1, D_2, \cdots, D_n\}$ is a chain of domains provided that $D_{j-1} \cap D_j \neq \varnothing$ $(2 \leqslant j \leqslant n)$. A collection (f_j, D_j) $(j = 1,2,\cdots,n)$ of meromorphic function elements is a meromorphic continuation along the chain of domains $\{D_1, D_2, \cdots, D_n\}$ if $f_{j-1}(z) \equiv f_j(z)$ in $D_{j-1} \cap D_j$ and in this case we say that (f_1, D_1) can be continued to (f_n, D_n) along the chain of domains and (f_n, D_n) can be obtained by a meromorphic continuation of (f_1, D_1) along the chain of domains. Let $\gamma : [0,1] \to \widehat{\mathbb{C}}$ be a path. If for each $t \in [0,1]$ there is a meromorphic function element (f_t, D_t) such that $\gamma(t) \in D_t$ and in a neighborhood I_t of $t \in [0,1]$ we have $D_s \cap D_t \neq \varnothing$, $\forall s \in I_t$ and

$$f_s(z) \equiv f_t(z), \ \forall z \in D_s \cap D_t, \tag{6.1.2}$$

then we say that one meromorphically continues (f_0, D_0) to (f_1, D_1) along the path γ and (f_1, D_1) is the meromorphic continuation of (f_0, D_0) along the path γ. It is obvious that (6.1.2) can be rewritten in terms of germs into

$$[f_s]_{\gamma(s)} = [f_t]_{\gamma(t)}.$$

Thus we understand the meaning of meromorphic continuation of a germ $[f]_a$ to another germ $[g]_b$ along a path connecting two points a and b. By that a germ $[f]_a$ or a meromorphic function element (f,D) can be continued to b along a path connecting a and b we mean that a germ $[g]_b$ can be obtained from $[f]_a$ or (f,D).

The continuation along a fixed path is unique in the sense of that $[g_1]_b$ and $[g_2]_b$ are meromorphic continuation of, respectively, $[f_1]_a$ and $[f_2]_a$ along a path connecting two points a and b, if $[f_1]_a = [f_2]_a$, then $[g_1]_b = [g_2]_b$.

That we say to continue meromorphically a germ or a function element in a domain without any restriction means that this germ or element is continued along any path in the considered domain along which we can do. And we say that we can continue meromorphically a germ or a function element in a domain without any restriction provided that this germ or element can be continued along any path in the considered domain. Thus we can state the monodromy theorem as follows. If a germ $[f]_a$ can be continued in a domain U without any restriction, then for any $b \in U$ we obtain the same germ along any two paths connecting a and b which are homotopic in U.

Now we go back to the construction of $\mathscr{L}_0(\widehat{\mathbb{C}})$ and the associated complete meromorphic function $\mathscr{F}$ over $\mathscr{L}_0(\widehat{\mathbb{C}})$. Notice that $\pi(\mathscr{L}_0(\widehat{\mathbb{C}}))$ is a domain on $\widehat{\mathbb{C}}$, denoted by U, and hence $\mathscr{L}_0(\widehat{\mathbb{C}})$ is a component of $\mathscr{L}(U)$. Take a point $[f]_a \in \mathscr{L}_0(\widehat{\mathbb{C}})$ and we have a corresponding meromorphic function element (f,D) with $a \in D \subseteq U$. Then we can meromorphically continued (f,D) in U without any restriction to obtain its complete meromorphic function $\mathscr{F}$ and the associated Riemann surface $\mathscr{L}_0(\widehat{\mathbb{C}})$. This is asserted in the following theorem.

Theorem 6.1.2. *$\mathfrak{M}$ is a component of $\mathscr{L}(\widehat{\mathbb{C}})$ if and only if arbitrarily choosing a fixed point $[f]_a \in \mathfrak{M}$, we have*

$$\mathfrak{M} = \{[g]_b : \ [g]_b \text{ is obtained from the continuation of } [f]_a \text{ along a curve}\}. \quad (6.1.3)$$

Proof. Assume that $\mathfrak{M}$ is a component of $\mathscr{L}(\widehat{\mathbb{C}})$ and therefore $\mathfrak{M}$ contains point $[g]_b$ which is a continuation of $[f]_a$ along a curve. On the other hand, for any point $[g]_b \in \mathfrak{M}$ there exists a path $\tilde{\gamma}: [0,1] \to \mathfrak{M}$ connecting $[f]_a$ and $[g]_b$. Write $\gamma = \pi(\tilde{\gamma})$ and it is a path in $\widehat{\mathbb{C}}$ connecting a and b. Then $\tilde{\gamma}(t) = [f_t]_{\gamma(t)}$ with $[f_0]_{\gamma(0)} = [f]_a$ and $[f_1]_{\gamma(1)} = [g]_b$. It is easy to see that $\{[f_t]_{\gamma(t)}\}$ satisfies (6.1.2) and so $[g]_b$ is a continuation of $[f]_a$ along γ. This immediately implies (6.1.3).

Now assume that (6.1.3) holds. This means that $\mathfrak{M}$ is arcwise connected and contains any connected open subset of $\mathscr{L}(\widehat{\mathbb{C}})$ containing $[f]_a$. Then it is clear that $\mathfrak{M}$ is a component of $\mathscr{L}(\widehat{\mathbb{C}})$. □

Below let us consider the boundary of $\mathscr{L}_0(\widehat{\mathbb{C}})$ and meromorphic extension of $\mathscr{F}$. Every boundary point of $\mathscr{L}_0(\widehat{\mathbb{C}})$ is produced by (f,D) corresponding to some germ and some path γ with an end point a such that (f,D) cannot be meromorphically continued to a along γ but can be done to the other points in γ, such boundary point will be denoted by notation $[f]_{a,\gamma}$.

For an isolated boundary point Q of $\mathscr{L}_0(\widehat{\mathbb{C}})$, if there exists a neighborhood W on $\mathscr{L}_0(\widehat{\mathbb{C}})$ such that $\pi(W \setminus \{Q\}) = B^*(a,\delta)$ with $0 < \delta \leqslant 1$ (here we assume $a \neq \infty$) and $\pi : W \setminus \{Q\} \to B^*(a,\delta)$ is finitely sheeted, then define $\pi(Q) = a$ and Q is called an algebraic singular point or branch point of the Riemann surface $\mathscr{L}_0(\widehat{\mathbb{C}})$. If π is m-sheeted, then $m-1$ is the order of this branch point. In this case, $\mathscr{F}$ is bounded in $W \setminus \{Q\}$, in view of the Riemann's Removable Singularity Theorem $\mathscr{F}$ can be meromorphically extended to Q. If $\pi : W \setminus \{Q\} \to B^*(a,\delta)$ is infinitely sheeted, then Q is called a transcendental singular point or branch point of the Riemann surface $\mathscr{L}_0(\widehat{\mathbb{C}})$.

In order to make these clear we observe two special examples: e^z and z^m. Take a branch of $\mathrm{Log}z$ and continue it in $\widehat{\mathbb{C}}$ without any restriction to produce a Riemann surface which is formed by gluing along their cutting lines remaining parts of infinitely many complex plane cut down the negative real axis. Obviously, this

Riemann surface is simply connected and 0 and ∞ are only two boundary points of it. Therefore both 0 and ∞ are the transcendental singular points of this Riemann surface. We continue a branch of $\sqrt[n]{z}$ to obtain a Riemann surface and 0 and ∞ are the algebraic singular points of this Riemann surface, which shall become simply connected after adding 0 and ∞ onto it.

In what follows we add all algebraic singular points onto $\mathscr{L}_0(\widehat{\mathbb{C}})$, namely $\mathscr{L}_0(\widehat{\mathbb{C}})$ is obtained through analytic continuation with algebraic character. Analytic continuation with algebraic character means that function $f(z)$ in analytic function element may have the following expansion

$$f(z) = \sum_{n=n_0}^{+\infty} a_n (z-z_0)^{n/p}, \tag{6.1.4}$$

where n_0 is an integer and p is a positive integer. When $p=1$, that is the analytic continuation mentioned previously. Therefore an algebraic singular point is expressed as $[f]_{z_0}$ for some $f(z)$ with the form (6.1.4), which is an algebraic singular point of order $p-1$.

Now let us give a complete description to the above process. Let $Q=[g]_{a,\gamma}$ be an algebraic singular point of $\mathscr{L}_0(\widehat{\mathbb{C}})$ and W is a vicinity of Q. Define $\pi(Q)=a$ so that $\pi(W)$ is a vicinity of a in $\widehat{\mathbb{C}}$ and take a $\delta>0$ such that $D=\{z:|z-a|<\delta\}\subset\pi(W)$. Set $D^*=D\setminus\{a\}$. When δ is chosen to be sufficiently small, $\pi: W_Q\to D^*$ is an unbranched holomorphic covering, where W_Q is the component of $\pi^{-1}(D^*)$ lying in W. Draw a closed curve $\widetilde{\iota}$ around Q in W and then $\iota=\pi(\widetilde{\iota})$ goes around a. It is clear that ι is homotopic to α^m in D^* where α is the circle $\{z:|z-a|=\varepsilon\}$ with $\varepsilon<\delta$. Hence $g(z)$ is continued meromorphically m times along α starting from a point b of $\alpha\cap\gamma$, the resulting germ coincides with $[g]_b$, namely it goes back to the starting germ. In view of Theorem 6.1.1, there exists a conformal mapping ψ_Q of W_Q onto $A^*=\{\zeta: 0<|\zeta|<\sqrt[p]{\delta}\}$ such that $\pi=a+(\psi_Q)^p$ for some positive integer p. Set

$$F_Q(\zeta)=\mathscr{F}\circ\psi_Q^{-1}(\zeta): A^*\to\mathscr{F}(W_Q)\subset\widehat{\mathbb{C}}.$$

It is an analytic function in A^*. Since $\widehat{\mathbb{C}}\setminus\mathscr{F}(W_Q)$ contains at least three points, in view of the Picard Theorem a is not an essential singular point of $F_Q(\zeta)$ and thus we have the Laurant series

$$F_Q(\zeta)=\sum_{n=n_0}^{\infty} a_n\zeta^n$$

for some integer n_0. Set $z=a+\zeta^p$ and $f_Q(z)=F_Q(\zeta)$ and then $f_Q(z)$ has the Puiseux series with the form (6.1.4). It is easy to see that $[g]_b$ is a germ from a branch of $f_Q(z)$. Set $Q=[f_Q]_a$, which is added onto $\mathscr{L}_0(\widehat{\mathbb{C}})$. Define $\pi([f_Q]_a)=a$ and $\mathscr{F}([f_Q]_a)=F_Q(0)$. Thus π and $\mathscr{F}$ are extended forward algebraic singular point of $\mathscr{L}_0(\widehat{\mathbb{C}})$.

We denote by $\widetilde{\mathscr{L}_0}(\widehat{\mathbb{C}})$ the union of $\mathscr{L}_0(\widehat{\mathbb{C}})$ and all its algebraic singular points. We want to show that $\widetilde{\mathscr{L}_0}(\widehat{\mathbb{C}})$ is a Riemann surface equipped with a complex structure such that $\mathscr{L}_0(\widehat{\mathbb{C}})$ becomes its Riemann sub-surface. We add all neighborhoods of algebraic singular points to the topology system of $\mathscr{L}_0(\widehat{\mathbb{C}})$ to obtain the topology system of $\widetilde{\mathscr{L}_0}(\widehat{\mathbb{C}})$ and make $\widetilde{\mathscr{L}_0}(\widehat{\mathbb{C}})$ become a topological space. Since all algebraic singular points are isolated, under the topology formed in the above way $\widetilde{\mathscr{L}_0}(\widehat{\mathbb{C}})$ is connected and Hausdorff and $\mathscr{L}_0(\widehat{\mathbb{C}})$ is an open subset of $\widetilde{\mathscr{L}_0}(\widehat{\mathbb{C}})$. As did in the above, π and $\mathscr{F}$ are well defined on $\widetilde{\mathscr{L}_0}(\widehat{\mathbb{C}})$ and it is easy to show that they are continuous.

Theorem 6.1.3. *$\widetilde{\mathscr{L}_0}(\widehat{\mathbb{C}})$ is a Riemann surface. Furthermore, if for each $[f]_a \in \mathscr{L}_0(\widehat{\mathbb{C}})$, $f(z)$ is injective in a vicinity of a and for each $Q \in \widetilde{\mathscr{L}_0}(\widehat{\mathbb{C}}) \setminus \mathscr{L}_0(\widehat{\mathbb{C}})$, F_Q is injective in a vicinity of 0, then*

$$\mathscr{F} : \widetilde{\mathscr{L}_0}(\widehat{\mathbb{C}}) \to \mathscr{F}(\widetilde{\mathscr{L}_0}(\widehat{\mathbb{C}})) \subseteq \widehat{\mathbb{C}}$$

is an unbranched holomorphic covering and if, in addition, $\mathscr{F}(\widetilde{\mathscr{L}_0}(\widehat{\mathbb{C}}))$ is simply-connected, then $\mathscr{F}$ is a conformal map of $\widetilde{\mathscr{L}_0}(\widehat{\mathbb{C}})$ onto $\mathscr{F}(\widetilde{\mathscr{L}_0}(\widehat{\mathbb{C}}))$ and so $\widetilde{\mathscr{L}_0}(\widehat{\mathbb{C}})$ is simply-connected.

Proof. Set

$$\widetilde{\Phi}_0 = \Phi_0 \cup \{(W_Q, \psi_Q) : \forall\, Q \in \widetilde{\mathscr{L}_0}(\widehat{\mathbb{C}}) \setminus \mathscr{L}_0(\widehat{\mathbb{C}})\}.$$

We want to show that $\widetilde{\Phi}_0$ is a complex atlas. Obviously it suffices to prove that $\pi|_{\mathscr{N}(f,D)}$ and ψ_Q are analytically compatible when W_Q intersects $\mathscr{N}(f,D)$. In fact,

$$\pi \circ \psi_Q^{-1} : \psi_Q(W_Q \cap \mathscr{N}(f,D)) \to \pi(W_Q \cap \mathscr{N}(f,D))$$

is with the form

$$\pi \circ \psi_Q^{-1}(z) = \pi(Q) + z^p$$

for some natural number p, which is certainly analytic. Consequently, $\widetilde{\Phi}_0$ is a complex atlas and produces a complex structure of $\widetilde{\mathscr{L}_0}(\widehat{\mathbb{C}})$ to make it be a Riemann surface.

For a point $[f]_a \in \mathscr{L}_0(\widehat{\mathbb{C}})$, we have an associated function element (f,D) and under the assumption of Theorem 6.1.3 $f(z)$ is required to be injective in D. It is obvious that $\mathscr{F}$ is injective of $\mathscr{N}(f,D)$ onto $f(D)$. For a point $Q \in \widetilde{\mathscr{L}_0}(\widehat{\mathbb{C}}) \setminus \mathscr{L}_0(\widehat{\mathbb{C}})$, then $\mathscr{F} = F_Q \circ \psi_Q$ is an injective map of W_Q onto $\mathscr{F}(W_Q)$. The reason for the result is that $\psi_Q : W_Q \to A$ is injective and $F_Q : A \to \mathscr{F}(W_Q)$ is injective under the assumption of Theorem 6.1.3 (if necessary, we reduce the radius of A) where $A = \{\zeta : |\zeta| < \sqrt[p]{\delta}\}$. Thus we have proved that $\mathscr{F} : \widetilde{\mathscr{L}_0}(\widehat{\mathbb{C}}) \to \mathscr{F}(\widetilde{\mathscr{L}_0}(\widehat{\mathbb{C}}))$ is a local homeomorphism. $\mathscr{F}$ has the curve lifting property and so it is a covering.

Thus Theorem 6.1.3 follows. □

Assume that $\widetilde{\mathscr{L}_0}(\widehat{\mathbb{C}})$ is simply-connected. Then there exists a conformal mapping

$$\Theta : \Omega \to \widetilde{\mathscr{L}_0}(\widehat{\mathbb{C}})$$

of Ω onto $\widetilde{\mathscr{L}_0}(\widehat{\mathbb{C}})$, where Ω is one of $\widehat{\mathbb{C}}$, $\mathbb{C}$ and Δ. Set

$$F(z) = \pi \circ \Theta(z) : \Omega \to \widehat{\mathbb{C}}$$

and it is a meromorphic function. We shall call $\widetilde{\mathscr{L}_0}(\widehat{\mathbb{C}})$ the Riemann surface associated with $F(z)$ and the boundary of Ω the natural boundary of $F(z)$. Actually, we can obtain $\widetilde{\mathscr{L}_0}(\widehat{\mathbb{C}})$ from the inverse of $F(z)$ with the help of meromorphic continuation with algebraic character, which will be explained later and we shall point out the connection of $\mathscr{F}$ and Θ.

Concerning $\widetilde{\mathscr{L}_0}(\widehat{\mathbb{C}})$ three possibilities occur: (1) $\Omega = \widehat{\mathbb{C}}$, $\widetilde{\mathscr{L}_0}(\widehat{\mathbb{C}})$ is elliptic, and then π is a m-fold mapping of $\widehat{\mathbb{C}}$ from $\widetilde{\mathscr{L}_0}(\widehat{\mathbb{C}})$, namely, $F(z)$ is a rational function with degree m; (2) $\Omega = \mathbb{C}$, $\widetilde{\mathscr{L}_0}(\widehat{\mathbb{C}})$ is parabolic, and then $F(z)$ cannot be extended to ∞ and it is a transcendental meromorphic function in the whole plane; (3) $\Omega = \Delta$, $\widetilde{\mathscr{L}_0}(\widehat{\mathbb{C}})$ is hyperbolic, and $F(z)$ has the natural boundary $\{z : |z| = 1\}$, namely, F can not be continued through $\{z : |z| = 1\}$ forward to outside of Δ.

In the final case (3), we may consider as an example

$$F(z) = \sum_{n=1}^{\infty} z^{2^n}.$$

It is well-known that the natural boundary of $F(z)$ is the unit circle and so the Riemann surface associated with it is hyperbolic.

In what follows, we confine our discussion to the case when $\widetilde{\mathscr{L}_0}(\widehat{\mathbb{C}})$ is simply-connected and parabolic. In this case, the associated function $F(z)$ is a transcendental meromorphic function in $\mathbb{C}$. Conversely, given a meromorphic function $w = F(z)$ in $\mathbb{C}$, we come obtain the Riemann surface associated with it. Actually, starting from a branch of the inverse $f(w)$ of $w = F(z)$ in the w-plane, we continue this branch meromorphically with algebraic character without any restriction in the extended w-plane to generate a Riemann surface $\widetilde{\mathscr{L}_0}(\widehat{\mathbb{C}})$. Then we have

Theorem 6.1.4. *$\widetilde{\mathscr{L}_0}(\widehat{\mathbb{C}})$ is simply-connected and parabolic if and only if the associated function is a transcendental meromorphic function in $\mathbb{C}$, namely, $\widetilde{\mathscr{L}_0}(\widehat{\mathbb{C}})$ is generated by the inverse of a transcendental meromorphic function in $\mathbb{C}$.*

Proof. It suffices to prove that $\widetilde{\mathscr{L}_0}(\widehat{\mathbb{C}})$ is simply-connected and parabolic if it is generated by the inverse of a transcendental meromorphic function $F(z)$ in $\mathbb{C}$ in the previous way. Set $V = F(\mathbb{C})$ and then $V = \widehat{C}$, $\mathbb{C}$ or $\mathbb{C} \setminus \{a\}$ for some $a \in \mathbb{C}$. Since $F^{-1}(V) = \mathbb{C}$, for $b \in \mathbb{C}$ we have a branch g of F^{-1} sending $F(b)$ to b and g has at most algebraic singularity over $F(b)$. We therefore have $[g]_{F(b)} \in \widetilde{\mathscr{L}_0}(\widehat{\mathbb{C}})$, $\mathscr{F}([g]_{F(b)}) = b$ and $\mathscr{F}(\widetilde{\mathscr{L}_0}(\widehat{\mathbb{C}})) = \mathbb{C}$. In view of Theorem 6.1.3, $\mathscr{F} : \widetilde{\mathscr{L}_0}(\widehat{\mathbb{C}}) \to \mathbb{C}$ is conformal and $\widetilde{\mathscr{L}_0}(\widehat{\mathbb{C}})$ is simply-connected. Set $\Theta = \mathscr{F}^{-1} : \mathbb{C} \to \widetilde{\mathscr{L}_0}(\widehat{\mathbb{C}})$ and hence $\widetilde{\mathscr{L}_0}(\widehat{\mathbb{C}})$ is parabolic. It is easy to see that the following diagram

$$\begin{array}{ccc}
\widetilde{\mathscr{L}_0}(\widehat{\mathbb{C}}) & \xrightarrow{\ \mathrm{id}\ } & \widetilde{\mathscr{L}_0}(\widehat{\mathbb{C}}) \\
{\scriptstyle \mathscr{F}}\big\downarrow & & \big\downarrow{\scriptstyle \pi} \\
\mathbb{C} & \xrightarrow{\ F\ } & \widehat{\mathbb{C}}
\end{array}$$

is commutative, that is, $F = \pi \circ \mathscr{F}^{-1} = \pi \circ \Theta$. □

Usually, one does not distinguish $F(z)$ from $\Theta(z)$ when no confusion may occur. Thus we say that $F(z)$ maps conformally $\mathbb{C}$ onto its associated Riemann surface $\widetilde{\mathscr{L}_0}(\widehat{\mathbb{C}}))$. Below by $\mathfrak{R}_F$ we denote the associated Riemann surface $\widetilde{\mathscr{L}_0}(\widehat{\mathbb{C}}))$ of a transcendental meromorphic function $F(z)$ in $\mathbb{C}$.

In above point of view, both of the Riemann surfaces associated with e^z and z^n are simply-connected and $\mathfrak{R}_{\mathrm{e}^z}$ is parabolic and $\mathfrak{R}_{z^n}$ is elliptic. Both of $\mathrm{e}^z : \mathbb{C} \to \mathfrak{R}_{\mathrm{e}^z}$ and $z^n : \widehat{\mathbb{C}} \to \mathfrak{R}_{z^n}$ are conformal.

According to a theorem of Iversen [14], all boundary points of the parabolic surface are accessible. This means that for every boundary point Q there exists a curve Γ in $\mathscr{L}_0(\widehat{\mathbb{C}})$ tending to Q and in the point of view of continuation, namely, there exist a curve γ in the complex plane with $\pi(Q)$ as an end point and a function element (f, D) such that $D \cap \gamma \neq \varnothing$ and f can be continued meromorphically with algebraic character forward to $\pi(Q)$. This will be used in the sequel.

Now we observe the singular points of $\mathfrak{R}_F$ from $F(z)$. A point z_0 in $\mathbb{C}$ is called a critical point of $F(z)$ if $F'(z_0) = 0$ or z_0 is a pole of $F(z)$ with multiplicity greater than 1. The value of $F(z)$ at a critical point is called a critical value of $F(z)$. A value $a \in \widehat{\mathbb{C}}$ is called an asymptotic value of $F(z)$ if there exists a curve Γ in $\mathbb{C}$ tending to ∞ such that $F(z) \to a$ as $z \in \Gamma \to \infty$ and in this case we call Γ the corresponding asymptotic curve. We call critical values together with asymptotic values of a transcendental meromorphic function its singular values. The following is reason of the names. We observe the behavior of the inverse in vicinities of singular values. Let z_0 be a critical point of $F(z)$ and then $a = F(z_0)$ is a critical value of $F(z)$. First of all we consider the case $a \neq \infty$. We can write

$$F(z) = a + (z - z_0)^p \phi(z)$$

for some $p > 1$ with $\phi(z) \neq 0$ in a neighborhood V_{z_0} of z_0, that is, $0 \notin \phi(V_{z_0})$. Then $\phi_0(z) = (z - z_0)\phi^{1/p}(z)$ is univalent on V_{z_0} where $\phi^{1/p}(z)$ is chosen to be an analytic branch. Let $\psi(\zeta)$ be the inverse of $\phi_0(z)$, namely there exists a disk A centered at 0 such that $\psi(\zeta) : A \to V_{z_0}$ is analytic and univalent with $\phi_0(\psi(\zeta)) = \zeta$. Then the inverse $F^{-1}(w)$ of $F(z)$ can be expanded into the Puiseux series

$$F^{-1}(w) = \psi \circ (w - a)^{1/p} = \sum_{n=0}^{\infty} a_n (w - a)^{n/p}, \ w \in B(a, \delta)$$

where $a_0 = z_0$ and $a_1 = \phi^{-1/p}(z_0) \neq 0$. When $a = \infty$, we consider $1/F(z)$ in the same method as above to obtain $\psi(\zeta)$ and hence we have the expansion in the Puiseux series from $F^{-1}(w) = \psi \circ w^{-1/p}$ with $a_1 \neq 0$. Therefore, there exists a branch of F^{-1} which cannot be continued up onto a meromorphically, but can be done with algebraic character. This implies that a produces an algebraic singular point of $\mathfrak{R}_F$. For sufficient small $r > 0$, thus $F^{-1}(B(a,r))$ has a bounded component $U(r)$ containing z_0 and for $r > r' > 0$, $U(r') \subset U(r)$ such that $\cap_{r>0} U(r) = \{z_0\}$.

Now we come to consider asymptotic values. Let a be an asymptotic value of $F(z)$ with asymptotic curve Γ. It is clear that for arbitrarily $r > 0$, $F^{-1}(B(a,r))$ has a component $U(r)$ containing tail of Γ and for $r > r' > 0$, $U(r') \subset U(r)$. Then $\cap_{r>0} U(r) = \varnothing$. Indeed, if we have a point $z_0 \in \cap_{r>0} U(r) \neq \varnothing$, $\cap_{r>0} U(r)$ is a continuum connecting z_0 and ∞ at which $F(z) = a$, but this is impossible. For any $r > 0$, $F : U(r) \to B(a,r)$ is not univalent. Suppose that it would not be true and then we have a branch G of F^{-1} which is an univalent analytic map from $B(a,r)$ onto $U(r)$ for all sufficient small $r > 0$ (Since $G(B^*(a,r)) \subseteq U(r)$, a cannot be a pole or essential singular point of G and thus G can be analytically extended to a so that we have $G(B(a,r)) = U(r)$.) This implies the existence of a point $z_r \in U(r)$ such that $F(z_r) = a$ and we can choose a $0 < r' < r$ such that $z_r \notin U(r')$ and so $z_{r'} \neq z_r$, which contradicts the univalence of $F(z)$ on $U(r)$. This implying process shows that $F : U(r) \to B(a,r)$ is ∞-to-one. Conversely, assume that there exists a component $U(r)$ of $F^{-1}(B(a,r))$ such that for $r > r' > 0$, $U(r') \subset U(r)$ and $\cap_{r>0} U(r) = \varnothing$. Take a sequence of decreasing positive numbers $\{r_n\}$ with $r_n \to 0$ and a sequence of points $\{z_n\}$ with $z_n \in U(r_n)$. We draw a curve γ_n from z_n to z_{n+1} in $U(r_n)$ and so $\Gamma = \cup_{n=1}^{\infty}$ is a curve tending to ∞. Since for each n, $F(\gamma_n) \subset B(a,r_n)$, we therefore have $F(z) \to a$ as $z \in \Gamma \to \infty$, that is, a is an asymptotic value of $F(z)$ and Γ is the corresponding asymptotic curve.

From the above discussion, we have known that for each asymptotic value of $F(z)$ there exists a branch of the inverse of $F(z)$ which cannot be continued up onto a meromorphically with algebraic character. Therefore the asymptotic value a will produce a boundary point of the Riemann surface $\mathfrak{R}_F$ associated with $F(z)$. Conversely, let Q be a boundary point of $\mathfrak{R}_F$. Draw a curve γ tending to Q on $\mathfrak{R}_F$ (The existence of γ is shown by a theorem of Iversen as mentioned above) and write $\tilde{\gamma} = \pi(\gamma)$, which tends to a point $a \in \widehat{\mathbb{C}}$. Set $a = \pi(Q)$. Therefore $\Gamma = \mathscr{F}(\gamma)$ is produced via a meromorphic continuation of a branch of F^{-1} along $\tilde{\gamma}$. If Γ tends to a finite complex point z_0 as going to Q along γ and clearly $F(z_0) = a$ in view of $F = \pi \circ \mathscr{F}^{-1}$, thus $F^{-1}(B(a,r))$ has a bounded component containing Γ so that the branch can be continued onto a, this is a contradiction. This implies that Γ tends to ∞ and $F(z) \to a$ as $z \in \Gamma \to \infty$, that is, a is an asymptotic value of F.

Up to now we have known that there exists at least a branch of F^{-1} which is not single-valued in any vicinity of a if and only if a is a singular value of $F(z)$. Below once the case takes place, we say that F^{-1} has singularity over a, precisely speaking, F^{-1} has algebraic singularity over a critical value and transcendental singularity over asymptotic value of $F(z)$. Thus we have proved

Theorem 6.1.5. *Let $F(z)$ be a transcendental meromorphic function. Then $a \in \widehat{\mathbb{C}}$ is an asymptotic value of $F(z)$ if and only if a corresponds to a boundary point of $\mathfrak{R}_F$ under π, in other words, there exists a transcendental singularity of F^{-1} over a.*

It is reasonable that we call $\mathscr{F}^{-1}(U(r))$ a vicinity of the boundary point of $\mathfrak{R}_F$ associated with a and sometimes we directly say $U(r)$ to be a neighborhood of the corresponding transcendental singularity of the inverse of $F(z)$. Iversen [14] is the first one to introduce the following classification of transcendental singularities. A transcendental singularity over a is called direct if for some $r > 0$, $F(z) \neq a$ in its neighborhood $U(r)$, namely the projection π misses $\pi(Q)$ in $\mathscr{F}^{-1}(U(r))$. If $F : U(r) \to B_0(a,r)$ is a universal covering, then we call this direct singularity logarithmic. Actually, in the case of logarithmic singularity, it follows from Theorem 6.1.1 that there exists a conformal mapping ψ of the half plane $H = \{z : \mathrm{Re} z < \log r\}$ from $U(r)$ such that $F = \exp \circ \psi - a$ and the singularity of the inverse of $F(z)$ is characterized by the singularity of $\mathrm{Log}(z-a)$. A direct transcendental singularity which is not logarithmic exists. Let us observe the function $f(z) = z \sin z$ and $g(z) = \frac{1}{z}\exp(-\mathrm{e}^z)$. It is clear that the singularities of f^{-1} over ∞ and g^{-1} over 0 are direct. A simple calculation implies that ∞ is a limit point of critical values of $f(z)$ and 0 that of critical values of $g(z)$. Actually, $g'(z) = -\frac{1}{z}(\frac{1}{z} + \mathrm{e}^z)\exp(-\mathrm{e}^z)$ and all critical points come from the roots of $-\mathrm{e}^z = \frac{1}{z}$ and thus the critical value of g at critical point z is $\frac{1}{z}\mathrm{e}^{1/z}$. This shows that 0 is unique limit point of sequence of the critical values. A transcendental singularity over a is called indirect if it is not direct, namely, for each r, $F(z)$ can take a in $U(r)$ and so takes a infinitely often. Let us observe the function $h(z) = \frac{\sin z}{z}$ to show the existence of indirect singularity. It is obvious that $\frac{\sin z}{z} \to 0$ as $z \to +\infty$ along the positive real axis and the $U(r)$ containing tail of positive real axis contains infinitely many zeros $n\pi$ of $h(z)$. And a simple calculation implies that 0 is a limit point of critical values of $h(z)$. These hint the existence of certain possible connection among singularities, which will be discussed in the sequel.

Transcendental singularities are determined not only by asymptotic values and also their associated asymptotic curves. However, asymptotic curves are obviously not unique. Actually two different asymptotic curves may determine the same singularity and there may exist several singularities over a fixed value. For example, the inverse of e^{z^2} has two distinct logarithmic singularities over ∞, one is decided by the positive real axis and the other by the negative real axis. Hence an equivalent relationship is necessary among asymptotic curves over a fixed value. Two asymptotic curves $\Gamma_j (j = 1,2)$ associated a are called equivalent if there exist a sequence of curves $\{\gamma_k\}$ connecting Γ_1 and Γ_2 such that

$$\lim_{k\to\infty} \mathrm{dist}(\gamma_k, 0) = \infty \text{ and } \lim_{z\in\cup_{k=1}^{\infty}\gamma_k \to \infty} F(z) = a.$$

In other words, the branch of F^{-1} can be meromorphically continued from Γ_1 to Γ_2 along every curve γ_k to produce a common branch and thus (a,Γ_1) and (a,Γ_2) correspond the same boundary point of $\mathfrak{R}_F$. These equivalent classes of asymptotic curves are in a bijective correspondence with transcendental singularities of F^{-1} over a.

6.2 Density of Singularities

Let $F(z)$ be a transcendental meromorphic function and $\mathfrak{R}_F$ its associated Riemann surface. Then the boundary of $\mathfrak{R}_F$ is totally disconnected, which can be proved by the following Gross Theorem.

Theorem 6.2.1. *For every point $a \in \widehat{\mathbb{C}}$, every single-valued analytic branch of F^{-1} can be analytically continued up to its antipodal point on the sphere along direction $\arg(z-a) = \theta$ ($\arg z = \theta$ for the case $a = \infty$) originated from a with exception of a set of θ of measure zero.*

In other words, Theorem 6.2.1 is to say that $\mathfrak{R}_F$ contains a geodesic ray from a to its antipodal point in almost every direction.

Let us observe the function $f(z) = z^2(z-1)e^z$. 0 is a critical value as well as an asymptotic value of $f(z)$. There exist an algebraic singularity and a logarithmic singularity over 0 and at the same time, $f(z)$ maps conformally a simply-connected domain containing 1 onto a vicinity of 0. There exist two distinct indirect singularities of the inverse of $\sin z/z$ over 0 and two distinct direct ones over ∞. We can construct an example whose inverse has logarithmic, non-logarithmic direct and indirect singularities over a fixed value a.

From Theorem 6.1.1 we have the following

Theorem 6.2.2. *If a is an isolated singular value of $F(z)$, then all singularities of F^{-1} over a are algebraic or logarithmic.*

Proof. Choose a $\delta > 0$ such that $B_0(a,\delta)$ does not contain any singular values of $F(z)$. For every component $U(\delta)$ of $F^{-1}(B_0(a,\delta))$, $F: U(\delta) \to B_0(a,\delta)$ is an unbranched holomorphic covering. Then Theorem 6.2.2 immediately follows from Theorem 6.1.1. □

In view of Theorem 6.2.2, an asymptotic value must be a limit point of other singular values if there exists at least one transcendental singularity over it which is not logarithmic. The following result is formulated from the proof of Theorem 1 of [3].

Theorem 6.2.3. *Assume that F^{-1} has an indirect singularity over a fixed value a on $\widehat{\mathbb{C}}$. If a is not a limit point of critical values, then there exists a sequence of asymptotic values $\{a_n\}$ of $F(z)$ satisfying*

$$|a_n - a| > |a_{n+1} - a| \to 0 \ (n \to \infty)$$

together with the property that there exists a sequence of disjoint unbounded simply-connected domains U_n and a sequence of asymptotic curves $\Gamma_n \subset U_n$ associated to a_n such that $F(z)$ is univalent in U_n, $D_n = F(U_n)$ is the disk $\{w : |(w-a) - \frac{2}{3}(a_n - a)| < \frac{|a_n - a|}{3}\}$.

Proof. Let $U(R)$ be a neighborhood of the indirect singularity over a such that no critical points are in $U(R)$ according to the assumption of Theorem 6.2.3. Assume that we have found a_n $(n=1,2,\cdots,m)$ with U_n, Γ_n and D_n mentioned as in Theorem 6.2.3. Now since $a \notin \overline{D_k}$, we can choose a R_0 with $0<R_0<|a_m-a|$ and $R_0<R$ such that $U(R_0)\cap U_k=\varnothing$ $(1\leqslant k\leqslant m)$ and $U(R_0)\subset U(R)$. Since $U(R_0)$ is also a neighborhood of the indirect singularity over a, there exists a point $z_{m+1}\in U(R_0)$ with $F(z_{m+1})=a$. We have an analytic branch ϕ of F^{-1} which sends a to z_{m+1} and we expand ϕ in the power series with the radius r_{m+1} of convergence.

Suppose that $r_{m+1}\geqslant R_0$. Then $\phi(\{w:|w-a|<R_0\})$ is a component of $F^{-1}(B(a,R_0))$ and noting that $z_{m+1}\in U(R_0)\cap\phi(\{w:|w-a|<R_0\})\neq\varnothing$, then $U(R_0)=\phi(\{w:|w-a|<R_0\})$. This implies that $F(z):U(R_0)\to B(a,R_0)$ is of one-to-one, a contradiction is derived. We have shown that $0<r_{m+1}<R_0$.

Let a_{m+1} be a singular point of ϕ on $|w-a|=r_{m+1}$, that is, ϕ cannot be analytically continued through a_{m+1} forward outside of the disk $\{w:|w-a|<r_{m+1}\}$. Thus a_{m+1} is a singular value of $F(z)$ and hence a_{m+1} is an asymptotic value of $F(z)$ by the assumption of Theorem 6.2.3. Set

$$D_{m+1}=\left\{w:|(w-a)-\frac{2}{3}(a_{m+1}-a)|<\frac{|a_{m+1}-a|}{3}\right\}$$

and write $U_{m+1}=\phi(D_{m+1})$ and $\Gamma_{m+1}=\phi(L_{m+1})$ where L_{m+1} is the radius of D_{m+1} terminal at a_{m+1}. It is obvious that U_{m+1} is unbounded and $\Gamma_{m+1}\to\infty$ and $F(z)\to a_{m+1}$ as $z\in\Gamma_{m+1}\to\infty$.

By induction, we have proved Theorem 6.2.3. □

Generally, direct singularities are rare, which is asserted by the following result due to Heins [13].

Theorem 6.2.4. *The set of asymptotic values over which there is at least one direct singularity is at most countable.*

The situation for a transcendental meromorphic function with finite order is few complicated. In order to make further discussion of singularities of the inverse of a meromorphic function with finite order, we first of all establish the following lemma, whose idea is essentially due to Ahlfors (see the proof of Theorem 4.19 of Zhang [22] and Page 305 of Nevanlinna [18]).

Lemma 6.2.1. *Let $F(z)$ be a transcendental meromorphic function. Assume that there exist p values a_j in $\widehat{\mathbb{C}}$, p non-negative integers k_j and p disjoint unbounded domains U_j $(j=1,2,\cdots,p)$ bounded by certain analytic curves such that $F^{(k_j)}(z)\neq a_j$ in $z\in U_j$ and*

$$\inf\{|F^{(k_j)}(z)-a_j|:z\in U_j\}<\min\{h_j,1\}$$

where $h_j=\inf\{|F^{(k_j)}(z)-a_j|:z\in\partial U_j\}$(here for $a_j=\infty$ we use $1/|F^{(k_j)}(z)|$ in the place of $|F^{(k_j)}(z)-a_j|$).

Then we have

$$\liminf_{r\to\infty}\frac{T(r,F)}{r^{p/2}}>0, \tag{6.2.1}$$

and hence $p\leqslant 2\mu(F)$.

Proof. Assume without any loss of generalities that each $a_j \neq \infty$. For each j we take a point z_j from U_j such that $|F^{(k_j)}(z_j) - a_j| < \min\{h_j, 1\} = b_j$(say). Then the circle $\{z : |z| = t\}$ for $t \geqslant |z_j|$ intersects U_j and set

$$\theta_j(t) = \mathrm{mes}\{\theta \in [0, 2\pi) : te^{\mathrm{i}\theta} \in U_j\},$$

$D_j(t) = U_j \cap \{z : |z| \leqslant t\}$ and $\Gamma_j(t) = U_j \cap \{z : |z| = t\}$. In view of Lemma 4.1.4 and two constant theorem on the harmonic measure, for all sufficiently large r we have

$$\begin{aligned}
\log \frac{1}{|F^{(k_j)}(z_j) - a_j|} &\leqslant \omega(z_j, \partial U_j, D_j(r)) \log \frac{1}{b_j} \\
&\quad + \omega(z_j, \Gamma_j(r), D_j(r)) \log M\left(r, U_j, \frac{1}{F^{(k_j)} - a_j}\right) \\
&\leqslant \log \frac{1}{b_j} + 9\sqrt{2} \exp\left(-\pi \int_{2|z_j|}^{\frac{1}{2}r} \frac{\mathrm{d}t}{t\theta_j(t)}\right) \\
&\quad \times \log M\left(r, U_j, \frac{1}{F^{(k_j)} - a_j}\right),
\end{aligned}$$

and equivalently

$$\pi \int_{2|z_j|}^{\frac{1}{2}r} \frac{\mathrm{d}t}{t\theta_j(t)} \leqslant \log\log M\left(r, U_j, \frac{1}{F^{(k_j)} - a_j}\right) + \log c_j, \tag{6.2.2}$$

where $c_j = 9\sqrt{2}\left(\log \frac{b_j}{|F^{(k_j)}(z_j) - a_j|}\right)^{-1} > 0$.

Next to complete our proof we estimate $\log M\left(r, U_j, \frac{1}{F^{(k_j)} - a_j}\right)$ from above in term of the characteristic $T(4r, F)$ and hence produce a lower bound for the characteristic. In virtue of Lemma 2.1.3 and (2.6.1) we have a $R_j \in [r, 2r)$ such that on the circle $\{z : |z| = R_j\}$

$$\begin{aligned}
\log \frac{1}{|F^{(k_j)}(z) - a_j|} &\leqslant d_j T\left(2r, \frac{1}{F^{(k_j)} - a_j}\right) \\
&\leqslant d_j T(2r, F^{(k_j)}) + O(1) \\
&\leqslant d_j K_j T(4r, F) + O(1)
\end{aligned}$$

for positive constants d_j and K_j. This implies that

$$\log M\left(R_j, U_j, \frac{1}{F^{(k_j)} - a_j}\right) \leqslant d_j K_j T(4r, F) + O(1). \tag{6.2.3}$$

Combining (6.2.2) and (6.2.3) yields that

$$\pi\int_{2|z_j|}^{\frac{1}{2}r}\frac{dt}{t\theta_j(t)} \leqslant \pi\int_{2|z_j|}^{\frac{1}{2}R_j}\frac{dt}{t\theta_j(t)} \leqslant \log T(4r,F)+O(1)$$

and thus

$$\pi\int_{r_0}^{\frac{1}{2}r}\sum_{j=1}^{p}\frac{1}{\theta_j(t)}\frac{dt}{t} \leqslant p\log T(4r,F)+O(1), \tag{6.2.4}$$

where $r_0=\max\{2|z_j| : 1\leqslant j\leqslant p\}$. Noting that $\sum_{j=1}^{p}\theta_j(t)\leqslant 2\pi$ because all U_j are disjoint and in view of the Schwarz inequality, thus we have

$$p^2=\left(\sum_{j=1}^{p}\sqrt{\theta_j(t)}\frac{1}{\sqrt{\theta_j(t)}}\right)^2 \leqslant \sum_{j=1}^{p}\theta_j(t)\sum_{j=1}^{p}\frac{1}{\theta_j(t)} \leqslant 2\pi\sum_{j=1}^{p}\frac{1}{\theta_j(t)}.$$

This is applied to (6.2.4) to obtain

$$\frac{p}{2}\log\frac{r}{2r_0} \leqslant \log T(4r,F)+O(1),$$

that is, for some positive constant K

$$\frac{T(r,F)}{r^{p/2}} \geqslant K(8r_0)^{-p/2}.$$

Lemma 6.2.1 follows. □

Denjoy [7] conjectured in 1907 that an entire function of order λ has at most 2λ distinct finite asymptotic values. When asymptotic curves are rays from the origin, Denjoy himself confirmed the conjecture. Carleman [5] proved in 1921 that the number of finite asymptotic values is less than 5λ. Finally, the conjecture was demonstrated in 1930 and extended to the following format in 1932 by Ahlfors.

Theorem 6.2.5. *Let $F(z)$ be a meromorphic function. If the inverse F^{-1} has p distinct direct singularities, then we have (6.2.1).*

Proof. Under the assumption of Theorem 6.2.5 we have p disjoint neighborhoods $U_j(r)$ of direct singularities and we can choose a suitable r such that the boundary of $U_j(r)$ consists of analytic curves, namely it does go through no critical points of $F(z)$. Thus Theorem 6.2.5 immediately follows from Lemma 6.2.1. □

Basically, we can deal with the derivatives of $F(z)$ in the condition of Theorem 6.2.5. Usually, this theorem is known as Denjoy-Carleman-Ahlfors Theorem. Now we use Theorem 6.2.5 to confirm the Denjoy conjecture. Let Γ_1 and Γ_2 be two asymptotic curves associated to two distinct finite asymptotic values a_1 and a_2 of entire function $F(z)$. We can assume that Γ_1 and Γ_2 divide the complex plane $\mathbb{C}$ into two simply-connected domain U_1 and U_2. In view of the Lindelöf Theorem, $F(z)$ is unbounded in both of U_1 and U_2 and for sufficiently large r, $F^{-1}(B_0(\infty,r))$ has unbounded components $U_1(r)$ and $U_2(r)$ with $U_1(r)\subset U_1$ and $U_2(r)\subset U_2$. Then $U_j(r)$

is a vicinity of direct singularity over ∞, that is, there exist two direct singularities over ∞. This yields that corresponding to p distinct finite asymptotic values are at least p direct singularities over ∞ and in view of Theorem 6.2.5 the lower order of $F(z)$ is at least $p/2$, namely $p \leqslant 2\mu(F)$. The Denjoy conjecture is proved. However, the situation is not simple for an entire function with infinite order. Actually, Gross [12] found an entire function of infinite order the set of whose asymptotic values is the extended complex plane. In view of Theorem 6.2.4, all but at most countable singularities for the Gross function are indirect.

We can say more for a neighborhood of a direct singularity. The following is a improving version of a result of W. Fuchs [11] and a result of Zhang [22].

Theorem 6.2.6. *Let $f(z)$ be a transcendental meromorphic function and U an unbounded domain in $\mathbb{C}$ whose boundary contains at least one unbounded component. If $f(z)$ is analytic in U and for each $\zeta \in \partial U \setminus \{\infty\}$,*

$$\limsup_{z \in U \to \zeta} |f(z)| \leqslant 1,$$

then we have either $|f(z)| \leqslant 1$ for $z \in U$ or a curve Γ in U tending to ∞ such that

$$\liminf_{z \in \Gamma \to \infty} \frac{\log\log|f(z)|}{\log|z|} \geqslant \frac{1}{2}.$$

Proof. Assume that there exists a point $z_0 \in U$ such that $|f(z_0)| > 1$. Take a real number d with $1 < d < |f(z_0)|$. The set $\{z : |f(z)| > d\}$ has a component $U(d)$ which contains z_0. It is clear that $U(d) \subseteq U$. We can choose d such that $\partial U(d)$ consists of analytic Jordan curves. The maximal principle yields that $U(d)$ is unbounded. As in the proof of Lemma 6.2.1, we have

$$\pi \int_{2|z_0|}^{\frac{1}{2}r} \frac{dt}{t\theta_j(t)} \leqslant \log\log M(r, U(d), f) + \log 9\sqrt{2} - \log\log \frac{|f(z_0)|}{d} \tag{6.2.5}$$

and by noting $\theta_j \leqslant 2\pi$, this implies that

$$\liminf_{r \to \infty} \frac{\log\log M(r, U, f)}{\log r} \geqslant \frac{1}{2}. \tag{6.2.6}$$

The main idea of the proof of the remainder comes from Zhang [22]. Write $\eta_n = \frac{1}{n}$ for $n \geqslant 7$ and we take a sequence of positive numbers $\{t_n\}$ such that

$$t_{n+1} = (36\sqrt{2})^{n+1} t_n^{1+\eta_n},$$

that is,

$$t_{n+1}^{\eta_{n+1}} = 36\sqrt{2} t_n^{\eta_n},$$

and

$$t_7 \geqslant (36\sqrt{2})^{1+2+3+4+5+6}$$

and in view of (6.2.6) t_7 can be chosen for the existence of a point z_7 in U with $|z_7| = t_7$ such that

$$\log|f(z_7)| \geqslant |z_7|^{\frac{1}{2}-\eta_7}.$$

We shall complete our proof in induction. Assume that we can take a point z_n in U with $t_n = |z_n|$ such that

$$\log|f(z_n)| \geqslant |z_n|^{\frac{1}{2}-\eta_n}.$$

Write $C_n = |z_n|^{\frac{1}{2}-\eta_n}$ and we can find an $A_n \in [\frac{1}{4}C_n, \frac{1}{2}C_n]$ such that the set $\{z : \log|f(z)| > A_n\}$ has a component U_n containing z_n whose boundary consists of Jordan analytic curves. It is clear that $U_n \subset U$ as $C_n > 4$. Then as in (6.2.5) we have

$$\frac{1}{2}|z_n|^{\frac{1}{2}-\eta_n} \leqslant \log|f(z_n)| - A_n \leqslant 9\sqrt{2}\left(\frac{r}{4|z_n|}\right)^{-1/2} \log M(r, U_n, f)$$

and equivalently

$$\frac{1}{36\sqrt{2}} r^{\frac{1}{2}} t_n^{-\eta_n} \leqslant \log M(r, U_n, f)$$

so that we have a point z_{n+1} in U_n with $t_{n+1} = |z_{n+1}|$ satisfying

$$\log|f(z_{n+1})| \geqslant \frac{1}{36\sqrt{2}} t_{n+1}^{\frac{1}{2}} t_n^{-\eta_n} = |z_{n+1}|^{\frac{1}{2}-\eta_{n+1}}.$$

Draw a curve L_n in U_n connecting z_n and z_{n+1} and $L_n \subset \{z : |z| \leqslant t_{n+1}\}$. Consider a point $z \in L_n$. If $|z| \leqslant t_n$, then we have

$$\log|f(z)| > A_n \geqslant \frac{1}{4}C_n \geqslant \frac{1}{4}|z|^{\frac{1}{2}-\eta_n};$$

If $|z| > t_n$, then we have

$$\begin{aligned}
\log|f(z)| > \frac{1}{4} t_n^{\frac{1}{2}-\eta_n} &= \frac{1}{4}|z|^{\frac{1}{2}-\eta_n}\left(\frac{t_n}{t_{n+1}}\right)^{1/2-\eta_n} \\
&= \frac{1}{4}|z|^{\frac{1}{2}-\eta_n}(36\sqrt{2})^{-(n+1)(n-2)/2n} t_n^{-\frac{n-2}{2n}\eta_n} \\
&\geqslant \frac{1}{4}|z|^{\frac{1}{2}-\eta_n}(36\sqrt{2})^{-n/2} t_n^{-\frac{1}{2}\eta_n}.
\end{aligned}$$

From the definition of t_n it is easily seen that $t_n > (36\sqrt{2})^{n+(n-1)+\cdots+1} > (36\sqrt{2})^{\frac{1}{2}n(n+1)}$ and hence $t_n^{\eta_n} > (36\sqrt{2})^{n/2}$. This deduces for $|z| > t_n$,

$$\log|f(z)| > \frac{1}{4}|z|^{\frac{1}{2}-3\eta_n}.$$

By induction, we have obtained a sequence of curves $\{L_n\}$ which satisfies the above properties. Set $L = \cup_{n=7}^{\infty} L_n$ and L is in U and tends to ∞. It is obvious that L is our desired curve mentioned in Theorem 6.2.6. □

Zhang G. H. [23] took all order derivatives into account and established the following

Theorem 6.2.7. *Let $F(z)$ be a meromorphic function with order λ and let p_i be the number of non-zero and finite distinct direct singularities of the inverse of ith order derivative $F^{(i)}$. Then we have*

$$\sum_{i=0}^{\infty} p_i \leqslant 2\lambda.$$

The following result is extracted from the proof of Theorem 1 of [3].

Lemma 6.2.2. *Let $f(z)$ be a meromorphic function and let U be a component of $f^{-1}(B(a,R))$ containing no critical points. For every $1 \leqslant n \leqslant 2p$, $p > 3$, there exists a sequence $\{z_{n,j}\}_{j=1}^{\infty}$ in U such that $z_{n,j} \to \infty$, $f(z_{n,j}) \to a_n \in B(a,R/2)$ $(j \to \infty)$ with $a_n \neq a_m$ for $n \neq m$, and*

$$|f'(z_{n,j})| \leqslant |z_{n,j}|^{-2p-1}.$$

Then $f(z)$ has the order at least $p-3$.

For meromorphic functions with finite order, Bergweiler and Eremenko [3] found connection between critical values and asymptotic values with indirect singularities and proved the following.

Theorem 6.2.8. *Let $F(z)$ be a meromorphic function with finite order. If a is a value on $\widehat{\mathbb{C}}$ over which a non-logarithmic singularity of F^{-1} exists, then there exists a sequence of critical values $\{a_n\}$ of F tending to a with $a_n \neq a$.*

Proof. Suppose that a is not a limit point of critical values of $F(z)$. Then in view of Theorem 6.2.2, a is a limit point of asymptotic values of $F(z)$ each of which is not a limit point of critical values of $F(z)$. From Theorem 6.2.5, then there exists an asymptotic value b over which an indirect singularity exists is not a limit point of critical values of $F(z)$. Then there exist $\{a_n\}$, $\{\Gamma_n\}$ and $\{U_n\}$ for b satisfying the properties stated in Theorem 6.2.3. Thus for any $p > 3$, the condition of Lemma 6.2.2 can be deduced and so $f(z)$ has the infinite order, a contradiction is derived. □

Combination of Theorem 6.2.5 and Theorem 6.2.8 immediately yields the following

Corollary 6.2.1. *Let $F(z)$ be a meromorphic function with finite order. If $F(z)$ has only finitely many critical values, then the inverse of $F(z)$ has only finitely many logarithmic singularities and algebraic singularities without others.*

We have known from Denjoy-Carleman-Ahlfors Theorem that an entire function of finite order has at most finitely many asymptotic values. The result is not true for a meromorphic function with finite order, which is deduced by a result of Valiron [21] which says that there exists a meromorphic function of finite order the set of whose asymptotic values has the cardinality of the continuum and by Eremenko [8] in 1978 who constructed such a meromorphic function which has every value on $\widehat{\mathbb{C}}$ as its asymptotic value.

In what follows, we introduce the result of Eremenko [8]. We write each number in $[0,1]$ in base seven excluding the expression in which all entries behind some position are six and thus the expression of the number in the base seven is unique. We denote by A_n, B_n, C_n and D_n the sets of number in $[0,1]$ in whose expression the n-th entries are, respectively, $0,2,4$ and 6, for example,

$$A_n = \{x = 0.a_1a_2\cdots a_n\cdots \in [0,1] : a_n = 0\}.$$

Set

$$E_n = \overline{A_n} \cup \overline{B_n} \cup \overline{C_n} \cup \overline{D_n}$$

and

$$F_n = E_n \cup [\pi - 1, \pi].$$

Then F_n can be expressed into union of a finite number of intervals. Actually, for instance, we have

$$\overline{A_n} = \bigcup_{\substack{a_j \in \{0,1,\cdots,6\} \\ 1 \leqslant j \leqslant n-1}} [0.a_1\cdots a_{n-1}, 0.a_1\cdots a_{n-1}1].$$

Lemma 6.2.3. *For each natural number n, there exists a meromorphic function $f_n(z)$ of order one satisfying*

$$|f_n(z)| \leqslant 2, \ \arg z \in F_n, \tag{6.2.7}$$

$$f_n(0) = 0, \tag{6.2.8}$$

and

$$\left.\begin{array}{ll} f_n(z) \to 1, & \arg z \in \overline{A_n}, \\ f_n(z) \to \mathrm{i}, & \arg z \in \overline{B_n}, \\ f_n(z) \to 1 + \mathrm{i}, & \arg z \in \overline{C_n}, \\ f_n(z) \to 0, & \arg z \in \overline{D_n} \cup [\pi - 1, \pi], \end{array}\right\} \tag{6.2.9}$$

uniformly in $\arg z$ *as* $|z| \to \infty$.

Proof. Let $\{\theta_j\}_{j=1}^N$ be the set of endpoints of all maximum intervals in F_n. Consider the function

$$g(z) = \frac{\sum_{j=1}^N a_j \exp(ze^{-\mathrm{i}\theta_j})}{\sum_{j=1}^N \exp(ze^{-\mathrm{i}\theta_j})},$$

where $a_j = 1$ for $\theta_j \in \overline{A_n}$; $a_j = \mathrm{i}$ for $\theta_j \in \overline{B_n}$; $a_j = 1 + \mathrm{i}$ for $\theta_j \in \overline{C_n}$; $a_j = 0$ for $\theta_j \in \overline{D_n} \cup [\pi - 1, \pi]$. We check that $g(z)$ satisfies the property (6.2.9). It suffices to treat the case when $\arg z \in \overline{A_n}$. Actually, in this case, there exists a j_0 such that $\arg z \in [\theta_{j_0}, \theta_{j_0+1}] \subset \overline{A_n}$ and then $0 \leqslant \arg z - \theta_{j_0} \leqslant 2\left(\frac{1}{7}\right)^n$ and $|\arg z - \theta_{j_0}| \leqslant |\theta_{j_0+1} - \theta_{j_0}| \leqslant |\arg z - \theta_j|$ for $j \neq j_0, j_0 + 1$ where the second "=" is possible only for $\arg z = \theta_{j_0+1}$.

This implies that for $j \neq j_0, j_0+1$,

$$\operatorname{Re}(ze^{-i\theta_{j_0}} - ze^{-i\theta_j}) = |z|(\cos(\arg z - \theta_{j_0}) - \cos(\arg z - \theta_j)) \to +\infty$$

or $\operatorname{Re}(ze^{-i\theta_{j_0+1}} - ze^{-i\theta_j}) \to +\infty$ uniformly in $\arg z \in [\theta_{j_0}, \theta_{j_0+1}]$ as $|z| \to +\infty$. It is clear that

$$\sum_{j=1}^{N} a_j \exp(ze^{-i\theta_j}) \sim \exp(ze^{-i\theta_{j_0}}) + \exp(ze^{-i\theta_{j_0+1}}) \to \infty$$

uniformly in $\arg z \in [\theta_{j_0}, \theta_{j_0+1}]$ as $|z| \to +\infty$ and furthermore actually it follows that $g(z) \to 1$ uniformly in $\arg z \in \overline{A_n}$ as $|z| \to +\infty$.

Now we modify $g(z)$ to satisfy the properties (6.2.7) and (6.2.8). Let $\{b_k\}_{k=1}^m$ be all poles of $g(z)$ on $U = \{z : \arg z \in F_n\}$ and in view of (6.2.9), $m < \infty$. Consider the function

$$h(z) = g(z)(z+i)^{-m} \prod_{k=1}^{m} (z - b_k).$$

Since $-i \notin U$, $h(z)$ has no poles in U and still satisfies the property (6.2.9). Then $h(z)$ is bounded in U, that is, for $z \in U$ and for some $M > 0$, we have $|h(z)| \leqslant M$; in view of (6.2.9), for $|z| > r_0$ and $z \in U$, $|h(z)| < 2$. Choose a positive number τ so small that for $|z| \leqslant r_0$, $|\tau z(\tau z + i)^{-1}| < 2/M$. Now define

$$f_n(z) = \frac{\tau z}{\tau z + i} h(z).$$

Obviously $f_n(z)$ satisfies (6.2.8) and (6.2.9). Noting that $|\tau z(\tau z + i)^{-1}| < 1$ in the upper half plane, for $-i$ is in the lower half plane, we have $|f_n(z)| < 2$ for $z \in U$, that is (6.2.7). And it is obvious that $f_n(z)$ is of order one. □

Actually, the restriction about "order one" produced in the above construction can be removed. For arbitrary positive increasing function $\psi(r)$ tending to ∞ as $r \to \infty$, Valiron [21] constructed a meromorphic function $g(z)$ with the property (6.2.9) and such that

$$T(r, g) = o(\psi(r)(\log r)^2), \ r \to \infty.$$

Then with the help of Valiron's function $g(z)$ in the place of $g(z)$ in the proof of Lemma 6.2.3, we obtain $f_n(z)$ with the properties mentioned in Lemma 6.2.3 and

$$T(r, f_n) = o(\psi(r)(\log r)^2), \ r \to \infty. \tag{6.2.10}$$

Theorem 6.2.9. *There exists a meromorphic function $F(z)$ satisfying (6.2.10) with $F(z)$ in the place of $f_n(z)$ and such that every value on $\widehat{\mathbb{C}}$ is its asymptotic value and the corresponding asymptotic curves are the rays on the upper half complex plane.*

Proof. In view of Lemma 6.2.3 and the above remark we have a sequence of meromorphic functions $\{f_n\}$ with the properties mentioned in Lemma 6.2.3 and (6.2.10). Set

$$\psi_n(r) = \sqrt{\frac{T(r,f_n)}{\psi(r)(\log r)^2}}.$$

Then $\psi_n(r) \to 0$ as $r \to \infty$ and we can take a sequence of positive numbers $\{r_n\}$ such that $r_{n+1} > r_n > n$ and for $r \geqslant r_n$

$$T(r,f_n) < 2^{-n}\psi_n(r)\psi(r)(\log r)^2, \tag{6.2.11}$$

and

$$\sum_{j=1}^{n} 2^{-j}\psi_j(r) < \frac{1}{n}.$$

Since $f_n(0) = 0$, we can take a δ_n, $0 < \delta_n < 1$ so small that

$$|f_n(\delta_n z)| < 1 \text{ for } |z| < r_n.$$

Thus for $r \leqslant r_n$, $T(r,f_n(\delta_n z)) = 0$ and noting that $T(r,f_n(\delta_n z)) \leqslant T(r,f_n)$ together with (6.2.11) hence we have for all r

$$T(r,f_n(\delta_n z)) < 2^{-n}\psi_n(r)\psi(r)(\log r)^2.$$

Define

$$f(z) = \sum_{n=1}^{\infty} 2^{-n} f_n(\delta_n z).$$

The series converges uniformly on any compact subset of the complex plane and hence $f(z)$ is a meromorphic function in $\mathbb{C}$. We shall obtain the desired function of Theorem 6.2.9 through $f(z)$. For this end, we check the properties of $f(z)$ we need in our purpose. First of all for any $r > r_1$ we have $r_N \leqslant r < r_{N+1}$ for some $N \geqslant 1$ and we have the following estimation that for $|z| \leqslant r$,

$$\left|\sum_{n=N+1}^{\infty} 2^{-n} f_n(\delta_n z)\right| \leqslant \sum_{n=N+1}^{\infty} 2^{-n}|f_n(\delta_n z)| < \sum_{n=N+1}^{\infty} 2^{-n} < 1.$$

This yields that

$$\begin{aligned}
T(r,f) &\leqslant T\left(r, \sum_{n=1}^{N} 2^{-n} f_n(\delta_n z)\right) + T\left(r, \sum_{n=N+1}^{\infty} 2^{-n} f_n(\delta_n z)\right) + \log 2 \\
&\leqslant \sum_{n=1}^{N} T(r, 2^{-n} f_n(\delta_n z)) + \log N + \log 2 \\
&\leqslant \left(\sum_{n=1}^{N} 2^{-n}\psi_n(r)\right)\psi(r)(\log r)^2 + \log r + \log 2 \\
&< \frac{1}{N}\psi(r)(\log r)^2 + \log r + \log 2 \\
&= o(\psi(r)(\log r)^2).
\end{aligned}$$

Next let us check that every complex number in the square

$$S = \{z:\ 0 \leqslant \mathrm{Re}z \leqslant 1, 0 \leqslant \mathrm{Im}z \leqslant 1\}$$

is an asymptotic value of $f(z)$. We can write every $a \in S$ into the form $a = \sum_{n=1}^{\infty} 2^{-n} b_n$ where each $b_n \in \{0,1,1+\mathrm{i},\mathrm{i}\}$. From the expansion of a we construct a number $\phi_a \in [0,1]$ in the base seven

$$\phi_a = 0.t_1 t_2 \cdots t_n \cdots$$

where $t_n = 0$ for $b_n = 1$; $t_n = 2$ for $b_n = \mathrm{i}$; $t_n = 4$ for $b_n = 1+\mathrm{i}$; $t_n = 6$ for $b_n = 0$. Let us check that a is an asymptotic value of $f(z)$ with asymptotic curve $\arg z = \phi_a$. Since $\phi_a \in \bigcap_{n=1}^{\infty} F_n$, in view of (6.2.7), for each n, $|f_n(z)| \leqslant 2$ on $\arg z = \phi_a$. Given $\varepsilon > 0$, we have

$$\sum_{n=N+1}^{\infty} 2^{-n} |f_n(\delta_n z)| < \varepsilon$$

and

$$\sum_{n=N+1}^{\infty} 2^{-n} |b_n| < \varepsilon$$

for some N. Now for the fixed N, in view of (6.2.9) for $\arg z = \phi_a$ and $|z| > r_0$ we have

$$\sum_{n=1}^{N} 2^{-n} |f_n(\delta_n z) - b_n| < \varepsilon.$$

Therefore it follows that

$$\begin{aligned} |f(z) - a| &= \Big|\sum_{n=1}^{\infty} 2^{-n} f_n(\delta_n z) - \sum_{n=1}^{\infty} 2^{-n} b_n\Big| \\ &\leqslant \sum_{n=N+1}^{\infty} 2^{-n} |f_n(\delta_n z)| + \sum_{n=N+1}^{\infty} 2^{-n} |b_n| + \sum_{n=1}^{N} 2^{-n} |f_n(\delta_n z) - b_n| \\ &< 3\varepsilon, \quad \text{for } \arg z = \phi_a \text{ and } |z| > r_0. \end{aligned}$$

It has been proved that $f(z) \to a$ as $z \to \infty$ along $\arg z = \phi_a$. The same argument implies that $f(z) \to 0$ as $z \to \infty$ uniformly in $\pi - 1 \leqslant \arg z \leqslant \pi$.

Consider the function $w = M(z) = z\left(z - \frac{1+\mathrm{i}}{2}\right)$. Set $\Omega = \{z : \left|z - \frac{1+\mathrm{i}}{2}\right| < \frac{1}{2}\} \subset S$ and hence $0 \in M(\Omega)$, namely for some $\tau > 0$, $\{z : |z| \leqslant \tau\} \subset M(\Omega)$. Thus $\{z : |z| \leqslant 1\} \subset \frac{1}{\tau} M(\Omega)$. Set $G_1(z) = \frac{1}{\tau} M(f(z))$. Then all values in the closed unit disk Δ are asymptotic values of $G_1(z)$ and the associated asymptotic curves lie in the angle $\{z : 0 \leqslant \arg z \leqslant 1\}$ and $G_1(z) \to 0$ as $z \to \infty$ uniformly in the angle $\{z : \pi - 1 \leqslant \arg z \leqslant \pi\}$.

The same method is available to construct a meromorphic function G_2 satisfying (6.2.10) and such that all values in the unit disk Δ are its asymptotic values and the associated asymptotic curves lie in the angle $\{z : \pi - 1 \leqslant \arg z \leqslant \pi\}$ and $G_2(z) \to 0$ as $z \to \infty$ uniformly in the angle $\{z : 0 \leqslant \arg z \leqslant 1\}$.

We consider the Möbius transformation

$$w = T(z) = \frac{z}{z - \frac{1}{2}}.$$

Then $T(\Delta)$ is a neighborhood of ∞, namely there exists a $R > 0$ such that $\{z : |z| \geqslant R\} \subset T(\Delta)$ and thus $\widehat{\mathbb{C}} \setminus \Delta \subset \frac{1}{R}T(\Delta)$.

Define

$$F(z) = G_1(z) + \frac{1}{R}T(G_2(z)).$$

A simple calculation yields that $F(z)$ satisfies (6.2.10) and each value on $\widehat{\mathbb{C}}$ is an asymptotic value of $F(z)$ by noting that $G_1(z) \to 0$ uniformly in the angle $\{z : \pi - 1 \leqslant \arg z \leqslant \pi\}$ and $\frac{1}{R}T(G_2(z)) \to 0$ in the angle $\{z : 0 \leqslant \arg z \leqslant 1\}$ as $z \to \infty$.

We complete the proof of Theorem 6.2.9. □

Thus in view of Theorem 6.2.9, for arbitrary $\lambda \in (0, +\infty)$, we can find a meromorphic function $F_1(z)$ with order $\lambda(F_1) < \min\{1, \lambda\}$ satisfying the result of Theorem 6.2.9. Now find a meromorphic function $F_2(z)$ with order λ which tends to zero as $|z| \to \infty$ uniformly on the upper half plane. Then $F(z) = F_1(z) + F_2(z)$ has the order λ and has every value on $\widehat{\mathbb{C}}$ as its asymptotic value. Namely, the following result have been obtained.

Theorem 6.2.10. *Given arbitrarily a $\lambda \in (0, +\infty)$, there exists a meromorphic function $F(z)$ with order λ such that every value on $\widehat{\mathbb{C}}$ is its asymptotic value.*

We know that every meromorphic function has an at most countable number of critical points and so of critical values. However, combining Theorem 6.2.10 and Theorem 6.2.8 yields directly the following

Theorem 6.2.11. *Given arbitrarily a $\lambda \in (0, +\infty)$, there exists a meromorphic function $F(z)$ with order λ such that critical values of F is dense on $\widehat{\mathbb{C}}$.*

Theorem 6.2.11 holds because a meromorphic function with finite order has only finitely many asymptotic values over which direct singularities exist.

6.3 Meromorphic Functions of Bounded Type

It is an interesting topic to study properties of meromorphic functions on which some restrictions on their singular values are imposed. In this section, we mainly discuss fixed points of meromorphic functions of bounded type. A transcendental meromorphic function is said to be of bounded type if the set of its finite singular values is bounded and of finite type if the set is finite. We denote by $\mathscr{B}$ the set of all meromorphic functions of bounded type and by $\mathscr{S}$ the set of all finite-type functions.

Consider the functions

$$f(z) = \frac{1}{z^2} + e^z \text{ and } g(z) = \frac{1}{z} + \tan z.$$

Both of $f(z)$ and $g(z)$ are of bounded type. Here we only check $g(z)$ is in $\mathscr{B}$. $g(z)$ has only two asymptotic values $\pm i$. Since $g'(z) = -z^{-2} + \cos^{-2} z$, the critical points of $g(z)$ are exactly the roots of $\cos^2 z = z^2$ and hence $\tan^2 z = -1 + z^{-2}$ at all critical points. This deduces that all limit points of the critical values are $\pm i$. Thus it has been proved that $g(z)$ is of bounded type.

In view of Theorem 6.2.2, for a function in $\mathscr{B}$, ∞ may only be its critical value and asymptotic value over which none but logarithmic singularities exist. The inverse of $f(z)$ has algebraic and logarithmic singularities over ∞ without others. However ∞ is neither critical values nor asymptotic values of $g(z)$, that is, ∞ is a normal point of the inverse of $g(z)$. A finite-type function may have only asymptotic values over which logarithmic singularities exist. Obviously for a non-zero polynomial $P(z)$ and a non-constant polynomial $Q(z)$, $\int^z P(z)e^{Q(z)}dz$ is in $\mathscr{S}$ and there exist elements in $\mathscr{S}$ with other forms. Actually, the composition of a finite-type meromorphic function and a finite-type entire function is of finite type.

In view of Theorem 6.2.5, a function $f(z)$ in $\mathscr{S}$ has lower order at least $p/2$ where p is the number of asymptotic values. Therefore if $f(z)$ is of lower order less than $1/2$, then it has no asymptotic values at all. In Langley and Zheng [17], the following result is proved.

Theorem 6.3.1. *Let $\phi(r)$ be an unbounded increasing positive function. Then there exist a transcendental entire function $g(z)$ and a transcendental and meromorphic function $f(z)$ such that $F = f \circ g(z)$ is in $\mathscr{S}$ and $T(r,F) = O(\phi(r)(\log r)^2)$ as $r \to \infty$.*

The proof of Theorem 6.3.1 needs the following lemma, which is Lemma 2 of [17].

Lemma 6.3.1. *Assume $\{w_n\}$ is a sequence of complex numbers such that for some fixed $R > 1$ and for all large r, the annulus $\{z : R^{-1}r \leqslant |w| \leqslant Rr\}$ contains at least one element of the sequence $\{w_n\}$. Then there exists a transcendental entire function $g(z)$ with $T(r,g) = O(\phi(r)(\log r))$ as $r \to \infty$ such that all but finitely many critical values of $g(z)$ are elements of $\{w_n\}$.*

The proof of Lemma 6.3.1 is omitted here and the reader is referred to the paper [17].

Proof of Theorem 6.3.1. Let $\mathfrak{p}$ be the Weierstrass Pe function with period 1 and $2\pi i$ such that

$$(\mathfrak{p}')^2 = 4(\mathfrak{p} - e_1)(\mathfrak{p} - e_2)(\mathfrak{p} - e_3)$$

for three distinct complex constants e_j, $j = 1,2,3$. Define the function as in [4]

$$f(z) = \mathfrak{p}(\log v), \ v + v^{-1} = z.$$

$f(z)$ is a meromorphic function satisfying the first order equation

$$(z^2 - 4)(f'(z))^2 = 4(f(z) - e_1)(f(z) - e_2)(f(z) - e_3)$$

and $T(r,f)=O(\log r)^2$, as $r\to\infty$. Thus $f(z)$ has finitely many critical values and in view of Corollary 6.2.1, $f(z)$ is in $\mathscr{S}$. It is well-known that the set of points where $\mathfrak{p}=e_j,\ j=1,2,3$, namely the set of critical points of $\mathfrak{p}$ is

$$\{a_j+m+2n\pi\mathrm{i}:\ m,n\in\mathbb{Z}\}$$

here $a_1=\frac{1}{2},\ a_2=\pi\mathrm{i},\ a_3=\frac{1}{2}+\pi\mathrm{i}$. Then for each $m\in\mathbb{Z}$,

$$w_m^j=\exp(a_j+m)+\exp(-a_j-m), j=1,2,3$$

are critical points of $f(z)$. For large $r>0$, the annulus $\{z:\mathrm{e}^{-1}r\leqslant|z|\leqslant \mathrm{e}r\}$ contains at least one element of $\{w_m^j\}$. In view of Lemma 6.3.1, there exists a $g(z)$ having the properties mentioned in Lemma 6.3.1 for $\{w_m^j\}$ and $\sqrt{\phi(r/2)}$. Set $F(z)=f\circ g(z)$. Obviously, $F(z)$ is in $\mathscr{S}$ and

$$\begin{aligned} T(r,F) &= T(r,f\circ g)\leqslant T(M(r,g),f) \\ &= O(\log M(r,g))^2 = O(T(2r,g))^2 \\ &= O(\sqrt{\phi(r)}\log 2r)^2 = O(\phi(r)(\log r)^2). \end{aligned}$$

Theorem 6.3.1 follows. □

Langley proved in [15] that a meromorphic function $f(z)$ with finitely many singular values must satisfy

$$\liminf_{r\to\infty}\frac{T(r,f)}{(\log r)^2}>0$$

and in [16] that for every $\varepsilon>0$ there exists a meromorphic function $g(z)$ with four singular values such that

$$\limsup_{r\to\infty}\frac{T(r,g)}{(\log r)^2}<\varepsilon$$

while a function $h(z)$ with three singular values satisfies

$$\liminf_{r\to\infty}\frac{T(r,g)}{(\log r)^2}\geqslant c$$

where c is an absolute constant, which has been precisely determined by Eremenko [9] to be equal to $\frac{\sqrt{3}}{\pi}$.

For a function $f\in\mathscr{B}$, ∞ is a normal point or a logarithmic singularity of its inverse. In view of this property, an inequality concerning the first order derivative can be established, which will be used in our later discussion of the existence of fixed-points of the function. And the case when the singular values do not distribute along a sequence of annuli will also be discussed as an extension of the above result.

To the end, we collect some basic knowledge about the hyperbolic metric. Let D be a hyperbolic domain on $\mathbb{C}$, that is, $\mathbb{C}\setminus D$ contains at least two points. In other words, the unite disk Δ is the universal covering space of D. Then there exists the hyperbolic metric on D whose hyperbolic density is denoted by λ_D. The hyperbolic density of Δ is

$$\lambda_\Delta(z) = \frac{2}{1-|z|^2}, \ \forall z \in \Delta$$

and λ_D can be found from the following equality

$$\lambda_D(p(z))|p'(z)| = \frac{2}{1-|z|^2}, \ \forall z \in \Delta,$$

where $p : \Delta \to D$ is an universal covering. Therefore the right-half plane H, $U = \mathbb{C} \setminus \overline{B(0,R)}$ and the annulus $A = \{z : r < |z| < R\}$ have in turn the hyperbolic densities:

$$\lambda_H(z) = (\text{Re} z)^{-1}, \ \ \lambda_U(z) = \frac{1}{|z|(\log|z| - \log R)}$$

and

$$\lambda_A(z) = \frac{\pi}{2|z|\text{mod}(A)\sin(\pi \log(R/|z|)/\text{mod}(A))},$$

where mod(A) is the modulus of A, i.e., $\log \frac{R}{r}$.

Using the Schwarz-Pick Lemma yields the following Principle of Hyperbolic Metric.

Lemma 6.3.2. *Let $f(z)$ be a holomorphic mapping from a hyperbolic domain D_1 into a hyperbolic domain D_2. Then we have*

$$\lambda_{D_2}(f(z))|f'(z)| \leqslant \lambda_{D_1}(z), \ z \in D_1.$$

Here the equality holds if and only if f is a covering from D_1 onto D_2.

Now we can establish a fundamental inequality for functions in $\mathscr{B}$.

Theorem 6.3.2. *Let $f(z)$ be in $\mathscr{B}$ and all its finite singular values are in $B(0,R)$ for some $R > 0$. Then for a with $|f(a)| < R$ we have*

$$|f'(z)| \geqslant \frac{|f(z)|(\log|f(z)| - \log R)}{4|z-a|}. \tag{6.3.1}$$

Proof. Clearly, it suffices to prove (6.3.1) for z with $|f(z)| > R$. Then there exists a component V of $f^{-1}(U), U = \mathbb{C} \setminus \{w : |w| \leqslant R\}$, containing z. Since the inverse of $f(z)$ has at most a logarithmic singularity over ∞, $f : V \to U$ is either a conformal map or a universal covering, and V is simply connected. In view of Lemma 6.3.2, we have

$$\lambda_V(z) = \lambda_U(f(z))|f'(z)| = \frac{|f'(z)|}{|f(z)|(\log|f(z)| - \log R)}, \ z \in V.$$

In order to obtain (6.3.1) we estimate $\lambda_V(z)$. Set $\delta_V(z) = \inf\{|z-c| : \ c \in \partial V\}$ and certainly $\delta_V(z) \leqslant |z-a|$. In view of the Koebe distortion theorem, it can be proved that $\delta_V(z)\lambda_V(z) \geqslant \frac{1}{4}$, and so $\lambda_V(z) \geqslant \frac{1}{4|z-a|}$. This implies immediately (6.3.1). □

Theorem 6.3.2 is essentially due to Eremenko and Lyubich [10] and Rippon and Stallard [20], while the above proof was offered by Zheng [26] in which a corresponding inequality to a finite isolated singular value instead is also established and used to study of the complex dynamics.

The case considered in Theorem 6.3.2 is essentially that in a disk punctured at the center. Recently, in [28] we consider the case of an annulus where we also establish a fundamental inequality. To the end, we need a lemma which is directly produced from Lemma 4.3.1 and Lemma 4.3.2 of Zheng [27].

Lemma 6.3.3. *Let U be a hyperbolic domain on $\mathbb{C}$. Then for $a \notin U \cup \{\infty\}$, we have*

$$\lambda_U(z)|z-a| \geqslant (\mathrm{Mod}(U)+2\kappa)^{-1},$$

where $\kappa = \Gamma(1/4)^4/(4\pi^2)$ and

$$\mathrm{Mod}(U) = \sup\{\mathrm{mod}(A) : \textit{A is a doubly connected domain in } U$$

$$\textit{separating the boundary of } U\}.$$

Theorem 6.3.3. *Let $f(z)$ be a transcendental meromorphic function. Let U be a component of $f^{-1}(W)$ where*

$$W = \{z : r < |z| < R\}$$

with $0 < r < R < +\infty$. Assume that U contains no critical points of $f(z)$. Then for $z \in U$ and $a \notin U \cup \{\infty\}$, one of the following statements holds:

(1) *if a is in an unbounded component of complement of U, then*

$$|f'(z)| \geqslant \frac{\mathrm{mod}(W)}{2\pi}\frac{|f(z)|}{|z-a|}\sin\frac{\pi(\log R - \log|f(z)|)}{\mathrm{mod}(W)}; \tag{6.3.2}$$

(2) *if a is in the bounded component of complement of U, then*

$$|f'(z)| \geqslant \frac{2\min\{\mathrm{mod}(W), m\}}{(2\kappa+1)\pi}\frac{|f(z)|}{|z-a|}\sin\frac{\pi(\log R - \log|f(z)|)}{\mathrm{mod}(W)}, \tag{6.3.3}$$

where m is the covering number of $f(z)$ from U onto W.

Proof. Since U does not contain any critical points of $f(z)$, f is a covering from U onto W and further, in terms of Lemma 6.3.2 and the formula of hyperbolic density of an annulus, we have

$$\lambda_U(z) = \lambda_W(f(z))|f'(z)| = \frac{\pi|f'(z)|}{2|f(z)|\mathrm{mod}(W)\sin(\pi\log(R/|f(z)|)/\mathrm{mod}(W))}.$$

Next step is to estimate $\lambda_U(z)$ from below. Since f is a covering, U is simply connected or doubly connected. We need to treat two cases.

(I) a is in an unbounded component of $\mathbb{C}\setminus U$. Then we can choose a curve Γ starting at a forward to ∞ such that $U \subset \mathbb{C}\setminus\Gamma$. We have

$$\lambda_U(z)|z-a| \geqslant \lambda_{\mathbb{C}\setminus\Gamma}(z)|z-a| \geqslant \frac{1}{4}$$

and therefore, combining the above inequalities yields (6.3.2).

(II) a is in the bounded component of $\mathbb{C}\setminus U$. Then U is doubly connected and separates a and ∞. In terms of Lemma 6.3.3, we have

$$\lambda_U(z)|z-a| \geqslant (\mathrm{mod}(U)+2\kappa)^{-1}. \tag{6.3.4}$$

There exists a conformal map $\phi : U \to \{w : r_0 < |w| < d\}$ with $r_0 = r^{1/m}$ and then $f\circ\phi^{-1} : \{w : r_0 < |w| < d\} \to W$ is proper and it has the form $f\circ\phi^{-1}(w) = w^m$ and further, $d^m = R$. Thus $\mathrm{mod}(U) = \mathrm{mod}(W)/m$. In terms of (6.3.4), we obtain

$$\begin{aligned}|f'(z)| &\geqslant \frac{2\mathrm{mod}(W)m}{(\mathrm{mod}(W)+2\kappa m)\pi}\frac{|f(z)|}{|z-a|}\sin\frac{\pi(\log R-\log|f(z)|)}{\mathrm{mod}(W)}\\ &\geqslant \frac{2\mathrm{mod}(W)m}{\max\{\mathrm{mod}(W),m\}(1+2\kappa)\pi}\frac{|f(z)|}{|z-a|}\sin\frac{\pi(\log R-\log|f(z)|)}{\mathrm{mod}(W)}\\ &= \frac{2\min\{\mathrm{mod}(W),m\}}{(1+2\kappa)\pi}\frac{|f(z)|}{|z-a|}\sin\frac{\pi(\log R-\log|f(z)|)}{\mathrm{mod}(W)}.\end{aligned}$$

This is our desired inequality (6.3.3). □

We discuss the number of fixed-points of meromorphic functions in terms of Theorem 6.3.2 and Theorem 6.3.3. A root of $f(z) = z$ is called a fixed-point of $f(z)$ and the first order derivative $f'(z)$ at fixed-point z is called multiplier of this fixed-point. Furthermore, a fixed-point z is said to be attracting, indifferent or repelling if the modulus of its multiplier is less than, equal to or greater than one. A meromorphic function $f(z)$ may have no fixed-points, for instant, $z+\mathrm{e}^z$ has no fixed-points. However, for a function whose singular values are restricted, especially in $\mathscr{B}$, we can say something about the number of fixed-points of the function. In what follows, we show the main results of [17] and [28].

Theorem 6.3.4. *Let $f(z)$ be a transcendental meromorphic function in $\mathscr{B}$. Then we have*

$$m\left(r, \frac{1}{f-z}\right) = O(\log rT(r,f)), \tag{6.3.5}$$

as $r\to\infty$ outside $E(f)$, and in particular, $\delta(0, f-z) = 0$.

Proof. Set $g(z) = f(z) - z$ and $E(r) = \{\theta\in[0,2\pi) : |g(re^{\mathrm{i}\theta})| \leqslant 1\}$. Then for all sufficiently large $r > 0$, in view of (6.3.1), for $z = re^{i\theta}$ with $\theta\in E(r)$, we have

$$\begin{aligned}|g'(z)| \geqslant |f'(z)| - 1 &\geqslant \frac{|f(z)|(\log|f(z)| - \log R)}{4|z-a|} - 1\\ &\geqslant \frac{(|z|-1)(\log(|z|-1) - \log R)}{4|z-a|} - 1 \geqslant 1.\end{aligned}$$

Thus

$$\begin{aligned} m\left(r,\frac{1}{g}\right) &= \frac{1}{2\pi}\int_{E(r)} \log\left|\frac{g'(re^{i\theta})}{g(re^{i\theta})}\frac{1}{g'(re^{i\theta})}\right| d\theta \\ &\leqslant \frac{1}{2\pi}\int_{E(r)} \log\left|\frac{g'(re^{i\theta})}{g(re^{i\theta})}\right| d\theta \\ &\leqslant m\left(r,\frac{g'}{g}\right) = O(\log rT(r,f)) \end{aligned}$$

for all r outside $E(f)$. □

Theorem 6.3.4 is proved in [17]. The following is an extension of the theorem with f being entire and essentially attained in [28].

Theorem 6.3.5. *Let $f(z)$ be a transcendental meromorphic function with finitely many poles. Assume that there exist a sequence of annuli $W_n = \{z : R_n < |z| < cR_n\}$ with $c > \exp\left(\frac{2(2\kappa+1)\pi}{\sqrt{3}}\right)$ and $R_n \to \infty$ as $n \to \infty$ such that for each n, W_n contains no singular values of f. Then for any $\widehat{R}_n \in (\sqrt[3]{c}R_n, \sqrt{c}R_n) \setminus E(f)$, we have*

$$m\left(\widehat{R}_n, \frac{1}{f-z}\right) = O(\log \widehat{R}_n T(\widehat{R}_n, f)), \tag{6.3.6}$$

as $n \to \infty$ and in particular, $\delta(0, f-z) = 0$.

Proof. Since $\sqrt{c}R_n - \sqrt[3]{c}R_n = (\sqrt{c} - \sqrt[3]{c})R_n \to \infty$ as $n \to \infty$ and $E(f)$ has finite measure, for all sufficiently large n, $\in (\sqrt[3]{c}R_n, \sqrt{c}R_n)\setminus E(f) \neq \varnothing$. Set $g(z) = f(z) - z$ and for an $\widehat{R}_n \in (\sqrt[3]{c}R_n, \sqrt{c}R_n) \setminus E(f)$, set

$$E_n = \{\theta \in [0, 2\pi) : |g(\widehat{R}_n e^{i\theta})| \leqslant 1\}.$$

Then for $z = \widehat{R}_n e^{i\theta}$ with $\theta \in E_n$, we have $\widehat{R}_n - 1 \leqslant |f(z)| \leqslant \widehat{R}_n + 1$ and hence $f(z) \in W_n$, that is to say, $z \in f^{-1}(W_n)$. We assume that $0 \notin f^{-1}(W_n)$. Actually, if $f(0) = \infty$, clearly $0 \notin f^{-1}(W_n)$; if $f(0)$ is finite, then we consider $R_n > |f(0)|$ and so $0 \notin f^{-1}(W_n)$. Now we claim that $|f'(z)| > K|f(z)|/|z|$ with a constant $K > 1$, which will be proved by using Theorem 6.3.3.

Noting that $\sqrt[3]{c}R_n - 1 \leqslant |f(z)| \leqslant \sqrt{c}R_n + 1$, we have

$$\pi\frac{\log cR_n - \log|f(z)|}{\log c} \geqslant \pi\frac{\log cR_n - \log(\sqrt{c}R_n + 1)}{\log c} \to \frac{\pi}{2}$$

and

$$\pi\frac{\log cR_n - \log|f(z)|}{\log c} \leqslant \pi\frac{\log cR_n - \log(\sqrt[3]{c}R_n - 1)}{\log c} \to \frac{2\pi}{3}$$

as $n \to \infty$. Therefore for all sufficiently large n, we have

$$\sin\frac{\pi(\log cR_n - \log|f(z)|)}{\log c} \geqslant \frac{\sqrt{3}}{4}.$$

Let U_n be the component of $f^{-1}(W_n)$ containing z. If U_n does not separate 0 and ∞, then from Theorem 6.3.3 we have

$$|f'(z)| \geqslant \frac{\log c}{2\pi}\frac{|f(z)|}{|z|}\sin\frac{\pi(\log cR_n - \log|f(z)|)}{\log c} \geqslant \frac{\sqrt{3}\log c}{8\pi}\frac{|f(z)|}{|z|}.$$

For our purpose, now we can assume that U_n separates 0 and ∞ for all sufficiently large n. Let Γ_n be the simple analytic curve in U_n around 0 which f maps onto the circle $\{z : |z| = \sqrt{c}R_n\}$. Obviously, $\mathrm{dist}(\Gamma_n, 0) \to \infty$ as $n \to \infty$. Let m_n be the covering number of f from U_n onto W_n and $n(\gamma, b)$ be the winding number of the closed curve γ going around b. Then by the argument principle, $n(f(\Gamma_n), 0)$ equals to the deference of the numbers of zeros and poles of f inside Γ_n. Since f has only finitely many poles, we can assume that f has infinitely many zeros, otherwise (6.3.5) holds in view of the Nevanlinna second fundamental theorem. Then we have $m_n \geqslant n(f(\Gamma_n), 0) \to \infty$ as $n \to \infty$. By Theorem 6.3.3, we have

$$|f'(z)| \geqslant \frac{2\log c}{(2\kappa+1)\pi}\frac{|f(z)|}{|z|}\sin\frac{\pi(\log cR_n - \log|f(z)|)}{\log c} \geqslant \frac{\sqrt{3}\log c}{2(2\kappa+1)\pi}\frac{|f(z)|}{|z|}.$$

Letting $K = \frac{\sqrt{3}\log c}{2(2\kappa+1)\pi} > 1$, we complete the proof of our claim.

For $z = \widehat{R}_n e^{i\theta}$ with $\theta \in E_n$, we have

$$|g'(z)| \geqslant |f'(z)| - 1 \geqslant K\frac{|f(z)|}{|z|} - 1 \geqslant (K-1) - \frac{K}{|z|} > \frac{K-1}{2}$$

so that

$$\frac{1}{|g(z)|} \leqslant \frac{2}{K-1}\left|\frac{g'(z)}{g(z)}\right|.$$

The same argument as in the proof of Theorem 6.3.4 yields (6.3.6). □

Actually, Theorem 6.3.4 provides a criterion of that a meromorphic function would not be in $\mathscr{B}$, that is, if $\delta(0, f - z) > 0$, then $f(z)$ is not in $\mathscr{B}$. An alternate version of Theorem 6.3.5 is that if $f(z)$ is a transcendental meromorphic function with finitely many poles such that $\delta(0, f - z) > 0$, then for all sufficiently large R and $c > \exp\left(\frac{2(2\kappa+1)\pi}{\sqrt{3}}\right)$, the annulus $\{z : R < |z| < cR\}$ contains singular values of $f(z)$.

From Theorem 6.3.2, we know that for a function $f \in \mathscr{B}$, all but finitely many fixed-points z of $f(z)$ satisfy the inequality $|f'(z)| > d\log|z| > 1$ for some $d > 0$ and so they are repelling. However, the following results were proved in [17].

Theorem 6.3.6. *Let $f(z)$ be in $\mathscr{B}$ with order $\lambda(f)$ such that $0 < \sigma < \lambda(f) \leqslant \infty$. Then $f(z)$ has infinitely many fixed-points z with*

$$|f'(z)| > |z|^{\sigma/2}.$$

In order to prove Theorem 6.3.6 and for application in the sequel, we need the following lemma, which directly comes from the Koebe distortion theorem.

Lemma 6.3.4. *Let U be a simply connected domain and $\psi(z)$ a conformal map from U onto $B(a,r)$ with $\psi(z_0)=a$ and $z_0\in U$. Then for each $z\in U$, we have*

$$\frac{1}{r^2}\frac{(r-|\psi(z)-a|)^3}{r+|\psi(z)-a|}|\psi'(z_0)|\leqslant|\psi'(z)|\leqslant\frac{1}{r^2}\frac{(r+|\psi(z)-a|)^3}{r-|\psi(z)-a|}|\psi'(z_0)| \tag{6.3.7}$$

and

$$r^2\frac{|\psi(z)-a|}{(r+|\psi(z)-a|)^2}\leqslant|z-z_0||\psi'(z_0)|\leqslant r^2\frac{|\psi(z)-a|}{(r-|\psi(z)-a|)^2}. \tag{6.3.8}$$

And furthermore, we have the inclusion

$$\psi^{-1}(B(a,r/2))\subset B(z_0,2r|\psi'(z_0)|^{-1}). \tag{6.3.9}$$

Proof. Using the Koebe distortion theorem to the inverse of $\psi(z)$ immediately yields (6.3.7) and (6.3.8), for $\psi^{-1}(w)$ is a conformal map of U from $B(a,r)$.

For each $z\in\psi^{-1}(B(a,r/2))$, we have $\psi(z)\in B(a,r/2)$. In view of (6.3.8), we have $|z-z_0||\psi'(z_0)|\leqslant 2r$ and this implies (6.3.9). □

Lemma 6.3.5. *Let $f(z)$ be in $\mathscr{B}$. Set*

$$h(z)=\frac{f(z)}{z}-1.$$

Then there exists a positive number ε such that all but at most finitely many components U of $h^{-1}(B(0,2\varepsilon))$ are simply-connected and $h(z)$ maps them conformally onto $B(0,2\varepsilon)$. Furthermore, $h(z)$ has only finitely many singular values in $B(0,2\varepsilon)$.

Proof. First of all we consider critical values of $h(z)$ near 0. Let b be a critical point of $h(z)$ such that $|h(b)|\leqslant\frac{1}{2}$. A simple calculation implies that

$$\begin{aligned}|h(b)+1|=|f'(b)|&\geqslant\frac{|f(b)|(\log|f(b)|-\log R)}{4|b-a|}\\&=\frac{|b||h(b)+1|(\log|f(b)|-\log R)}{4|b-a|},\end{aligned}$$

where we have used the equality $h'(b)=0$ and (6.3.1) and furthermore we have

$$2R\exp(4(1+|a||b|^{-1}))\geqslant 2R\exp\left(4\frac{|b-a|}{|b|}\right)\geqslant 2|f(b)|=2|b||h(b)+1|\geqslant|b|,$$

so that $|b|\leqslant|a|$ or $|b|\leqslant 2e^8R$. Thus there exist no critical points of $h(z)$ in $\{z:|z|>\max\{|a|,2e^8R\}\}$ such that its critical values are in the disk $\{z:|z|<\frac{1}{2}\}$ and $h(z)$ has only finitely many critical values in $\{z:|z|<\frac{1}{2}\}$.

Obviously, z_0 is a zero of $h(z)$ if and only if z_0 is a fixed-point of $f(z)$ and in this case, z_0 is a critical point of $h(z)$ only when $f'(z_0) = 1$. We have known that $f(z)$ has only finitely many non-repelling fixed-points. Thus all but at most finitely many zeros of $h(z)$ are not its critical points. Choose two positive numbers R_1 and R_2 with $R_2 > 2R_1 > \max\{|a|, (2e^8 R)^2\}$ such that every fixed-point of $f(z)$ lying in $|z| > R_1$ satisfies $|f'(z)| > d\log|z| > 1$ and on $|z| = R_2$, $h(z) \neq 0$ and furthermore we have ε with $0 < \varepsilon < \frac{1}{4}$ such that $|h(z)| > 2\varepsilon$ on $|z| = R_2$. For every fixed-point z_1 of $f(z)$ with $|z_1| > R_2$, we have an analytic branch $\psi(w)$ of $h^{-1}(w)$ sending 0 to z_1. Let r_1 be the radius of convergence of the power series of $\psi(w)$ at 0 and then there exists a point $w_1 = r_1 e^{i\theta_1}$ for some real θ_1 such that $\psi(w)$ can not be analytically continued to w_1 along the path $\gamma : te^{i\theta_1}, 0 \leqslant t \leqslant r_1$.

Suppose $r_1 \leqslant 2\varepsilon$. Let U be the component of $h^{-1}(B(0,r_1))$ containing z_1 and hence $U \subset \{z : |z| > R_2\}$ so that $\psi(\gamma) \subset \{z : |z| > R_2\}$. Since $h(z)$ maps U into $B(0,2\varepsilon)$, that is, on U, $|f(z)| = |z||h(z)+1| > (1-2\varepsilon)|z| > \frac{1}{2}|z| > \frac{1}{2}R_2$, we therefore have

$$\begin{aligned}
\left|\frac{zh'(z)}{h(z)+1}\right| &= \left|\frac{zf'(z)}{f(z)} - 1\right| \\
&\geqslant \frac{|z|(\log|f(z)| - \log R)}{4|z-a|} - 1 \\
&\geqslant \frac{1}{8}\log\frac{|f(z)|}{R} \\
&\geqslant \frac{1}{8}\log\frac{R_2^{1/2}}{2R}R_2^{1/2} \\
&> \frac{1}{16}\log R_2,
\end{aligned}$$

and so

$$|zh'(z)| > |h(z)+1|\frac{1}{16}\log R_2 \geqslant (1-2\varepsilon)\frac{1}{16}\log R_2 \geqslant \frac{1}{32}\log R_2.$$

On γ, for $w = h(z)$ we have $\frac{\psi'(w)}{\psi(w)} = \frac{1}{zh'(z)}$, and it follows that

$$\left|\log\frac{\psi(w)}{z_1}\right| \leqslant \int_0^{|w|}\left|\frac{\psi'(te^{i\theta_1})}{\psi(te^{i\theta_1})}\right| dt \leqslant |w|\frac{32}{\log R_2} < \frac{64\varepsilon}{\log R_2}.$$

This implies that $\psi(\gamma)$ is bounded and then w_1 is a critical value of $h(z)$, while this contradicts the result obtained in the first paragraph of this proof. Hence we have proved $r_1 > 2\varepsilon$ and the component of $h^{-1}(B(0,2\varepsilon))$ containing z_1 is simply-connected and $h(z)$ maps it conformally onto $B(0,2\varepsilon)$. The same argument as above yields that $h(z)$ has no asymptotic values in $B(0,2\varepsilon)$. □

Now we are in position to prove Theorem 6.3.6.

Proof of Theorem 6.3.6. Consider the function $h(z) = \frac{f(z)}{z} - 1$. In view of Lemma 6.3.5, there exists a positive number ε such that $h(z)$ maps conformally all but finitely many components of $h^{-1}(B(0,2\varepsilon))$ onto $B(0,2\varepsilon)$. Set

$$g(z)=\frac{1}{\varepsilon}\left(\frac{f(z)}{z}-1\right).$$

Then $g(z)$ maps conformally every component U of $g^{-1}(B(0,2))$ containing a fixed-point of $f(z)$ in $\{z:|z|>R\}$ for some $R>0$ onto $B(0,2)$.

In view of Theorem 6.3.4, we can choose arbitrarily large $r>4R$ such that there exist

$$N(r)=n(r,f=z)-n(r/2,f=z)(\neq 0)$$

fixed-points of f lying in the annulus $I(r)=\{z:r/2\leqslant |z|\leqslant r\}$. Consider a fixed-point c of $f(z)$ in $I(r)$ and then c is a zero of $g(z)$. There exists a single-valued analytic branch of g^{-1} in $B(0,2)$ sending 0 to c, namely $g(z)$ conformally maps a simply connected domain containing c onto $B(0,2)$ and hence in view of Lemma 6.3.4, the component V_c containing c of $g^{-1}(B(0,1))$ satisfies

$$V_c\subset B(c,4|g'(c)|^{-1})\subset B(c,\frac{1}{2}|c|)\subset\left\{z:R\leqslant |z|\leqslant \frac{3}{2}r\right\}=J(r),$$

where we have used the inequality

$$|g'(c)|=\frac{1}{\varepsilon}\left|\frac{f'(c)-1}{c}\right|\geqslant\frac{|f'(c)|-1}{\varepsilon|c|}>\frac{d\log|c|-1}{\varepsilon|c|}>\frac{8}{|c|}$$

and R is chosen such that $R>\exp(9/d)$. Obviously V_c does not intersect each other and $\mathrm{Area}(J(r))<\frac{9}{4}\pi r^2$. Then

$$\sum_{c\in I(r)}\mathrm{Area}(V_c)\leqslant\frac{9}{4}\pi r^2.$$

Set

$$E(r)=\left\{c:\mathrm{Area}(V_c)\leqslant\frac{9\pi r^2}{N(r)}\right\}$$

and so

$$(N(r)-\#E(r))\frac{9\pi r^2}{N(r)}\leqslant\sum_{c\in I(r)\setminus E(r)}\mathrm{Area}(V_c)\leqslant\frac{9}{4}\pi r^2.$$

This yields $\#E(r)\geqslant\frac{3}{4}N(r)$. For each $c\in E(r)$, we have

$$\pi=\int_{V_c}|g'(z)|^2\mathrm{d}\sigma\leqslant|g'(z_c)|^2\mathrm{Area}(V_c)$$

for some $z_c\in V_c$, so that

$$|g'(z_c)|\geqslant\sqrt{\frac{\pi}{\mathrm{Area}(V_c)}}\geqslant\sqrt{\frac{1}{9r^2}N(r)}=\frac{1}{3}r^{-1}N(r)^{1/2}.$$

In view of Lemma 6.3.4 with $\psi = g$, $z_0 = c$, $a = 0$ and $r = 2$, we have

$$|g'(c)| \geqslant \frac{4}{27}|g'(z_c)| \geqslant \frac{4}{81}r^{-1}N(r)^{1/2}$$

and so

$$|f'(c)| \geqslant \varepsilon|c||g'(c)| - 1 \geqslant \frac{2\varepsilon}{81}N(r)^{1/2} - 1.$$

In view of Theorem 1.1.3 and Lemma 1.1.5, there exist a sequence of positive numbers $\{r_n\}$ tending to $+\infty$ such that $N(r_n) > r_n^{\rho}$ with $\sigma < \rho < \lambda$, as in view of Theorem 6.3.4, $n(r, f = z)$ has the order λ. Theorem 6.3.6 follows. □

From the proof of Theorem 6.3.6 it is easily seen that for all sufficiently large r there exist at least $\frac{1}{2}n(r, f = z)$ fixed-points of $f(z)$ at which

$$|f'(c)| \geqslant C_0\frac{|c|}{r}N(r)^{1/2} - 1,$$

where $N(r) = n(r, f = z) - n(2R, f = z)$ and C_0 is a positive constant. Indeed, we can obtain the result by using the same argument as in the proof of Theorem 6.3.6 to the annulus $I(r) = \{z : 2R \leqslant |z| \leqslant r\}$ instead. Thus if $\lambda(f) > 2$, there exists a sequence of positive numbers $\{r_n\}$ tending to $+\infty$ such that for each r_n, $f(z)$ has at least $\frac{1}{2}n(r_n, f = z)$ fixed-points of $f(z)$ at which $|f'(z)| \geqslant |z|^{\sigma/2-1}$; If $\mu(f) = \infty$, then for arbitrarily large $K > 0$ and for all sufficiently large r there exist at least $\frac{1}{2}n(r, f = z)$ fixed-points of $f(z)$ at which $|f'(z)| > |z|^K$. When an entire function $f(z)$ is considered, the following further result is proved in [17].

Theorem 6.3.7. *If $f(z)$ is a transcendental entire function in $\mathscr{B}$, then for $0 < \alpha < 1$, $f(z)$ has infinitely many fixed-points z with*

$$|f'(z)| > \varepsilon c \log M(\alpha|z|, f),$$

where c is an absolute constant and ε is one in Lemma 6.3.5.

In order to prove Theorem 6.3.7, we need the following result which is due to Pommerenke [19].

Lemma 6.3.6. *Let $g(z)$ be a transcendental entire function. Then the set $g^{-1}(B(0,1))$ contains infinitely many points z_n tending to ∞ such that*

$$|z_n g'(z_n)| \geqslant K \log M(|z_n|, g), \tag{6.3.10}$$

where K is an absolute constant.

Now we come to

The proof of Theorem 6.3.7. Assume without any loss of generalities that $f(0) = 0$. As in the proof of Theorem 6.3.6, consider the inverse of $g(z)$ on $B(0,2)$ and $g(z)$ is a transcendental entire function. The set $g^{-1}(B(0,2))$ consists of infinitely many components which contain a fixed-point of $f(z)$ and in which $g(z)$ is univalent possibly except finitely many members of them. Let $\{z_n\}$ be a sequence of points for $g(z)$ determined by Lemma 6.3.6. We can assume with any loss of generalities that each z_n is in a component $\{U_n\}$ of $g^{-1}(B(0,1))$ which contains a fixed-point of $f(z)$ written into c_n. Thus

$$|g'(c_n)| \geqslant \frac{4}{27}|g'(z_n)| \geqslant \frac{4}{27}|z_n|^{-1}K\log M(|z_n|,g)$$

and $z_n \in B(c_n,(1-\alpha)|c_n|)$ and so $\alpha \leqslant \left|\frac{z_n}{c_n}\right| \leqslant 2-\alpha$.

Since

$$|g'(c_n)| = \frac{1}{\varepsilon}\left|\frac{f'(c_n)-1}{c_n}\right|,$$

we have

$$\begin{aligned}|f'(c_n)| &\geqslant \frac{4}{27}\varepsilon K\frac{|c_n|}{|z_n|}\log M(|z_n|,g)-1\\ &\geqslant \frac{4\varepsilon K}{27(2-\alpha)}(\log M(|z_n|,f/z-1)-\log\varepsilon)-1\\ &\geqslant \frac{2\varepsilon K}{27}(\log M(|z_n|,f)-\log|z_n|-\log 2-\log\varepsilon)-1\\ &\geqslant \varepsilon c\log M(\alpha|c_n|,f)\end{aligned}$$

for sufficiently large n and some absolute positive constant c.

The proof of Theorem 6.3.9 is complete. □

We remark that Theorem 6.3.6 and Theorem 6.3.7 were extended by Clifford [6] to the roots of $f(z)-Q(z)$ for a rational function $Q(z)$ with $Q(z)\sim az^p$, $p\geqslant 1$, as $|z|\to\infty$.

Finally we simply discuss periodic points of an entire function. Let $f(z)$ be a transcendental entire function. A point $z_0\in\mathbb{C}$ is a periodic point of f if for a positive integer n, $f^n(z_0)=z_0$, where f^n means the nth iterate of f, namely $f^n(z)=f^{n-1}(f(z))$ and $f^1(z)=f(z)$. The smallest n such that $f^n(z_0)=z_0$ is called the order of periodic point z_0. Bergweiler [1] in 1991 proved the existence of infinitely many periodic points of order $n\geqslant 2$ for a transcendental entire function, which confirms the conjecture of I. N. Baker. And Zheng [25] in 1999 gave a quantity estimate of the number of periodic points of order $n\geqslant 2$ in $\{z:|z|<r\}$ in terms of logarithmic of maximum module of the function.

Theorem 6.3.8. *Let $f(z)$ be a transcendental entire function. For $m\geqslant 2$, we have*

$$\overline{n}_m\left(r,\frac{1}{f^m(z)-z}\right)\geqslant K\log M(r^d,f^m)$$

for an unbounded sequence of r, where $\overline{n}_m(r,1/(f^m(z)-z))$ is the number of distinct periodic points of order m and K is an absolute constant and $d>(1500)^{-2}$.

The proof of Theorem 6.3.8 can be found in [27]. Bergweiler [2] in 1997 shown furthermore that a transcendental entire function has infinitely many repelling periodic points of period n which do not lie on a given straight line. There exists an example of infinite growth order all but a finite number of whose periodic points lie in a small angular domain. However, this phenomenon cannot happen to an entire function with finite growth lower order (see [29]) (Note: if all but a finite number of periodic points of an entire function $f(z)$ lie in a small angular domain, then the Julia set of $f(z)$ lie in the small angular domain). It stimulates us to ask whether or not all but finitely many periodic points with the fixed order lie on finitely many straight lines which are not parallel or on finitely many angles whose openings are very small for an entire function with the finite lower order. However, it is well-known that $\tan z$ has all periodic points on the real axis.

If $f(z)$ is an entire function in $\mathscr{B}$, then for $n \geqslant 1$, f^n is in $\mathscr{B}$. In view of Theorem 6.3.4, we have $\delta(0, f^n - z) = 0$ and actually, from this we can attain a precise estimate of the number of periodic points of order n in $\{z : |z| < r\}$.

References

1. Bergweiler, W., Periodic points of entire functions: proof of a conjecture of Baker, Complex Varaibles Theory Appl., **17**(1991), 57-72.
2. Bergweiler, W., Non-real periodic points of entire functions, Canad. Math. Bull., **40**(1997), 271-275.
3. Bergweiler, W. and Eremenko, A., On the singularities of the inverse to a meromorphic function of finite order, Revista Math. Iberoamericana, (2)**11**(1995), 355-373.
4. Bank, S. and Kaufman, R., On meromorphic solutions of first-order differential equations, Comment. Math. Helv., **51**(1976), 289-299.
5. Carleman, T., Sur les fonctions inverses des fonctions entiéres, Ark. Mat. Astron. Fys., (10)**15**(1921).
6. Clifford, E. F., Extending two fixpoint theorems of Langley and Zheng, Results Math., (1)-(2)**47**(2005), 45-54.
7. Denjoy, A., Sur les fonctions entiéres de genre fini, C. R. Acad. Sci. Paris, **145**(1907).
8. Eremenko, A. E., Set of asymptotic values of a meromorphic function of finite order, Matematicheskie Zametki, (6)**24**(1978), 779-783.
9. Eremenko, A. E., Transcendental meromorphic functions with three singular values, Illinois J. Math., **48**(2004), 701-709.
10. Eremenko A. E. and Lyubich, M. Yu, Dynamical properties of some classes of entire functions, Ann. Inst. Fourier, **42**(1992), 989-1020.
11. Fuchs, W. H. J., A Phragmén-Lindelöf theorem conjectured by D. J. Newman, Trans. Amer. Math. Soc., **267**(1981), 285-293.
12. Gross, W., Eine ganze Funktion für die jede komplexe Zahl Konvergenzwert ist, Math. Ann., **79**(1918), 201-208.
13. Heins, M., Asymptotic spots of entire and meromorphic functions, Ann. of Math., **66**(1957), 430-439.
14. Iversen, F., Recherches sur les fonctions inverses des fonctions méromorphes, Thése, Helsingfors, 1914.
15. Langley, J. K., On differential polynomials, fixpoints and critical values of meromorphic functions, Result Math., **35**(1999), 284-309.
16. Langley, J. K., Critical values of slowly growing meromorphic functions, Comput. Methods Funct. Theory, (2)**2**(2002), 537-547.
17. Langley, J. K. and Zheng, J. H., On the fixpoints, multiplers and value distribution of certain classes of meromorphic functions, Ann. Acad. Sci. Fenn. Math., **23**(1998), 133-150.
18. Nevanlinna, R., Analytic functions, Springer, Berlin, Heidelberg, New York, 1970.

19. Pommerenke, Ch., Normal functions.- Proc. of the NRL Conference on Classical Function Theory, Naval Research Laboratory, Washington, 1970.
20. Rippon, P. J. and Stallard, G. M., Iteration of a class of hyperbolic meromorphic functions, Proc. Amer. Math. Soc., (11)**127**(1999), 3251-3258.
21. Valiron, G., Recherches sur le théoréme de M. Borel dans la théorie des fonctions méromorphes, Acta Math., **52**(1928), 67-92.
22. Zhang, G. H., Theory of entire and meromorphic functions (in Chinese) , Science Press Sinica, 1986.
23. Zhang, G. H., On direct transcendental singularities of the inverses of meromorphic functions and their derivatives, Science in China, Series A, **4**(1982), 289-297.
24. Zheng, J. H., On value distribution of meromorphic functions with respect to argument I, Complex Variables and Elliptic Equations, 2009.
25. Zheng, J. H., On the value ditribution of iterated entire functions, J. Austral. Math. Soc., Series A, **66**(1999), 18-31.
26. Zheng J. H., Singularities and limit functions in iteration of meromorphic functions. J. London Math. Soc., (2)**67**(2003), 195-207.
27. Zheng, J. H., Dynamics of meromorphic functions, Mongraph of Tsinghua University, Tsinghua University Press, 2006 (in Chinese).
28. Zheng, J. H., On fixed-points and singular values of meromorphic functions, to appear in Science in China, 2010.
29. Zheng, J. H., Wang, S and Huang, Z., Some properties of Fatou and Julia sets of transcendental meromorphic functions. Bull. Austral. Math. Soc., **66**(2002), 1-8.

Chapter 7
The Potential Theory in Value Distribution

Jianhua Zheng
Department of Mathematical Sciences, Tsinghua University, Beijing 100084, P. R. China
jzheng@math.tsinghua.edu.cn

Abstract: This chapter is mainly devoted to introducing the proof of the Nevanlinna's conjecture which Eremenko provided in terms of the potential theory. This conjecture proposed by F. Nevanlinna in 1929 had been at an important and special position in the Nevanlinna's value distribution theory. It was proved first by D. Drasin in 1987, but the Drasin's proof is very complicated. In our attempt to help readers easily grasp the Eremenko proof, we begin with the basic knowledge about subharmonic functions and discuss especially the normality of family and the Nevanlinna theory of δ-subharmonic functions. This reveals an approach of that some problems of value distribution of meromorphic functions are transferred to those of subharmonic functions. Finally, we make a simple survey on recent development and some related results of the Nevanlinna's conjecture.

Key words: Subharmonic functions, Subharmonic Normality, Subharmonic Nevanlinna theory, Nevanlinna conjecture, Deficiency

The potential theory is itself an important theory which studies mainly subharmonic functions and related problems. The subharmonic functions seem to be functions which are equipped with some distribution mass, while harmonic functions have zero distribution mass. The potential theory has proved powerful in study of value distribution of meromorphic functions, which was heavily stressed by the celebrated works of A. Baernstein, J. M. Anderson, A. É. Eremenko, M. L. Sodin, M. Tsuji and others. In fact, the logarithm of module of a meromorphic function is a defence of two subharmonic functions, called δ-subharmonic function, and its distribution mass on a domain D is defence of the number of its $\pm\infty$ valued points, i.e., zeros and poles of the meromorphic function lying in this D. Therefore, some problems (of value distribution) of meromorphic functions can be transferred toward those of subharmonic functions. This is basically a natural idea, but all the concrete approaches are not simple, direct or smooth. Thus remarkable approaches have been revealed in order to solve some important problems.

In this chapter, we will mainly introduce the proof of the Nevanlinna conjecture which Eremenko gave in 1987 and 1993 in terms of the potential theory. This conjecture stood at an important and special position in the Nevanlinna's value distribution theory and have attracted great interests. It was proposed by F. Nevanlinna in 1929 and proved first by D. Drasin in 1987. In order that the content of this chapter can be self-contained, we will collect some basic knowledge about the potential theory before we carry out our center purpose of this chapter.

7.1 Signed Measure and Distributions

Let $(X,\mathfrak{A})$ be a measurable space. An extended real-valued function μ defined on $\mathfrak{A}$ is called a signed measure if (i) $\mu(\varnothing)=0$ and (ii) $\mu(\cup_{j=1}^{\infty}E_j)=\sum_{j=1}^{\infty}\mu(E_j)$ for all pairwise disjoint sequences $\{E_j\}$ of elements of $\mathfrak{A}$. Obviously, a signed measure can assume at most one of the values $+\infty$ and $-\infty$. A complex-valued function on $\mathfrak{A}$ is a complex measure if it satisfies (i) and (ii). Let $\mu_1(E)=\mathrm{Re}\mu(E)$ and $\mu_2(E)=\mathrm{Im}\mu(E)$ for all $E\in\mathfrak{A}$ and then μ_1 and μ_2 are signed measures on $\mathfrak{A}$ and $\mu=\mu_1+\mathrm{i}\mu_2$. If $\mu\geqslant 0$, that is to say, $\mu(E)\geqslant 0$ for all $E\in\mathfrak{A}$, then μ is a measure on $\mathfrak{A}$.

Let μ be a signed measure on a measurable space $(X,\mathfrak{A})$. A set $P\in\mathfrak{A}$ is said to be non-negative for μ if $\mu(P\cap E)\geqslant 0$ for all $E\in\mathfrak{A}$ and non-positive if $\mu(P\cap E)\leqslant 0$ for all $E\in\mathfrak{A}$. There exists a non-negative set P for μ such that the complement set P^c of P in X is non-positive. The ordered pair (P,P^c) is called a Hahn decomposition of X for μ. The Hahn decomposition of X for μ is unique in the sense of that if (P,P^c) and (Q,Q^c) are both Hahn decompositions, then for every $E\in\mathfrak{A}$, we have $\mu(P\cap E)=\mu(Q\cap E)$ and $\mu(P^c\cap E)=\mu(Q^c\cap E)$. In terms of the Hahn decomposition (P,P^c) of X, we can produce the Jordan decomposition of μ, that is,

$$\mu=\mu^+-\mu^-$$

where μ^+ and μ^- are defined by

$$\mu^+(E)=\mu(E\cap P)\text{ and }\mu^-(E)=-\mu(E\cap P^c)$$

for all $E\in\mathfrak{A}$. Define $|\mu|=\mu^++\mu^-$ on $\mathfrak{A}$ by $|\mu|(E)=\mu^+(E)+\mu^-(E)$ for all $E\in\mathfrak{A}$, and then $|\mu|$, μ^+ and μ^- are well-defined measures on $(X,\mathfrak{A})$, which are called in turn the total variation of μ, the positive variation of μ and the negative variation of μ.

Now we take into account the case when X is an open subset of $\mathbb{R}^n$ (for $n=2$ we consider the complex plane $\mathbb{C}$.). Let $C_0(X)$ be the set of all bounded complex-valued continuous functions $f(x)$ on X such that for every positive number ε, there is a compact subset F of X satisfying for all $x\in X\cap F^c$, $|f(x)|<\varepsilon$. $C_c(X)$ is the subset of $C_0(X)$ consisting of functions $f(x)$ with compact support, that is, the closure of the set of points at which it does not vanish, denoted by $\mathrm{supp}f$, is compact. $C_0(X)^*$ is the dual space of $C_0(X)$, that is, the set of all bounded linear functionals from $C_0(X)$ into $\mathbb{C}$. Let $\mathscr{M}(X)$ be the set of all complex-valued regular Borel measures on X. For $T\in C_0(X)^*$, we write $\langle T,f\rangle$ for the value of T at $f\in C_0(X)$ and $\langle\mu,f\rangle=\int_X f\mathrm{d}\mu$

for $\mu \in \mathscr{M}(X)$. The Riesz Representation Theorem says that $T \in C_0(X)^*$ if and only if there exists a $\mu \in \mathscr{M}(X)$ such that for all $f \in C_0(X)$, $\langle T, f\rangle = \langle \mu, f\rangle$, and writing L_μ for T determined by μ, then the mapping Γ defined by $\Gamma(\mu) = L_\mu$ is a norm-preserving linear mapping of $\mathscr{M}(X)$ onto $C_0(X)^*$. Thus $C_0(X)^*$ and $\mathscr{M}(X)$ are isomorphic as Banach spaces.

Let $\mathscr{B}$ be a Banach space and $\mathscr{B}^*$ the dual space of $\mathscr{B}$. A sequence of elements $\{f_n\}$ in $\mathscr{B}^*$ is said to weakly* converge to $f \in \mathscr{B}^*$, write $w^*\text{-}\lim_{n\to\infty} f_n = f$, if for all $x \in \mathscr{B}$, we have $\lim_{n\to\infty} f_n(x) = f(x)$. The Alaoglu Theorem asserts that every closed bounded ball of $\mathscr{B}^*$ of a separable Banach space $\mathscr{B}$ is weakly* sequently compact, that is to say, every bounded sequence of $\mathscr{B}^*$ contains a weakly* convergent subsequence. This immediately deduces the following result.

Proposition 7.1.1. *Let $\{\mu_n\}$ be a sequence of elements in $\mathscr{M}(X)$. If $\{\mu_n\}$ is uniformly bounded, that is, for some positive constant M, $|\mu_n|(X) \leqslant M$ for all n, then there exist a subsequence of $\{\mu_n\}$ which weakly* converges to a $\mu \in \mathscr{M}(X)$.*

By $C^\infty(X)$ (resp. $C^m(X)$) we mean the family of all complex-valued functions which have continuous partial derivatives of all orders (resp. orders of 1 to m) and $C_c^\infty(X)$ is the subset of $C^\infty(X)$ consisting of functions $f(x)$ with compact support. $C_c^\infty(X)$ is a vector space and the members of $C_c^\infty(X)$ are usually called test functions.

For $f \in C^\infty(X)$, we write the partial derivatives as $\partial_j f = \partial f/\partial x_j$, $j = 1, 2, \cdots, n$ and when $X \subseteq \mathbb{C}$, we have

$$\partial f = \frac{\partial f}{\partial z} = \frac{1}{2}\left(\frac{\partial f}{\partial x} - \mathrm{i}\frac{\partial f}{\partial y}\right)$$

and

$$\bar{\partial} f = \frac{\partial f}{\partial \bar{z}} = \frac{1}{2}\left(\frac{\partial f}{\partial x} + \mathrm{i}\frac{\partial f}{\partial y}\right).$$

Thus the Laplacian $\Delta = 4\partial\bar{\partial}$. We consider derivatives of high order in terms of the multi-index. A multi-index is an n-tuple $\alpha = (\alpha_1, \alpha_2, \cdots, \alpha_n)$ of non-negative integers; its length is $|\alpha| = \alpha_1 + \cdots + \alpha_n$. In the natural way, define the addition of two multi-indices and

$$\partial^\alpha f = \frac{\partial^{|\alpha|} f}{\partial^{\alpha_1} x_1 \cdots \partial^{\alpha_n} x_n}.$$

A linear functional $L : C_c^\infty(X) \to \mathbb{C}$ is called a distribution if for every compact subset K of X there exists a $C \geqslant 0$ and a non-negative N such that

$$|\langle L, \phi\rangle| \leqslant C \sum_{|\alpha| \leqslant N} \sup |\partial^\alpha \phi|, \tag{7.1.1}$$

for all $\phi \in C_c^\infty(X)$ with $\operatorname{supp}\phi \subset K$. By $\mathscr{D}'(X)$ we denote the set of all distributions on X and then $\mathscr{D}'(X)$ is a vector space over $\mathbb{C}$ whose addition and scalar multiplication are defined as follows, for $I, J \in \mathscr{D}'(X)$ and $c \in \mathbb{C}$,

$$\langle I + J, \phi\rangle = \langle I, \phi\rangle + \langle J, \phi\rangle \text{ and } \langle cI, \phi\rangle = c\langle I, \phi\rangle,\ \phi \in C_c^\infty(X).$$

There is an equivalent definition of distributions. A linear functional $L : C_c^\infty(X) \to \mathbb{C}$ is a distribution if and only if it is sequential continuity in $C_c^\infty(X)$, that is, for any sequence $\{\phi_j\}$ in $C_c^\infty(X)$, if $\mathrm{supp}\phi_j (j = 1, 2, \cdots)$ are contained in a fixed compact subset of X and for each multi-index α, $\partial^\alpha \phi_j \to 0$ uniformly as $j \to \infty$, then $\langle L, \phi_j \rangle \to 0$ as $j \to \infty$.

There are two kinds of simple but important distributions.

(1) For each $\mu \in \mathscr{M}(X)$, the restriction of the linear functional on $C_0(X)$ determined by μ to $C_c^\infty(X)$ is a distribution on X ((7.1.1) is easily shown for this case), and hence sometimes below we consider the complex measure as a distribution, which won't be emphasized if no confusion occurs.

(2) We say a complex-valued function $f(x)$ on X to be locally integrable on X, denoted by $f \in \mathscr{L}^1_{\mathrm{loc}}(X)$, if $\int_K |f| dm < \infty$ for all compact sets $K \subset X$ where m is the Lebesgue measure on X. Then each $f \in \mathscr{L}^1_{\mathrm{loc}}(X)$ determines a distribution L_f defined by

$$\langle L_f, \phi \rangle = \int_X f\phi \; \mathrm{d}m, \; \phi \in C_c^\infty(X).$$

Obviously, if $f, g \in \mathscr{L}^1_{\mathrm{loc}}(X)$ with $f = g$ almost everywhere with respect to m, then $L_f = L_g$ and hence we often write f for the distribution L_f.

Now we introduce the notion of convergence of distributions. Let $\{L_j\}$ be a sequence of distributions on X. The sequence is said to converge in $\mathscr{D}'(X)$ to $L \in \mathscr{D}'(X)$, if

$$\lim_{j\to\infty} \langle L_j, \phi \rangle = \langle L, \phi \rangle, \; \forall \phi \in C_c^\infty(X).$$

Actually, it is a natural extension of the notion of weak* convergence.

The idea of distributions is remarkable. We can consider the derivatives of a distribution and therefore talk about the derivatives of a complex measure and a locally integrable function in the sense of distribution, which is a natural extension of the original derivatives of a smooth function, that is to say, for a smooth function, its derivative coincides with its derivative as a distribution. Let $L \in \mathscr{D}'(X)$. We define $\partial_i L \; (i = 1, 2, \cdots, n)$ by

$$\langle \partial_i L, \phi \rangle = -\langle L, \partial_i \phi \rangle, \; \phi \in C_c^\infty(X) \tag{7.1.2}$$

in terms of the Green formula for smooth functions. It is easily seen from a simple calculation that $\partial_i L$ is a distribution on X, and it is called the first order derivative of L. Then for any multi-index α, from (7.1.2), we have

$$\langle \partial^\alpha L, \phi \rangle = (-1)^{|\alpha|} \langle L, \partial^\alpha \phi \rangle, \; \phi \in C_c^\infty(X).$$

7.2 δ-Subharmonic Functions

In this section, we will first of all state some basic knowledge about subharmonic functions and then some results which themselves are important in this theory and will serve the main purpose of this chapter. The proofs of the basic results concerning subharmonic functions we collect below will be omitted when they can be easily found in some textbooks. The reader is referred to Conway's book [6], Hayman's book [17], Ransford's book [24] and Tsuji's book [26].

7.2.1 Basic Results Concerning δ-Subharmonic Functions

We take into account functions with range in $[-\infty,\infty)$ defined on a complex domain D of $\mathbb{C}$ and by $\mathscr{R}(D)$ denote the set of all such functions $u : D \to [-\infty,+\infty)$. There is a natural partial order "$\geqslant$" on $\mathscr{R}(D)$: $u \geqslant v$ means $u-v \geqslant 0$. Then for a subfamily $\mathscr{F}$ of $R(D)$, define $\vee_{u\in\mathscr{F}} u = \sup\{u : u \in \mathscr{F}\}$ and $\wedge_{u\in\mathscr{F}} u = \inf\{u : u \in \mathscr{F}\}$. It is obvious that $\wedge_{u\in\mathscr{F}} u$ is always in $\mathscr{R}(D)$, while $\vee_{u\in\mathscr{F}} u$ may not be. Thus $\wedge_{u\in\mathscr{F}} u$ is the greatest lower bound of $\mathscr{F}$ under the order and when $\vee_{u\in\mathscr{F}} u \in \mathscr{R}(D)$, it is the least upper bound of $\mathscr{F}$. For $u, v \in \mathscr{R}(D)$, we have $u \vee v \in \mathscr{R}(D)$ and $u \wedge v \in \mathscr{R}(D)$ and indeed, $u \vee v = \max\{u,v\}$ and $u \wedge v = \min\{u,v\}$.

A function $u \in \mathscr{R}(D)$ is upper semi-continuous (abbreviated usc), provided that for every c in $[-\infty,+\infty)$, the set $\{z : u(z) < c\}$ is an open subset of D. That u is usc is equivalent to that for each $z \in D$, $u(z) \geqslant \limsup_{\zeta\to z} u(\zeta)$. An important property of a usc function is that it is bounded above on any compact subset of its domain and assumes its smallest bound at some point.

A function u in $\mathscr{R}(D)$ is subharmonic on D if it is usc and satisfies the submean inequality, that is, for every closed disk $\overline{B}(a,r)$ contained in D, the inequality

$$u(a) \leqslant \frac{1}{2\pi}\int_0^{2\pi} u(a+r\mathrm{e}^{\mathrm{i}\theta})\mathrm{d}\theta \tag{7.2.1}$$

holds. For a usc function u on D, u is subharmonic if and only if u satisfies the Maximum Principle, that is to say, for every compact set K contained in D and every harmonic function h on K, $u \leqslant h$ on K whenever $u \leqslant h$ on ∂K. It is easily seen that a subharmonic function u on D will be a constant if it attains the global maximum on D at a point of D and that $u \leqslant 0$ on D if $\limsup_{\zeta\to z} u(\zeta) \leqslant 0$ for $z \in \partial D$. The subharmonic function is very flexible, which can be illustrated by the following result.

Lemma 7.2.1. *Let v be an usc function on D that is subharmonic on an open subset U of D. Then v is subharmonic on D, if $v \geqslant u$ on U and $v = u$ on $D\setminus U$ for a subharmonic function u on D.*

This brings us convenience to construct some subharmonic functions according to our requirement.

By $\mathscr{S}(D)$ we mean the set of all subharmonic functions on D. Then for $u(z) \in \mathscr{S}(D)$ with $u \not\equiv -\infty$, we have $u \in \mathscr{L}^1_{\mathrm{loc}}(D)$ and $\int_0^{2\pi} u(a+r\mathrm{e}^{\mathrm{i}\theta})\mathrm{d}\theta > -\infty$ for any $\overline{B}(a,r) \subset D$. For $u, v \in \mathscr{S}(D)$, it is easy to see that $u+v \in \mathscr{S}(D)$ and $u \vee v \in \mathscr{S}(D)$. Let $\mathscr{F}$ be a subfamily of $\mathscr{S}(D)$ which is locally bounded above. Define $u(z) = \limsup_{\zeta\to z} \bigvee_{v\in\mathscr{F}} v(\zeta)$ and then $u(z)$ is subharmonic on D. If $\bigvee_{v\in\mathscr{F}} v(z)$ is usc, then

$u(z) = \bigvee_{v \in \mathscr{F}} v(z)$. And $\psi \circ u$ is subharmonic for an increasing convex function ψ on $\mathbb{R}$ and $u \in \mathscr{S}(D)$.

Let $u(z)$ be a subharmonic function on the disk $B(0,R)$ with $u \not\equiv -\infty$. For $0 < r < R$, define

$$M_u(r) = \sup_{|z|=r} u(z),$$

$$J(r,u) = \frac{1}{2\pi} \int_0^{2\pi} u(re^{i\theta}) d\theta.$$

Then we have the following basic results:

(1) both $M_u(r)$ and $J(r,u)$ are increasing convex functions of $\log r$;

(2) $u(0) \leqslant J(r,u) \leqslant M_u(r)$;

(3) $\lim_{r \to 0^+} M_u(r) = \lim_{r \to 0^+} J(r,u) = u(0)$.

There is a close relation between subharmonic functions and Borel measures, which was revealed by F. Riesz. Since a subharmonic function $u(z)$ is locally integrable, it therefore determines a distribution and Δu is well defined in the sense of distribution, which is usually known as the generalized Laplacian of u. Indeed, Δu is a Radon measure, that is, a Borel measure such that for every compact subset K, $\Delta u(K) < \infty$. The Weyl Theorem asserts that for $u, v \in \mathscr{S}(D)$, if $\Delta u = \Delta v$, then $u = v + h$ where h is harmonic on D.

We shall write $\mathscr{M}_c^r(\mathbb{C})$ ($\mathscr{M}_c^+(\mathbb{C})$, resp.) for the set of all real-valued (positive, resp.) elements of $\mathscr{M}(\mathbb{C})$ with compact support on $\mathbb{C}$. Let $\mu \in \mathscr{M}_c^r(\mathbb{C})$. Define the logarithmic potential of μ to be the function

$$p_\mu(z) = \int \log|z-w| d\mu(w) \ (z \in \mathbb{C}).$$

Then $p_\mu(z)$ has the following properties:

(4) $p_\mu(z)$ is a locally integrable function and harmonic on the complement of $\operatorname{supp}\mu$ on $\mathbb{C}$;

(5) $p_\mu(z) = \mu(\mathbb{C}) \log|z| + O(|z|^{-1})$ as $z \to \infty$.

(6) $\Delta p_\mu = 2\pi\mu$ in the sense of distribution. Moreover, $p_\mu(z)$ is the unique solution of this equation, that is to say, if $u \in \mathscr{L}^1_{\text{loc}}(\mathbb{C})$ with $u(z) - \mu(\mathbb{C})\log|z| \to 0$ as $z \to \infty$ and such that $\Delta u = 2\pi\mu$, then $u = p_\mu$ a.e..

(7) If $\mu \in \mathscr{M}_c^+(\mathbb{C})$, then $p_\mu(z)$ is subharmonic on $\mathbb{C}$.

An extended real-valued function $u \in \mathscr{L}^1_{\text{loc}}(D)$ is a δ-subharmonic function if there exist two subharmonic functions u_1 and u_2 on D such that $u = u_1 - u_2$ a.e.. Then we have the Riesz Decomposition Theorem which says that, for arbitrary relatively compact open subset G of D, we can decompose u as

$$u = h + p_\mu \text{ on } G$$

where h is harmonic on G and $\mu \equiv \frac{1}{2\pi}\Delta u|_G$. Since $\Delta u = \Delta u_1 - \Delta u_2$, Δu therefore is a signed measure and it is called the Riesz charge of u, denoted by $\mu[u]$. Thus $\mu[u]^+ = \Delta u_1$ and $\mu[u]^- = \Delta u_2$. From here it follows that a δ-subharmonic function is subharmonic if and only if its Riesz charge is positive, in other words, its charge is a Borel measure, and it is equivalent to $\mu[u]^- = 0$.

A set E is a polar set if there is a non-constant subharmonic function u on $\mathbb{C}$ such that $E \subset \{z : u(z) = -\infty\}$. Then if u is a subharmonic function on a domain D with $u \not\equiv -\infty$, then $E = \{z \in D : u(z) = -\infty\}$ is a polar set. Let D be a domain on $\widehat{\mathbb{C}}$ and let $\phi : \partial D_\infty \to \mathbb{R}$ be an extended real-valued function. Define

$$\mathscr{P}(\phi, D) = \{u : u \text{ is subharmonic, bounded above on } D \text{ and } \limsup_{z \to a} u(z) \leqslant \phi(a) \text{ for every } a \text{ in } \partial_\infty D\}.$$

The associated Perron function $H_D\phi$ of ϕ is defined by

$$H_D\phi = \sup\{u : u \in \mathscr{P}(\phi, D)\}.$$

Then $H_D\phi$ is harmonic on D if it is not identically $\pm\infty$. Specially, if ϕ is bounded, then $H_D\phi$ is harmonic on D and

$$\sup_D |H_D\phi| \leqslant \sup_{\partial D} |\phi|.$$

Now let us introduce the harmonic measure. Let $\mathfrak{B}(\partial D)$ be the σ-algebra of Borel subsets of ∂D. A harmonic measure for D is a function $\omega_D : D \times \mathfrak{B}(\partial D) \to [0,1]$ such that

(1) for each fixed $z \in D$, $\omega_D(z, *)$ is a Borel probability measure on $\mathfrak{B}(\partial D)$;

(2) for each continuous function $\phi : \partial D \to \mathbb{R}$, we have

$$H_D\phi(z) = \int_{\partial D} \phi(\zeta) \mathrm{d}\omega_D(z, \zeta), \ z \in D.$$

The integral in the right side of the above equality is usually called the generalized Poisson integral of ϕ on D, denoted by $P_D\phi$. When ∂D is non-polar, a unique harmonic measure ω_D for D exists and for a fixed Borel subset α of ∂D, $\omega_D(z, \alpha)$ is harmonic and $0 \leqslant \omega_D \leqslant 1$ on D because $\omega_D(z, \alpha) = H_D\chi_\alpha(z)$. There exists following integral relation between the Green function and the harmonic measure for a bounded domain D:

$$G_D(z, w) = \int_{\partial D} \log|\zeta - w| \mathrm{d}\omega_D(z, \zeta) - \log|z - w|, \ z, w \in D.$$

A harmonic majorant of a subharmonic function u on a domain D is a harmonic function h on D such that $h \geqslant u$ there and h is called the least harmonic majorant of u if $h \leqslant H$ for every other harmonic majorant H of u.

The following is also known as the Riesz Decomposition Theorem.

Theorem 7.2.1. *Let D be a domain on $\mathbb{C}$ with non-polar ∂D and let u be a subharmonic function on D with $u \not\equiv -\infty$ and a harmonic majorant. Then*

$$u(z) = h(z) - \frac{1}{2\pi}\int_D G_D(z,w)\Delta u(w), z \in D,$$

where $h(z)$ is the least harmonic majorant of u.

In discussion of our center purpose of this chapter made in next section, we need the following results.

Lemma 7.2.2. *If $u_1 \geqslant u_2$ are two δ-subharmonic functions and $u_1(z) = u_2(z), z \in E$, for some Borel set E, then $\mu[u_1]|_E \leqslant \mu[u_2]|_E$.*

The lemma can be found in Grishin [16]. The next is main lemma of Eremenko [11] and also a modified version of the main lemma of Eremenko [10].

Lemma 7.2.3. *Let u_k $(k = 1, 2, \cdots, q)$ be non-negative subharmonic functions in a simply connected domain D with the Riesz measures μ_k $(k = 1, 2, \cdots, q)$ and have disjoint connected supports. Assume that*

$$\sum_{k=1}^{q} \mu_k \geqslant 2 \bigvee_{k=1}^{q} \mu_k. \tag{7.2.2}$$

Then there exist a Riemann surface Σ with a two-sheeted ramified covering $p : \Sigma \to D$ and a function h harmonic on Σ such that $u \circ p = |h|$, where $u = \sum_{k=1}^{q} u_k$. Furthermore, the covering p is ramified over at most $q-2$ points in D and each ramification point of p is a zero of h of order at least 3.

We know a subharmonic function is usc, but may not be continuous. Consideration of continuity of subharmonic functions leads to the introduction of another topology, the fine topology. This topology was suggested by H. Cartan in 1944 and has been found to have many pathologic properties and applications in the potential theory.

Definition 7.2.1. *The fine topology is the smallest topology on $\mathbb{C}$ under which each subharmonic function becomes a continuous function from $\mathbb{C}$ to $[-\infty, \infty)$. The fine topology will be denoted by $\mathfrak{F}$ and the usual topology by $\mathfrak{U}$.*

We will use terminologies such as finely open, finely closed and finely continuous, etc., when we consider topological phenomena relative to the fine topology. We will mean notations under the usual topology when we do not use the adjective "fine" before them. In what follows, we collect some basic properties related to the fine topology. The fine topology contains the usual topology, that is, $\mathfrak{U} \subset \mathfrak{F}$ and $\mathfrak{U}$ is a proper subset of $\mathfrak{F}$. The fine topology is a Hausdorff topology, which can follow from the following result. All sets with the form

$$W \cap \bigcap_{k=1}^{n} \{z : u_k(z) > c_k\}$$

for $W \in \mathfrak{U}$ and subharmonic functions u_k and constants c_k $(k = 1, 2, \cdots, n)$ composes a base for the fine topology. Any polar set has no fine limit points, all its points are finely isolated, and a finely compact set is finite. Thus it is not a locally compact topology. A fine domain, i.e., finely connected open set, is polygonally connected (see B. Fugleds [15]). If $D \in \mathfrak{F}$ and $z_0 \in D$, then the set of all $r > 0$ such that $\{z : |z - z_0| = r\}$ is contained in D has positive linear measure and thus a finely open set has positive area (see M. Brelot [5], Proposition IX.2 and Proposition IX.10). This implies that a finely open set has at most countable fine components. For a function u and a set E, define a function u_E on $\mathbb{C}$ as follows:

$$u_E(z) = u(z),\ z \in E \text{ and } u_E(z) = 0,\ z \in \mathbb{C} \setminus E.$$

Lemma 7.2.4. *Let $u(z)$ be a non-negative subharmonic function in a simply connected domain D on $\mathbb{C}$. Then*

$$u = \sum u_E,$$

where the sum is taken over all fine components E of $\{z \in D : u(z) > 0\}$. Furthermore, $u_E(z)$ is subharmonic on $\mathbb{C}$ and has disjoint support for different E's.

Proof. Let E be a fine component of $\{z : u(z) > 0\}$ and so E is a fine domain. We denote by E' the union of all discs whose boundaries are in E and then E' is open, indeed it is a domain as E is polygonally connected. It is clear that $E \subset E' \subset D$. Let F be other fine component of $\{z : u(z) > 0\}$ than E. Suppose that $F \cap E' \neq \varnothing$ and $z_0 \in F \cap E'$. Then for some $r > 0$ and some z_1, we have $z_0 \in \{z : |z - z_1| < r\} \subset E'$ and $\{z : |z - z_1| = r\} \subset E$. This together with $F \cap E = \varnothing$ implies that $F \subset \{z : |z - z_1| < r\} \subset D$, but from the Maximum Principle F cannot be relatively compact in D, otherwise, $u \equiv 0$ on F. Therefore $F \cap E' = \varnothing$. We have proved that $u_E(z) = u(z),\ z \in E'$ and $u_E(z) = 0,\ z \in \mathbb{C} \setminus E'$, that is, $u_E(z) = u_{E'}(z)$. Since 0 is subharmonic on $\mathbb{C}$, in terms of Lemma 7.2.1 it follows that u_E is subharmonic. □

Consider two δ-subharmonic functions u_1 and u_2 and then the set $\{z : u_1(z) > u_2(z)\}$ is finely open. The following result is often used in the sequel (see Doob [7]).

Lemma 7.2.5. *Let u_1 and u_2 be two δ-subharmonic functions. Assume that $u_1(z) = u_2(z)$ in some finely open set E. Then the restrictions of their Riesz charges to E coincide.*

Since an open set is also finely open, Lemma 7.2.5 holds for an open set and this strengthens the result of Lemma 7.2.2 for E being open.

7.2.2 Normality of Family of δ-Subharmonic Functions

Let D be a domain in $\mathbb{C}$. For a sequence of functions $\{f_n\}$ in $C^m(D)$ with $0 \leqslant m \leqslant \infty$ (here $C^0 = C$), we say that $\{f_n\}$ converges to $f \in C^m(D)$ in C^m_{loc} if for any compact subset K of D and any multi-index α with $|\alpha| \leqslant m$, we have $||\partial^\alpha f_n - \partial^\alpha f||_K = \max\{|\partial^\alpha f_n(z) - \partial^\alpha f(z)| : z \in K\} \to 0\ (n \to \infty)$. Actually, $||\partial^\alpha f_n - \partial^\alpha f||_K \to 0\ (n \to \infty)$ is equivalently that $\{\partial^\alpha f_n\}$ converges to $\partial^\alpha f$ in $C(K)$, i.e., uniformly on K with respect to the Euclidean metric. If each $f_n(z)$ is analytic or harmonic on D, then

$f_n(z)$ has derivatives of all orders on D and if $\{f_n\}$ converges in C_{loc}, then for any multi-index α, $\{\partial^\alpha f_n\}$ converges in C_{loc}.

When we put ∞ in our consideration, we use the spherical metric d_∞ instead of the Euclidean metric, that is to say, $d_\infty(f_n(z), f(z)) \to \infty$ as $n \to \infty$ uniformly on any compact subset of D, which we mean by saying that $\{f_n\}$ converges to f in $C_{\text{loc},s}$, and it is equivalent to $||f_n - f||_{K,s} = \sup\{d_\infty(f_n(z), f(z)) : z \in K\} \to 0$ $(n \to \infty)$ for any compact subset K of D. In the same way, we define that $\{f_n\}$ converges to a $f \in C^m(D)$ in $C^m_{\text{loc},s}$ and thus we can consider the uniform convergence of functions meromorphic on a domain D.

Definition 7.2.2. *A family $\mathscr{F}$ of functions in $C^m(D)$ for $0 \leqslant m \leqslant \infty$ is called normal in C^m (resp. in C^m_s) on a domain D if any sequence of its elements contains a subsequence which converges in C^m_{loc} (resp. in $C^m_{\text{loc},s}$). And we say $\mathscr{F}$ to be normal in C^m (resp. in C^m_s) at a point a if it is normal in C^m (resp. in C^m_s) at a neighborhood of a.*

For an analytic or meromorphic or harmonic function family $\mathfrak{F}$ on D, in the sequel whenever no confusion occurs we mean $\mathfrak{F}$ is normal in C or in C_s for short by simply saying it to be normal.

A large number of results about normality of analytic or meromorphic function family were established and a great number of connections between normality and value distribution were revealed. Some of them applies the harmonic function family. For example, a family of harmonic functions on D is normal if it is locally uniformly bounded. The Montel Theorem asserts that an analytic function family is normal on a domain D if its member does not take two finite fixed values on D. From this we easily get that a harmonic function family is normal on D if its member does not take a fixed finite value on D. And we can also prove the following version of Zalcman's Lemma for harmonic function family.

Theorem 7.2.2. *Let $\mathscr{F}$ be a family of harmonic functions on D. If $\mathscr{F}$ is not normal at $z_0 \in D$, then there exist a sequence of elements $\{h_n\} \subset \mathscr{F}$, a sequence of complex numbers $\{z_n\}$ in D and a sequence of positive numbers $\{\rho_n\}$ and a non-constant harmonic function h on $\mathbb{C}$ such that as $n \to \infty$, we have $z_n \to z_0$, $\rho_n \to 0$ and*

$$h_n(z_n + \rho_n z) \to h(z) \tag{7.2.3}$$

uniformly on any compact subset of $\mathbb{C}$.

Proof. We can find a harmonic function $v(z)$ with $v(z_0) = 0$ for each $u \in \mathscr{F}$ in a fixed neighborhood $U \subset D$ of z_0 such that $u + \mathrm{i}v$ is analytic on U. Then the family $\{u + \mathrm{i}v : u \in \mathscr{F}\}$ is not normal at $z_0 \in U$. Theorem 7.2.2 follows from the Zalcman's Lemma about analytic function family. □

However, the usual criterion of normality on families of harmonic or analytic functions may not be valid in ensuring normality of families of subharmonic functions. We do not know if Theorem 7.2.2 holds for subharmonic function family. However, we guess that it would be this case if we suitably and uniformly restrict in the local sense the total variation of the Riesz charges related to the family as Arsove did in [2].

In this section, we discuss a weaker version of normality than the usual normality mentioned above, that is, normality in the sense of $\mathscr{L}^1_{\mathrm{loc}}$, i.e., under the metric of $\mathscr{L}^1_{\mathrm{loc}}$ instead of the metric of C_{loc}. Let D be a Borel set on $\mathbb{R}$ or $\mathbb{R}^2$. We say that a sequence of functions $\{u_n\}$ in $\mathscr{L}^1_{\mathrm{loc}}(D)$ converges in $\mathscr{L}^1_{\mathrm{loc}}$ to a function $u \in \mathscr{L}^1_{\mathrm{loc}}(D)$ on D, if for any compact subset K of D, $\|u_n - u\|_{\mathscr{L}^1(K)} = \int_K |u_n(x) - u(x)| \mathrm{d}m(x) \to 0$ as $n \to \infty$ where m is the Lebesgue measure over D. $\{u_n\}$ in $\mathscr{L}^1_{\mathrm{loc}}(D)$ is said to converge to $u \in \mathscr{L}^1_{\mathrm{loc}}(D)$ weakly in $\mathscr{L}^1_{\mathrm{loc}}$ if for every locally bounded measurable function g on D, we have

$$\lim_{n\to\infty} \int_K u_n g \, \mathrm{d}m = \int_K ug \, \mathrm{d}m$$

for every compact subset of D. Obviously, if $\{u_n\}$ converges in $\mathscr{L}^1_{\mathrm{loc}}$, then it converges weakly in $\mathscr{L}^1_{\mathrm{loc}}$.

Definition 7.2.3. *Let $\mathscr{F}$ be a family of functions in $\mathscr{L}^1_{\mathrm{loc}}(D)$. $\mathscr{F}$ is called normal in $\mathscr{L}^1_{\mathrm{loc}}$ on D, if every sequence of elements in $\mathscr{F}$ contains a subsequence which converges on D in $\mathscr{L}^1_{\mathrm{loc}}$ to a function in $\mathscr{L}^1_{\mathrm{loc}}(D)$. And we say $\mathscr{F}$ to be normal in $\mathscr{L}^1_{\mathrm{loc}}$ at a point a if it is normal in $\mathscr{L}^1_{\mathrm{loc}}$ at a neighborhood of a.*

The following result is obvious.

Proposition 7.2.1. *Let $\mathscr{F}$ be a family of functions in $\mathscr{L}^1_{\mathrm{loc}}(D)$ for a domain D. $\mathscr{F}$ is normal in $\mathscr{L}^1_{\mathrm{loc}}$ on D if and only if it is normal in $\mathscr{L}^1_{\mathrm{loc}}$ at every point of D.*

If $h_n(z)$ is a sequence of harmonic functions on D and tends to a harmonic function h in $\mathscr{L}^1_{\mathrm{loc}}$, then $h_n \to h$ uniformly on any compact subset of D. This result is not always true when h_n and h are subharmonic functions, while in this case we have that for every compact subset K of D, $h_n \to h$ on K in the Lebesgue measure m, that is, for arbitrarily small $\varepsilon > 0$, we have

$$m(\{z \in K : |h_n(z) - h(z)| > \varepsilon\}) \to 0 \ (n \to \infty);$$

In view of Riesz Theorem (Theorem III.11.26 of [18]) and Egorov Theorem (Theorem III.11.32 of [18]), $\{h_n\}$ contains a subsequence which converges uniformly on K with respect to the Lebesgue measure m, that is, for any small $\varepsilon > 0$, uniformly on $K \setminus E$ for some set E with $m(E) < \varepsilon$. This is also true for the Carleson measure, which was proved by Azarin [3] (see below Theorem 7.2.4). Let E be a bounded subset of $\mathbb{C}$. For a fixed real number $\alpha > 0$, define

$$\alpha\_\mathrm{mes}E = \inf \sum r_j^{\alpha},$$

where the inf is taken over the disks which forms a covering of E and r_j's are the radius of these disks. Here α_mes is the α-Carleson measure. For a subset E of $\mathbb{C}$ which is allowed to be unbounded, define

$$\alpha\_\overline{\mathrm{mes}}(E) = \limsup_{r\to\infty} \{\alpha\_\mathrm{mes}(E \cap B(0,r))\} r^{-\alpha}.$$

It is obvious that when $\alpha_1 > \alpha_2$, we have $\alpha_1\_\overline{\text{mes}}(E) \leqslant \alpha_2\_\overline{\text{mes}}(E)$. A set E is called a C_0^α set if $\alpha\_\overline{\text{mes}}(E) = 0$ and a C_0^0 set if $\alpha\_\overline{\text{mes}}(E) = 0$ for all $\alpha > 0$. Obviously, a bounded set E is a C_0^0 set.

According to the Arzela-Ascoli Theorem, a family $\mathscr{F}$ is normal in $\mathscr{L}^1_{\text{loc}}$ on D, then it is locally bounded in $\mathscr{L}^1_{\text{loc}}$ on D. The main purpose of this section is to investigate possibility of the inverse of the result for a subharmonic function family.

The Riesz decomposition of subharmonic function leads us to begin our below discussion of normality of subharmonic function family with normality of family of the logarithmic potentials.

Lemma 7.2.6. *Let $\{\mu_n\}$ be a subsequence of elements in $\mathscr{M}_c(\mathbb{C})$ with $\text{supp}\mu_n \subset X$ for some fixed compact subset X of $\mathbb{C}$, which weakly* converges to a $\mu \in \mathscr{M}_c(\mathbb{C})$. Then*

(1) *for $0 < r < \infty$, we have*

$$\lim_{n\to\infty} \int_0^{2\pi} |p_{\mu_n}(re^{i\theta}) - p_\mu(re^{i\theta})| d\theta = 0; \tag{7.2.4}$$

(2) *p_{μ_n} converges to p_μ in $\mathscr{L}^1_{\text{loc}}$ on $\mathbb{C}$;*

(3) *for $\alpha > 0$, p_{μ_n} converges to p_μ in the α-Carleson measure and furthermore contains a subsequence which converges to p_μ uniformly with respect to the α-Carleson measure.*

Proof. To prove (1), we are given a $r \in (0,\infty)$. Define a linear operator $T : \mathscr{L}^\infty(0,2\pi) \to C(X)$ by the formula

$$(Tf)(z) = \int_0^{2\pi} \log|re^{i\theta} - z| f(\theta) d\theta, \quad z \in X.$$

Noting that $\sup_{z\in X} \int_0^{2\pi} |\log|re^{i\theta} - z|| d\theta < \infty$ and $\int_0^{2\pi} |\log|re^{i\theta} - z| - \log|re^{i\theta} - w|| d\theta \to 0$ as $|z-w| \to 0$, it is easy to see that

$$S(r) = \{Tf : f \in \mathscr{L}^\infty(0,2\pi), \|f\|_\infty \leqslant 1\}$$

is uniformly bounded and equicontinuous on X, and therefore $S(r)$ is relatively compact in $C(X)$. We have

$$\begin{aligned}\int_0^{2\pi} |p_{\mu_n}(re^{i\theta}) - p_\mu(re^{i\theta})| d\theta &= \sup_f \int_0^{2\pi} (p_{\mu_n}(re^{i\theta}) - p_\mu(re^{i\theta})) f(\theta) d\theta \\ &= \sup_f \int_X (Tf)(z) d(\mu_n(z) - \mu(z)),\end{aligned} \tag{7.2.5}$$

where the supremum is taken over all f on the unit ball of $\mathscr{L}^\infty(0,2\pi)$ and so Tf goes over $S(r)$. Since $\int_X (Tf)(z) d(\mu_n(z) - \mu(z)) \to 0$ as $n \to \infty$ for each $Tf \in S(r)$, by noting the relatively compactness of $S(r)$ we deduce that the final quantity in (7.2.5) will tend to zero as $n \to \infty$. In another word, we have shown that $\int_X (Tf)(z) d(\mu_n(z) -$

$\mu(z)) \to 0$ uniformly in f on the unit ball of $\mathscr{L}^\infty(0,2\pi)$ as $n \to \infty$ (The observations will be used in other form in the sequel.)

To prove (2), we consider sufficiently any closed disk on which $p_{\mu_n} \to p_\mu$ in $\mathscr{L}_{\text{loc}}$ and then the same argument as above yields our desired result (2).

To prove (3). Set

$$[\log x]_\varepsilon = \max\{\log x, \log \varepsilon\}$$

and for a point $z \in \mathbb{C}$

$$J_{1n}(z) = \int [\log|z-\zeta|]_\varepsilon \mathrm{d}(\mu - \mu_n)$$

and

$$J_{2n}(z) = \int_{B(z,\varepsilon)} \{\log|z-\zeta| - \log\varepsilon\}\mathrm{d}(\mu-\mu_n).$$

We write

$$\begin{aligned}\int \log|z-\zeta|\mathrm{d}\mu(\zeta) - \int \log|z-\zeta|\mathrm{d}\mu_n(\zeta) &= \int \log|z-\zeta|\mathrm{d}(\mu-\mu_n)(\zeta)\\ &= \int_{\mathbb{C}\setminus B(z,\varepsilon)} \log|z-\zeta|\mathrm{d}(\mu-\mu_n)(\zeta) + \int_{B(z,\varepsilon)} \log|z-\zeta|\mathrm{d}(\mu-\mu_n)(\zeta)\\ &= J_{1n}(z) + J_{2n}(z).\end{aligned}$$

Since $[\log|z-\zeta|]_\varepsilon$ is continuous in (z,ζ), it follows from the weak* convergence of μ_n to μ that $J_{1n}(z) \to 0$ uniformly on X as $n \to \infty$. Therefore, to complete our proof it suffices to prove that $J_{2n}(z)$ converges to zero in the α-Carleson measure. Set $\mu_{z,n}(t) = |\mu - \mu_n|(\{\zeta : |\zeta - z| < t\})$ for $0 < t < \varepsilon$, and for $\beta < \alpha$, $E_\beta^n = \{z : \mu_{z,n}(t) < \varepsilon^{-\beta}t^\alpha\}$. For $z \in E_\beta^n$ we have

$$\begin{aligned}0 \leqslant |J_{2n}(z)| &= -\int_0^\varepsilon [\log t - \log\varepsilon]\mathrm{d}\mu_{z,n}(t)\\ &= -\mu_{z,n}(t)\log\frac{t}{\varepsilon}\Big|_0^\varepsilon + \int_0^\varepsilon \mu_{z,n}(t)\frac{\mathrm{d}t}{t}\\ &\leqslant \varepsilon^{-\beta}\int_0^\varepsilon t^{\alpha-1}\mathrm{d}t = \frac{1}{\alpha}\varepsilon^{\alpha-\beta}.\end{aligned}$$

We need to estimate the α-Carleson measure of the complement $E_\beta^{n,c}$ of E_β^n. For $z \in E_\beta^{c,n}$ we have a t_z such that $\mu_{z,n}(t_z) > \varepsilon^{-\beta}t_z^\alpha$. The disks $B(z,t_z), \forall z \in E_\beta^{c,n}$ form a covering of $E_\beta^{n,c}$. In terms of Lemma 3.2 of Ahlfors and Landkof in Section 4 of Chapter III [21], from the covering we can extract a subcovering $\{B(z_j,t_{z_j})\}_{j=1}^N$ $(1 \leqslant N \leqslant \infty)$ with absolutely finite multiplicity ν. Then

$$
\begin{aligned}
\alpha\_\mathrm{mes}(E_\beta^{n,c}) &\leqslant \sum_{j=1}^{N} t_{z_j}^{\alpha} \leqslant \sum_{j=1}^{N} \varepsilon^\beta \mu_{z_j,n}(t_{z_j}) \\
&\leqslant \varepsilon^\beta \nu |\mu - \mu_n|(\cup_{j=1}^{N} B(z_j, t_{z_j})) \\
&\leqslant \nu\varepsilon^\beta |\mu - \mu_n|(X).
\end{aligned}
$$

Obviously, there exists a $M > 0$ such that $|\mu - \mu_n|(X) \leqslant M$ for $n \in \mathbb{N}$.

Given arbitrarily two small τ and δ, we choose ε so that $\alpha^{-1}\varepsilon^{\alpha-\beta} < \tau/2$ and $\nu\varepsilon^\beta M < \delta$. Set

$$W_{\tau,n} = \{z : |J_{1n}(z) + J_{2n}(z)| > \tau\}.$$

Take a $n_0 = n_0(\varepsilon)$ such that $|J_{1n}(z)| < \tau/2$ when $n > n_0$ and for $z \in W_{\tau,n}$, we have $|J_{2n}(z)| > \tau/2$ for $n > n_0$. This implies that $W_{\tau,n} \subset E_\beta^{n,c}$ and thus

$$\alpha\_\mathrm{mes}(W_{\tau,n}) \leqslant \alpha\_\mathrm{mes}(E_\beta^{n,c}) < \delta$$

so that $\alpha\_\mathrm{mes}(W_{\tau,n}) \to 0$ as $n \to \infty$.

Thus Lemma 7.2.6 follows. □

We remark that using a translation we can transfer the result (1) on the circle $\{z : |z| = r\}$ to a circle centered any point. Actually, $p_\mu(z+a) = p_\nu(z)$ for $\nu = \mu \circ T$ and $T(w) = w + a$. Result (1) in Lemma 7.2.6 was proved in [1] and Result (3) in [3]. The following result immediately follows from Lemma 7.2.6 and Proposition 7.1.1.

Theorem 7.2.3. *Let $\mathscr{A}$ be a subfamily of $\mathscr{M}(X)$ for a compact subset X of $\mathbb{C}$. If $\mathscr{A}$ is uniformly bounded, then $\mathscr{P} = \{p_\mu : \mu \in \mathscr{A}\}$ is normal in $\mathscr{L}^1_{\mathrm{loc}}$ on $\mathbb{C}$ and every sequence in $\mathscr{A}$ contains a subsequence $\{\mu_n\}$ such that for some $\mu \in \mathscr{M}(X)$, the results in Lemma 7.2.6 hold.*

The following is Theorem 4.4.1 of Azarin [3], which establish a relation between the convergency in $\mathscr{L}^1_{\mathrm{loc}}$ and in α-Carleson measure.

Theorem 7.2.4. *Let u_n be a sequence of subharmonic functions on D. Assume that $u_n \to u$ in $\mathscr{D}'(D)$. Then on each compact subset of D, for arbitrary $\alpha > 0$, $u_n \to u$ in the α-Carleson measure and furthermore contains a subsequence which converges to p_μ uniformly with respect to the α-Carleson measure.*

Proof. Given arbitrarily a closed disk K of D, $\mu_n = \frac{1}{2\pi}\Delta u_n|_K \to \frac{1}{2\pi}\Delta u|_K = \mu$ in $\mathscr{D}'(K)$, equivalently week* convergence. We have the Rieze decomposition

$$u_n(z) = h_n(z) + p_{\mu_n}(z), \; u(z) = h(z) + p_\mu(z), \; z \in K,$$

where $h_n(z)$ and $h(z)$ are harmonic in K. In terms of the result (2) of Theorem 7.2.6, h_n converges in $\mathscr{D}'(K)$ to h and then this implies uniform convergence in any disk $K_0 \Subset K$. This together with the result (3) of Theorem 7.2.6 yields that $u_n \to u$ uniformly on K with respect to the α-Carleson measure. □

We have the following corollary of Theorem 7.2.4 by noticing that the convergence in $\mathscr{L}^1_{\text{loc}}$ implies that in $\mathscr{D}'$, which has be used in the above proof.

Corollary 7.2.1. *Let u_n and D be given as in above. Assume instead that $u_n \to u$ in $\mathscr{L}^1_{\text{loc}}$ on D. Then the result of Theorem 7.2.4 holds.*

Below we establish a basic criterion of normality of subharmonic function family in the sense of $\mathscr{L}^1_{\text{loc}}$. To the end, we need the following

Lemma 7.2.7. *Let $u(z)$ be a subharmonic function on the disk $\overline{B}(0,R)$ with $u(0) \not\equiv -\infty$. Then we have*

$$\mu(\overline{B}(0,r)) \leqslant \frac{63(R+r)}{R(R-r)^2}\|u\|_{\mathscr{L}^1(\overline{B}(0,R))} \tag{7.2.6}$$

and

$$\mu(\overline{B}(0,r)) \leqslant \frac{3(R+r)}{R-r}(J(R,u)-u(0)), \tag{7.2.7}$$

where $\mu = \frac{1}{2\pi}\Delta u$.

Proof. First of all we estimate $J(r,|u|)$ in terms of $\|u\|_{\mathscr{L}^1(\overline{B}(0,R))}$. It is easily seen that

$$\begin{aligned}\|u\|_{\mathscr{L}^1(\overline{B}(0,R))} &= \int_{\overline{B}(0,R)} |u(z)|\mathrm{d}m(z)\\ &= \int_0^{2\pi}\int_0^R |u(te^{\mathrm{i}\theta})|t\mathrm{d}t\mathrm{d}\theta\\ &= \int_0^R tJ(t,|u|)\mathrm{d}t.\end{aligned}$$

Since $J(t,|u|)$ is continuous in t, we have for some $r \leqslant r_1 < r+\frac{1}{3}(R-r)$

$$\begin{aligned}\int_r^{r+\frac{1}{3}(R-r)} tJ(t,|u|)\mathrm{d}t &= J(r_1,|u|)\frac{1}{2}\left[\left(r+\frac{1}{3}(R-r)\right)^2 - r^2\right]\\ &> \frac{R(R-r)}{18}J(r_1,|u|)\end{aligned}$$

and for some $r+\frac{1}{2}(R-r) \leqslant r_2 \leqslant R$

$$\begin{aligned}\int_{r+\frac{1}{2}(R-r)}^R tJ(t,|u|)\mathrm{d}t &= J(r_2,|u|)\frac{1}{2}\left[R^2 - \left(r+\frac{1}{2}(R-r)\right)^2\right]\\ &> \frac{3R(R-r)}{8}J(r_2,|u|).\end{aligned}$$

Thus

$$J(r_2,|u|)+J(r_1,|u|) \leqslant \frac{21}{R(R-r)}\|u\|_{\mathscr{L}^1(\overline{B}(0,R))}.$$

In terms of the Poisson-Jensen formula (7.2.9) for subharmonic functions, we have

$$J(r,u) = u(0) + \int_0^r \log\frac{r}{t}\, \mathrm{d}\mu(\overline{D}_t)$$

where $\overline{D}_t = B(0,t)$ and so

$$\begin{aligned} J(r_2,u) - J(r_1,u) &= \mu(\overline{D}_{r_2})\log r_2 - \mu(\overline{D}_{r_1})\log r_1 + \int_{r_1}^{r_2} \log\frac{1}{t}\mathrm{d}\mu(\overline{D}_t) \\ &= \int_{r_1}^{r_2} \frac{\mu(\overline{D}_t)}{t}\, \mathrm{d}t \\ &\geqslant \mu(\overline{D}_{r_1})\log\frac{r_2}{r_1} \\ &\geqslant \mu(\overline{D}_r)\log\frac{3(R+r)}{2(R+2r)} \\ &\geqslant \mu(\overline{D}_r)\frac{R-r}{3(R+r)}. \end{aligned}$$

Combining the above inequalities implies (7.2.6) and by noting $u(0) \leqslant J(r_1,u) \leqslant J(r_2,u) \leqslant J(R,u)$, the final inequality yields (7.2.7). □

Lemma 7.2.8. *Let μ be a Radon measure on $\mathbb{C}$. Then for any compact subset K, we have*

$$\frac{1}{2\pi}\int_0^{2\pi} |p_\mu(re^{\mathrm{i}\theta};K)|\mathrm{d}\theta \leqslant C_K(r)\mu(K),$$

where

$$p_\mu(z;K) = \int_K \log|z-w|\mathrm{d}\mu(w) = p_{\mu_K}(z),$$

$$C_K(r) = \sup_{w\in K}\frac{1}{2\pi}\int_0^{2\pi} |\log|re^{\mathrm{i}\theta} - w||\mathrm{d}\theta$$

and μ_K is the restriction of μ to K. Furthermore, we have

$$\|p_{\mu_K}\|_{\mathscr{L}^1(B(0,R))} \leqslant \mu(K)\int_0^R rC_K(r)\mathrm{d}r.$$

Proof. Lemma 7.2.8 follows from the following implication:

$$\begin{aligned} \frac{1}{2\pi}\int_0^{2\pi} |p_\mu(re^{\mathrm{i}\theta})|\mathrm{d}\theta &= \frac{1}{2\pi}\int_0^{2\pi} |\int_K \log|z-w|\mathrm{d}\mu(w)|\mathrm{d}\theta \\ &\leqslant \int_K \mathrm{d}\mu(w)\frac{1}{2\pi}\int_0^{2\pi} |\log|re^{\mathrm{i}\theta} - w||\mathrm{d}\theta \\ &\leqslant \mu(K)\sup_{w\in K}\frac{1}{2\pi}\int_0^{2\pi} |\log|re^{\mathrm{i}\theta} - w||\mathrm{d}\theta \end{aligned}$$

with

$$\begin{aligned}C_K(r) &= \sup_{w\in K}\frac{1}{2\pi}\int_0^{2\pi}|\log|re^{i\theta}-w||d\theta\\ &= \sup_{w\in K}\left\{m(r,z-w)+m\left(r,\frac{1}{z-w}\right)\right\}\\ &\leqslant \sup_{w\in K}\left\{\log^+(r+|w|)+\frac{1}{2}\log^+\frac{1}{r^2+|w|^2}\right\}-\frac{1}{\pi}\int_0^{\pi/2}\log(1-\cos\theta)d\theta,\end{aligned}$$

by noting that $|re^{i\theta}-w|^2 = r^2+|w|^2-2r|w|\cos(\arg w-\theta)\geqslant (r^2+|w|^2)(1-|\cos(\arg w-\theta)|)$. □

Under previous preliminary discussion, now we give out a sufficient condition of a subharmonic function family being normal in $\mathscr{L}^1_{\rm loc}$.

Theorem 7.2.5. *Let $\mathscr{F}$ be a family of δ-subharmonic functions on D. Assume that $\mathscr{F}$ in $\mathscr{L}^1_{\rm loc}(D)$ and $\mathscr{M}(\mathscr{F})^- = \{\mu[u]^- : u\in\mathscr{F}\}$ on D are locally uniformly bounded from above, that is to say, for every compact subset K of D, there exists a positive number $M(K)$ such that for each $u\in\mathscr{F}$, we have*

$$\|u\|_{\mathscr{L}^1(K)}\leqslant M(K) \text{ and } (\mu[u])^-(K)\leqslant M(K).$$

Then $\mathscr{F}$ is normal on D in $\mathscr{L}^1_{\rm loc}$.

Proof. In terms of Proposition 7.2.1, it suffices to establish the result of Theorem 7.2.5 on the closed disk $\overline{B}(a,R)$ with $\overline{B}(a,2R)\subset D$. And we can assume $a=0$ for simplicity.

First of all we prove that $\mu[u]^+$ for $u\in\mathscr{F}$ is uniformly bounded from above on $\overline{B}(0,R)$. According to the Riesz Decomposition of subharmonic functions, each $u\in\mathscr{F}$ can be written into

$$u = u_1 - p_\nu$$

where u_1 is subharmonic on $B(0,R)$ and $\nu = \mu[u]^-|_{B(0,R)}$. Applying Lemma 7.2.8 yields that

$$\|u_1\|_{\mathscr{L}^1(\overline{B}(0,2R))}\leqslant \|u\|_{\mathscr{L}^1(\overline{B}(0,2R))}+\|p_\nu\|_{\mathscr{L}^1(\overline{B}(0,2R))}\leqslant M,$$

where M is a positive constant only depending on R. From Lemma 7.2.7 it follows that $\mu[u]^+(B(0,R)) = \mu[u_1](B(0,R))\leqslant 95M$. Now we write $u = h+p_\kappa - p_\nu$ for some harmonic function h on $B(0,R)$ and here $\kappa = \mu[u]^+|_{B(0,R)}$. The previous argument also deduces that $\|h\|_{\mathscr{L}^1(\overline{B}(0,R))}$ is uniformly bounded. Then the family consisting of h is normal in $\mathscr{C}$. Theorem 7.2.5 follows from Theorem 7.2.3. □

We consider the case of the whole complex plane.

Theorem 7.2.6. *Let $A(r)$ be a positive real-valued function on $(0,\infty)$ and $\mathscr{F}$ be a family of δ-subharmonic functions on $\mathbb{C}$. Assume that for all $r\in(0,\infty)$, we have*

$$\|u\|_{\mathscr{L}^1(B(0,r))}+(\mu[u])^-(B(0,r))\leqslant A(r). \tag{7.2.8}$$

Then (1) $\mathscr{F}$ *is normal on* $\mathbb{C}$ *in* $\mathscr{L}^1_{\mathrm{loc}}$; (2) *for each sequence* $\{u_n\}$ *in* $\mathscr{F}$, *there exist a* δ*-subharmonic function u on* $\mathbb{C}$ *and a subsequence* $\{u_{n_k}\}$ *of* $\{u_n\}$ *such that for all* $r \in (0,\infty)$

$$\lim_{k\to\infty} J(r,|u_{n_k}-u|) = 0 \text{ and } \lim_{k\to\infty} J(r,u_{n_k}) = J(r,u).$$

Result (2) in Theorem 7.2.6 is essentially Theorem 4 of Anderson and Baernstein II [1]. Their assumption is $J(r,|u|)+(\mu[u])^-(B(0,r)) \leqslant A(r)$ instead of (7.2.8), while this inequality implies (7.2.8).

Proof. Under the assumption (7.2.8), applying Theorem 7.2.5 we conclude the result (1) in Theorem 7.2.6.

Below we prove the result (2). As in the proof of Theorem 7.2.5, we have $|\mu|[u_n]$ is locally uniformly bounded. Take a sequence of positive numbers $\{r_m\}$ such that $r_m < r_{m+1} \to +\infty$ as $m\to\infty$. Then the usual diagonal argument implies the existence of subsequence of $\{\mu[u_n]\}$, which we still denote by $\{\mu[u_n]\}$, such that as $n\to\infty$, $\mu[u_n]^{\pm} \to \mu_m^{\pm}$ weak* respectively on each $B(0,r_m)$. Set $\kappa_{n,m} = \mu[u_n]^+|_{B(0,r_m)}$ and $\nu_{n,m} = \mu[u_n]^-|_{B(0,r_m)}$. We can write

$$u_n = h_{n,m} + p_{\kappa_{n,m}} - p_{\nu_{n,m}},$$

where $h_{n,m}$ is harmonic on $B(0,r_m)$. Obviously, in view of Lemma 7.2.8 and (7.2.8), $\{h_{n,m}\}$ is uniformly bounded from above in $\mathscr{L}^1$ on $B(0,r_m)$ and in view of Theorem 7.2.5, $\{h_{n,m}\}$ contains a subsequence which converges uniformly on $\overline{B}(0,r_{m-1})$.

Thus for $m=2$, we extract a subsequence from $\{h_{n,2}\}$ which converges to a harmonic function h_2 uniformly on $B(0,r_1)$. Let $\{u_n^{(2)}\}$ be the corresponding subsequence of u_n and set $u^{(2)} = h_2 + p_{\mu_2}$. In view of Lemma 7.2.6, for $r\in(0,r_1]$ we have $J(r,|u_n^{(2)}-u^{(2)}|)\to 0$ as $n\to\infty$. We can extract a subsequence from $\{u_n^{(2)}\}$, denoted by $\{u_n^{(3)}\}$ such that for a harmonic function h_3 on $B(0,r_3)$, for $r\in(0,r_2]$ we have $J(r,|u_n^{(3)}-u^{(3)}|)\to 0$ as $n\to\infty$, where $u^{(3)} = h_3 + p_{\mu_3}$. Since $u^{(2)}$ and $u^{(3)}$ are subharmonic on $B(0,r_1)$, we have $u^{(2)}(z) = u^{(3)}(z)$ on $B(0,r_1)$. Thus we extract successive convergent subsequences on each $B(0,r_m)$ and the limit functions coincide on their common domains and hence produce uniquely a function u subharmonic on $\mathbb{C}$. Now by the usual diagonal argument we have a subsequence of $\{u_n\}$ such that the result (2) of Theorem 7.2.6 holds for the u.

We have completed the proof of Theorem 7.2.6. □

The final is a version of the Zalcman Lemma for normality of subharmonic function family in the sense of $\mathscr{L}^1_{\mathrm{loc}}$.

Theorem 7.2.7. *Let* $\mathscr{F}$ *be a family of subharmonic functions on D and* $\{\mu[u] : u\in\mathscr{F}\}$ *locally uniformly bounded at a neighborhood of* $z_0 \in D$. *Then the result of Theorem 7.2.2 holds with (7.2.3) in* $\mathscr{L}^1_{\mathrm{loc}}$.

7.2.3 The Nevanlinna Theory of δ-Subharmonic Functions

Let us start from the equality in Theorem 7.2.1. When D is a bounded regular domain and u is subharmonic on a neighborhood of $\overline{D}$, there we have $h(z) = P_D u(z)$ and thus we attain the Poisson-Jensen Formula for subharmonic functions

$$u(z) = \int_{\partial D} u(\zeta)\mathrm{d}\omega_D(z,\zeta) - \frac{1}{2\pi}\int_D G_D(z,w)\Delta u(w),\ z \in D, \tag{7.2.9}$$

which is a generalization of Theorem 2.1.1. This formula is true for a δ-subharmonic function u.

For a δ-subharmonic function u on a neighborhood of the closure of a bounded regular domain D and a point $a \in D$, define

$$m(D,a,u) = \int_{\partial D} u^+(\zeta)\mathrm{d}\omega_D(a,\zeta),$$

$$N(D,a,u) = \frac{1}{2\pi}\int_D G_D(a,w)(\Delta u)^-(w)$$

and

$$T(D,a,u) = m(D,a,u) + N(D,a,u).$$

It follows from (7.2.9) that

$$T(D,a,u) = T(D,a,-u) + u(a). \tag{7.2.10}$$

When $D = B(0,r)$, we write $m(r,u)$ for $m(D,0,u)$, $N(r,u)$ for $N(D,0,u)$ and $T(r,u)$ for $T(D,0,u)$.

For $z \in D$ with $u(z) \neq \infty$, we have an analogy of (2.1.28):

$$u(z) \leqslant m(D,z,u) + N(D,z,u).$$

If D is a finitely connected Jordan domain and $\Gamma = \partial D$ consists of analytic curves, then $\mathrm{d}\omega_D(z,\zeta) = \frac{1}{2\pi}\frac{\partial G_D}{\partial n}(z,\zeta)\mathrm{d}s$. Therefore, when $u(z)$ is subharmonic and $D = B(0,R)$, we have

$$u(z) \leqslant m(D,z,u) = \frac{1}{2\pi}\int_0^{2\pi} \frac{R^2 - |z|^2}{|Re^{\mathrm{i}\theta} - z|^2} u^+(Re^{\mathrm{i}\theta})\mathrm{d}\theta \leqslant \frac{R+r}{R-r} m(R,u).$$

This shows that

$$u(z) \leqslant \frac{R+r}{R-r} T(R,u). \tag{7.2.11}$$

For a δ-subharmonic function u, in view of (7.2.9), we have

$$u(a) = \frac{1}{2\pi}\int_{\partial D} u(a + re^{\mathrm{i}\theta})\mathrm{d}\theta - \frac{1}{2\pi}\int_D \log\frac{r}{|w-a|}\Delta u(w), \tag{7.2.12}$$

where $D = \{z : |z-a| < r\}$ and specially, we have

$$N(r,u) = J(r,u_2) - u_2(0).$$

Noting $G_D(0,w) = \log \frac{r}{|w|}$ for $D = B(0,r)$, we have

$$N(r,u) = \int_D \log \frac{r}{|w|} d\mu^-(w)$$

where $\mu = \frac{1}{2\pi}\Delta u$. In terms of the formula for integration by parts, setting $n(t,u) = \mu^-(B(0,t))$, we have

$$\begin{aligned} N(r,u) &= \int_0^r \log \frac{r}{t} dn(t,u) \\ &= \lim_{\varepsilon\to 0}\left(-n(\varepsilon,u)\log\frac{r}{\varepsilon} + \int_\varepsilon^r \frac{n(t,u)}{t}dt\right) \\ &= \lim_{\varepsilon\to 0}\int_\varepsilon^r \frac{n(t,u)-n(\varepsilon,u)}{t}dt. \end{aligned}$$

Since for arbitrary fixed $r_0 \in (\varepsilon, r)$,

$$\int_{r_0}^r \frac{n(t,u)-n(\varepsilon,u)}{t}dt \leqslant \int_\varepsilon^r \frac{n(t,u)-n(\varepsilon,u)}{t}dt \leqslant \int_0^r \frac{n(t,u)-n(0,u)}{t}dt,$$

the Lebesgue Theorem implies that

$$\int_{r_0}^r \frac{n(t,u)-n(0,u)}{t}dt \leqslant N(r,u) \leqslant \int_0^r \frac{n(t,u)-n(0,u)}{t}dt,$$

where $n(0,u) = \lim_{\varepsilon\to 0} n(\varepsilon,u)$. Thus we attain

$$N(r,u) = \int_0^r \frac{n(t,u)-n(0,u)}{t}dt. \tag{7.2.13}$$

If $u(0) \neq +\infty$, then μ^- has no mass at 0, and write $u = u_1 - u_2$ with $u_2(0) \neq -\infty$. It is easy to see that $N(r,u) = N(r,u_2)$ and $n(0,u) = n(0,u_2) = 0$.

$T(r,u)$ and $N(r,u)$ is a non-negative increasing function in r and $T(r,u) \to \infty$ as $r \to \infty$ if u is not a constant. Define the order and lower order of u by those of $T(r,u)$ and

$$\delta(\infty,u) = \liminf_{r\to\infty} \frac{m(r,u)}{T(r,u)} = 1 - \limsup_{r\to\infty} \frac{N(r,u)}{T(r,u)}$$

and $\delta(0,u) = \delta(\infty,-u)$.

The following is a modified format of Theorem 5 of Anderson and Baernstein [1].

Theorem 7.2.8. *Let $u = u_1 - u_2$ be a δ-subharmonic function on $\mathbb{C}$. Assume that $\{r_n\}$, $\{\tau_n\}$ and $\{\varepsilon_n\}$ are three sequences of positive numbers such that $r_n \to \infty$, $\tau_n \to 0$ and $\varepsilon_n \to 0$ and for $\eta > 0$ and $\tau > 0$, we have*

$$T(r,u) \leqslant (1+\tau_n)\left(\frac{r}{r_n}\right)^{\alpha} T(r_n,u), \ \varepsilon_n r_n < r < \eta r_n; \tag{7.2.14}$$

$$T(r,u) \leqslant (1+\tau_n)\left(\frac{r}{r_n}\right)^{\beta} T(r_n,u), \ \tau r_n < r < \varepsilon_n^{-1} r_n. \tag{7.2.15}$$

Set

$$v_n(z) = \frac{u(zr_n)}{T(r_n,u)}.$$

Then there exist a subsequence of $\{r_n\}$, which is denoted by the same notation, such that for a δ-subharmonic function $\widehat{u} = \widehat{u}_1 - \widehat{u}_2$ on $\mathbb{C}$ with $\widehat{u}_2(0) = 0$, we have

(1) $v_n(z) \to \widehat{u}$ in $\mathscr{L}^1_{\mathrm{loc}}$ as $n \to \infty$;

(2) for each $r \in (0,\infty)$,

$$\lim_{n\to\infty} \int_0^{2\pi} \left| v_n(re^{i\theta}) - \widehat{u}(re^{i\theta}) \right| d\theta = 0;$$

(3) $T(r,\widehat{u}) \leqslant r^{\alpha}$, for $0 < r < \eta$ and $T(r,\widehat{u}) \leqslant r^{\beta}$, for $\tau < r < \infty$;

(4) if, in addition, $\widehat{u}$ is subharmonic, then for some $c > 0$, we have $\widehat{u}(z) \leqslant c|z|^{\alpha}$ for $0 < |z| < \eta/2$ and $\widehat{u}(z) \leqslant c|z|^{\beta}$ for $\tau/2 < |z| < \infty$.

(5) $T(1,\widehat{u}) = 1$;

(6) $N(r,\widehat{u}) \leqslant (1-\delta)T(r,\widehat{u})$ for $0 < r < \infty$, where $\delta = \delta(\infty,u)$.

Proof. We write $v_n(z) = v_n^{(1)}(z) - v_n^{(2)}(z)$, where $v_n^{(1)}(z) = u_1(zr_n)/T(r_n,u)$ and $v_n^{(2)}(z) = u_2(zr_n)/T(r_n,u)$. We want to prove that $\{v_n(z)\}$ satisfies the assumption of Theorem 7.2.6.

Take a fixed $r \in (0,\infty)$. Since $rr_n \to \infty$, it is obvious that we may assume u_1 and u_2 are harmonic in $|z| < 1$ and $u_1(0) \geqslant 0$ and $u_2(0) \geqslant 0$, otherwise we replace u_1 and u_2 with their Poisson integrals along $|z| = 1$ and add suitable constants. Thus $J(r,u_i) \geqslant u_i(0) \geqslant 0 (i = 1,2)$ and in terms of (7.2.12), we have

$$N(r,-u) = J(r,u_1) - u_1(0)$$

and further in terms of (7.2.10), we have

$$\begin{aligned} J(r,|u|) &= m(r,u) + m(r,-u) \\ &= m(r,u) + T(r,u) - N(r,-u) - u(0) \\ &= m(r,u) + T(r,u) - J(r,u_1) + u_2(0) \\ &\leqslant 2T(r,u) + u_2(0), \ \forall\, r \in (0,\infty). \end{aligned}$$

In terms of (7.2.15), for all large n, we have

$$J(r,|v_n|) = \frac{J(rr_n,|u|)}{T(r_n,u)} \leqslant \frac{2T(rr_n,u)+u_2(0)}{T(r_n,u)}$$
$$\leqslant \frac{2T((r+\tau)r_n,u)+u_2(0)}{T(r_n,u)}$$
$$\leqslant 4(r+\tau)^{\beta}.$$

Thus $\sup_n J(r,|v_n|) < +\infty$. Similarly, we have

$$J(r,v_n^{(2)}) = \frac{J(rr_n,u_2)}{T(r_n,u)} \leqslant \frac{T(rr_n,u)+u_2(0)}{T(r_n,u)} \leqslant 2(r+\tau)^{\beta}$$

and so $\sup_n J(r,v_n^{(2)}) < +\infty$. Set $A(r) = \sup_n J(r,|v_n|) + \sup_n J(r,v_n^{(2)}) < +\infty$. In terms of (7.2.7), we have shown that $\{v_n(z)\}$ satisfies the assumption of Theorem 7.2.6.

In terms of Theorem 7.2.6, for a subsequence of $\{r_n\}$, there exists a δ-subharmonic function $\widehat{u} = \widehat{u}_1 - \widehat{u}_2$ on $\mathbb{C}$ with $\widehat{u}_2(0) = 0$ such that (1) and (2) hold and in view of (7.2.12)

$$\lim_{n\to\infty} N(r,v_n) = \lim_{n\to\infty} (J(r,v_n^{(2)}) - v_n^{(2)}(0)) = J(r,\widehat{u}_2) = N(r,\widehat{u}).$$

Noting $|v_n^+ - \widehat{u}^+| \leqslant |v_n - \widehat{u}|$, from (2) it follows that

$$m(r,v_n) = J(r,v_n^+) \to J(r,\widehat{u}^+) = m(r,\widehat{u})$$

as $n \to \infty$ and therefore $T(r,v_n) \to T(r,\widehat{u})$ as $n \to \infty$.

In terms of (7.2.14) and (7.2.15), we have

$$T(r,v_n) = \frac{T(rr_n,u)}{T(r_n,u)} \leqslant (1+\tau_n)r^{\alpha}, \text{ for } \varepsilon_n < r < \eta,$$

$$T(r,v_n) = \frac{T(rr_n,u)}{T(r_n,u)} \leqslant (1+\tau_n)r^{\beta}, \text{ for } \tau < r < \varepsilon_n^{-1},$$

and $T(1,v_n) = 1$. This immediately implies that (3) and (5) hold. Since

$$N(r,v_n) = J(r,v_n^{(2)}) - v_n^{(2)}(0) = \frac{J(rr_n,u_2) - u_2(0)}{T(r_n,u)}$$
$$= \frac{N(rr_n,u)}{T(r_n,u)} \leqslant \frac{(1-\delta+o(1))T(rr_n,u)}{T(r_n,u)}$$
$$= (1-\delta+o(1))T(r,v_n),$$

we immediately deduce $N(r,\widehat{u}) \leqslant (1-\delta)T(r,\widehat{u})$ and (6) follows.

Now assume that $\widehat{u}(z)$ is subharmonic. In terms of (7.2.11) and (3) which we have proved, we have

$$\widehat{u}(z) \leqslant 3T(2r,\widehat{u}) \leqslant 3 \times 2^{\beta}|z|^{\beta}, \ \forall\ |z| > \tau/2$$

and $\widehat{u}(z) \leqslant 3 \times 2^{\alpha}|z|^{\alpha}$, $\forall\ |z| < \eta/2$ and hence (4) follows.

We have completed the proof of Theorem 7.2.8. □

For a δ-subharmonic function u of the finite order, we consider its type function $U(r)$ replacing its characteristic $T(r,u)$ in Theorem 7.2.8 as V. S. Azarin [3] did. That is to consider the function

$$u_r(z) = u(rz)r^{-\lambda(r)},$$

where $\lambda(r)$ is a proximate order of u. Noting that for all sufficiently large r, $T(r,u) \leqslant 2U(r)$ and for any fixed $d > 0$, $U(dr) = (1+o(1))U(r)$, by the same argument as in the proof of Theorem 7.2.8, then we can verify that for an arbitrary sequence $r_n \to \infty$, the results (1), (2) of Theorem 7.2.8 hold and $T(r,\widehat{u}) \leqslant r^{\lambda}$, $0 < r < \infty$ and if $\widehat{u}(z)$ is subharmonic, then $\widehat{u}(z) \leqslant c|z|^{\lambda}$, $0 < |z| < \infty$, where λ is the order of u.

In what follows, we consider the special case when $u(z) = \log|f(z)|$ for some meromorphic function $f(z)$ on D. Obviously, $u(z)$ is a δ-subharmonic function. It is easy to see from their definitions that $m(D,a,u) = m(D,a,f)$ where $m(D,a,f)$ is the approximated function of f in the sense of meromorphic functions. Noting the following basic equalities

$$\partial_z \log|z-w| = [2(z-w)]^{-1},\ \overline{\partial}_z(z-w)^{-1} = \pi\delta_w$$

and

$$\Delta_z \log|z-w| = 2\pi\delta_w,$$

where δ_w is the unit point mass at w, that is, the Dirac measure at w, we have

$$\mu^- = \sum_a \delta_a,$$

where the sum is taken over all poles of $f(z)$ in D counted with their multiplicities and thus

$$N(D,a,u) = \int_D G_D(a,w)\mathrm{d}\mu^-(w) = \sum_{b_n \in D} G_D(b_n,a) = N(D,a,f).$$

This implies that the characteristic $T(D,a,u)$ of u coincides with the characteristic $T(D,a,f)$ of $f(z)$.

When $D = B(0,r)$, $n(r,u) = \mu^-(B(0,r)) = n(r,f)$, where $n(r,f)$ is the number of poles of $f(z)$ in the disk $\{z : |z| < r\}$. If 0 is not a pole of $f(z)$, then $N(r,u) = N(r,f)$.

Below we establish a following analogy of the lemma on logarithmic derivative of meromorphic functions, which is Lemma 2 of Eremenko [11].

Theorem 7.2.9. *Let $\{f_j\}$ be a sequence of meromorphic functions in D and $t_j \to 0$ be a sequence of positive numbers. Assume that the sequences of $t_j \log|f_j|$ and $t_j \log|f_j'|$ are normal in $\mathscr{L}^1_{\mathrm{loc}}$ on D and converge to u_1 and u_2 respectively in $\mathscr{L}^1_{\mathrm{loc}}$ on D. Then $u_2(z) \leqslant u_1(z)$ on D and on each fine component B of the set $\{z : u_2(z) < u_1(z)\}$, u_1 is identically equal to some constant t and B is precisely a fine component of the set $\{z : u_2(z) < t\}$.*

Proof. Given arbitrarily a fixed point $a \in D$, take a $R > 0$ such that $\overline{B}(a,R) \subset D$. We may assume $a = 0$ for simple statement. In terms of (2.1.2), we have

$$\log f_j(z) = \frac{1}{2\pi}\int_0^{2\pi} \log|f_j(Re^{i\theta})| \frac{Re^{i\theta}+z}{Re^{i\theta}-z} d\theta - \sum_{a_m} \log \frac{R^2 - \overline{a}_m z}{R(z-a_m)} + \sum_{b_n} \log \frac{R^2 - \overline{b}_n z}{R(z-b_n)} + iC,$$

where C is a real number, and differentiating both of the sides yields

$$\frac{f_j'(z)}{f_j(z)} = \frac{1}{2\pi}\int_0^{2\pi} \log|f_j(Re^{i\theta})| \frac{2Re^{i\theta}}{(Re^{i\theta}-z)^2} d\theta + \sum_{a_m}\left(\frac{\overline{a}_m}{R^2-\overline{a}_m z} - \frac{1}{a_m - z}\right) - \sum_{b_m}\left(\frac{\overline{b}_m}{R^2-\overline{b}_m z} - \frac{1}{b_m - z}\right),$$

where the sums are taken over all zeros a_m and poles b_n of f_j in $B(0,R)$. Thus

$$\frac{|f_j'(z)|}{|f_j(z)|} \leqslant \frac{R}{\pi(R-r)^2}\int_0^{2\pi} |\log|f_j(Re^{i\theta})||d\theta + \sum_{a_m}\frac{1}{|a_m - z|} + \sum_{b_n}\frac{1}{|b_n - z|} + \frac{n_j}{R-r},$$

where $n_j = n(R,f) + n(R,1/f)$. Since $\{t_j \log|f_j|\}$ is normal in $\mathscr{L}^1_{\text{loc}}$, we have

$$\int_0^{2\pi} |\log|f_j(Re^{i\theta})||d\theta = O\left(\frac{1}{t_j}\right), \ n_j = \mu_j(B(0,R)) = O\left(\frac{1}{t_j}\right), j \to \infty,$$

where μ_j is the measure associated with $\log|f_j(z)|$. In terms of Lemma 1.2.3, for a fixed $0 < d < 1$ it follows that

$$\begin{aligned} t_j \int_{B(0,r)} \log^+ \frac{|f_j'(z)|}{|f_j(z)|} dm(z) &\leqslant t_j d^{-1}\pi r^2 \log^+ \left((\pi r^2)^{-1} \int_{B(0,r)} \left|\frac{f_j'(z)}{f_j(z)}\right|^d dm(z)\right) \\ &\quad + t_j \log 2 \\ &\leqslant O(t_j \log t_j) + t_j d^{-1}\pi r^2 \log^+ \sum_{a_m} \int_{B(0,r)} \frac{1}{|a_m - z|^d} dm(z) \\ &\quad + t_j d^{-1}\pi r^2 \log^+ \sum_{b_n} \int_{B(0,r)} \frac{1}{|b_n - z|^d} dm(z) \\ &= O(t_j \log t_j) \to 0, \text{ as } t_j \to 0, \end{aligned}$$

so that

$$\int_{B(0,r)} (u_2 - u_1)^+ dm(z) \leqslant 0, \ \forall\, 0 < r < R.$$

This yields $u_2(0) \leqslant u_1(0)$ and furthermore, $u_2(z) \leqslant u_1(z)$ in D.

Now we prove that u_1 is a constant on B. For a fixed point $z_0 \in B$, we can choose an arbitrary small r such that $C_r = \{z : |z - z_0| = r\} \subset B$ and

$$t_j \log |f_j'| \to u_2, \ t_j \log |f_j| \to u_1$$

hold uniformly on C_r as j passes a subsequence of positive integers to ∞ by means of Theorem 7.2.4. Take two real numbers τ and τ' such that

$$u_2(z) < \tau' < \tau < u_1(z), \ \forall z \in C_r.$$

For a fixed point $z_1 \in C_r$ with $|f_j(z)| \geqslant |f_j(z_1)|, \forall z \in C_r$, we have

$$\begin{aligned}\left|\left|\frac{f_j(z)}{f_j(z_1)}\right| - 1\right| &\leqslant |f_j(z_1)|^{-1} \int_{\widehat{z_1 z}} |f_j'(z)||\mathrm{d}z| \\ &\leqslant \exp\left(-\frac{\tau}{t_j}\right) \exp\left(\frac{\tau'}{t_j}\right) 2\pi r \\ &= 2\pi r \exp\left(-\frac{\tau - \tau'}{t_j}\right),\end{aligned}$$

where $\widehat{z_1 z}$ is a part of C_r from z_1 to z, so that

$$\begin{aligned}|t_j \log |f_j(z)| - t_j \log |f_j(z_1)|| &\leqslant t_j \log\left(1 + \left|\left|\frac{f_j(z)}{f_j(z_1)}\right| - 1\right|\right) \\ &\leqslant t_j \left|\left|\frac{f_j(z)}{f_j(z_1)}\right| - 1\right| \\ &\leqslant 2\pi r t_j \exp\left(-\frac{\tau - \tau'}{t_j}\right) \to 0,\end{aligned}$$

as $j \to \infty$. This shows that $u_1(z) = u_1(z_1)$ on C_r. We have proved that for every point $z_0 \in B$, we can take arbitrary small circles around it on which u_1 is a constant.

Now consider two arbitrary points a and b in B. Since B is polygonally connected, we can have a polygonal curve $\Gamma \subset B$ to connect a and b. Noting that

$$\lim_{r \to 0} \frac{1}{2\pi} \int_0^{2\pi} u_1(a + r\mathrm{e}^{\mathrm{i}\theta}) \mathrm{d}\theta = u_1(a),$$

given an arbitrarily small $\varepsilon > 0$ we have a sufficiently small r such that on $C^1 = \{z : |z - a| = r\}$, u_1 is a constant and on C^1

$$|u_1(z) - u_1(a)| = |\frac{1}{2\pi} \int_0^{2\pi} u_1(a + r\mathrm{e}^{\mathrm{i}\theta}) \mathrm{d}\theta - u_1(a)| < \varepsilon.$$

The same argument implies the existence of a small circle C^2 centered at b such that on C^2, u_1 is a constant and

$$|u_1(z) - u_1(b)| < \varepsilon.$$

From the previous result we have attained, it follows that each point of Γ is the center of some arbitrarily small circles on which u_1 is a constant. We can choose a finite collection of these circles together with C^1 and C^2 such that their interiors cover Γ and the union of these circles is connected. Therefore, u_1 is a constant on the connected union so that $|u_1(a) - u_1(b)| < 2\varepsilon$ and this implies $u_1(a) = u_1(b)$, that is to say, $u_1(z)$ is a constant on B.

Finally, set $u_1(z) \equiv t$ on B and t is a real number. Then $B \subset \{z : u_2(z) < t\}$. Let B' be the fine component of $\{z : u_2(z) < t\}$ containing B. The same argument as in the preceding paragraph yields that $u_1(z) \equiv t$ on B', otherwise, we can find a circle $C' \subset B'$, but $C' \not\subset B$ and $C' \cap B \neq \varnothing$ such that on C', $u_2(z) < t - \delta$ for some $\delta > 0$, while as in above we have $u_1 \equiv t$ on C', a contradiction is derived. This implies $B = B'$.

We complete the proof of Theorem 7.2.9. □

7.3 Eremenko's Proof of the Nevanlinna Conjecture

In 1929, F. Nevanlinna raised the following famous conjecture which attracted a great of interests.

Nevanlinna's Conjecture *Let $f(z)$ be a meromorphic function with finite order λ such that*

$$\sum_{a \in \widehat{C}} \delta(a, f) = 2. \tag{7.3.1}$$

Then all of the following statements hold:

(1) 2λ is a natural number $\geqslant 2$;

(2) $\delta(a, f) = p(a)/\lambda$, where $p(a)$ is a non-negative integer and $\nu(f) \leqslant 2\lambda$ where $\nu(f)$ is the number of deficient values of f;

(3) all deficient values are asymptotic.

In order to solve this conjecture, many remarkable methods of analysis, geometry and potential theory as well have been introduced into the study of value distribution of meromorphic functions. This conjecture for entire functions was proved in 1946 by A. Pfluger [23] with the order λ being a natural number instead and moreover, he proved that for an entire function f with finite non-integer order λ, $\sum_a \delta(a, f) \leqslant 2 - k(\lambda$, where $k(\lambda) = 1$, $0 \leqslant \lambda \leqslant 1/2$; $k(\lambda) = \sin(\pi\lambda)$, $1/2 \leqslant \lambda \leqslant 1$; $k(\lambda) \geqslant (q + 1 - \lambda)(\lambda - q)/\{2\lambda(q + \lambda)[2 + \log q]\}$, $1 < \lambda, q = [\lambda]$. A substantial step for the proof of the conjecture was walked by A. Weitsman [27] in 1969 who proved that under the assumption of the conjecture, $\nu(f) \leqslant 2\mu(f)$, $\mu(f)$ is the lower order. Up to 1983, it is D. Drasin who gave a complete proof of this conjecture in [8]. However, his proof is very complicated and his paper has about 100 pages. A simple proof of using the potential theory was found by A. Eremenko, which is what we will introduce in this section.

Recall the quantity

$$N_1(r,f) = 2N(r,f) - N(r,f') + N\left(r, \frac{1}{f'}\right)$$

appearing in Theorem 2.1.4. Set

$$n_1(r,f) = 2n(r,f) - n(r,f') + n\left(r, \frac{1}{f'}\right),$$

which counts the multiple points of f including the multiple poles. Then $N_1(r,f)$ is the integrated counting function for $n_1(r,f)$ and so it is known as the ramification term. In terms of the basic inequality (2.1.13), for arbitrary finite number of points $a_j \in \widehat{\mathbb{C}}(1 \leqslant j \leqslant q)$ we have

$$\sum_{j=1}^{q} \delta(a_j, f) \leqslant 2 - \limsup_{r\to\infty} \frac{N_1(r,f)}{T(r,f)}.$$

If (7.3.1) holds, then

$$N_1(r,f) = o(T(r,f)), \text{ as } r \to \infty. \tag{7.3.2}$$

It is natural to ask if (7.3.2) implies (7.3.1) as well as the results stated in the Nevanlinna's conjecture. D. F. Shea [25] proved that if the lower order $\mu(f) < 1/2$, then

$$\inf_{A} \limsup_{r\in A\to\infty} \frac{N_1(r,f)}{T(r,f)} \geqslant \cos(\pi\mu),$$

where the inf is taken over all the set A with density one.

The following is theorem Eremenko [11] proved, which is main result introduced in this section.

Theorem 7.3.1. *Let $f(z)$ be a meromorphic function with finite lower order satisfying (7.3.2). Then the results in the Nevanlinna's conjecture hold. If we normalize such that $\delta(\infty, f) = 0$, then we have*

$$\log \frac{1}{|f'(re^{i\theta})|} = \pi r^{\lambda} h(r) |\cos \lambda(\theta - \phi(r))| + o(r^{\lambda} h(r)), \text{ as } r \to \infty, \tag{7.3.3}$$

uniformly with respect to θ outside a C_0^1 set; On

$$V_k = \{z = re^{i\theta} : \frac{\pi}{2\lambda}(2k-1) \leqslant \theta - \phi(r) \leqslant \frac{\pi}{2\lambda}(2k+1)\} \ (k = 1, 2, \cdots, q),$$

there exists a $a_k \in \widehat{\mathbb{C}}$ such that

$$\log \frac{1}{|f(re^{i\theta}) - a_k|} = \pi r^{\lambda} h(r) |\cos \lambda(\theta - \phi(r))| + o(r^{\lambda} h(r)), \text{ as } r \to \infty, \tag{7.3.4}$$

uniformly with respect to θ outside a C_0^1 set, while $h(r)$ and $\phi(r)$ are continuous with $h(cr) \sim h(r)$ and $\phi(cr) = \phi(r) + o(1)$ as $r \to \infty$ uniformly with respect to $c \in [1,2]$.

Moreover,

$$T(r,f) \sim r^{\lambda} h(r), \text{ as } r \to \infty. \tag{7.3.5}$$

To prove Theorem 7.3.1, we need the following

Lemma 7.3.1. *Let U be a subharmonic function on $\mathbb{C}$ such that for arbitrarily small $\delta > 0$*

$$0 \leqslant U(z) \leqslant c|z|^{\lambda-\delta}, |z| < \eta,$$

$$0 \leqslant U(z) \leqslant c|z|^{\lambda+\delta}, \tau < |z|,$$

where η and τ depend on δ and the set $\{z \in \mathbb{C} : U(z) > 0\}$ has precisely q fine components. Assume that there is a Riemann surface Σ with a two-sheeted ramified covering $p : \Sigma \to \mathbb{C}$ and a function h harmonic on Σ such that $U \circ p = |h|$ and p is ramified over at most $q-2$ points on $\mathbb{C}$ and each ramification point of p is a zero of h of order at least 3. Then we have $q = 2\lambda$ and

$$U(re^{i\theta}) = |a| r^{\lambda} |\cos\lambda(\theta - \theta_0)| \tag{7.3.6}$$

for some $\theta_0 \in [0, 2\pi]$ and a complex number a.

Proof. First of all, we prove that $q = 2\lambda$. Consider a disk $D_0 = B(0,r)$ so small that p is unramified over $D_0 \setminus \{0\}$. Then we have a multi-valued analytic function $H(z)$ on D_0 which has the expansion of the Puiseux series

$$H(z) = \sum_{k=s}^{\infty} c_k z^{k/2}$$

with $c_s \neq 0$ such that $U(z) = |\mathrm{Re} H(z)|$ in D_0. Thus we obtain an asymptotic representation of $U(z)$

$$U(z) = |\mathrm{Re}(c_s z^{s/2}(1 + o(1)))|, \; z \to 0. \tag{7.3.7}$$

Using the inequality $0 \leqslant U(z) \leqslant c|z|^{\lambda-\delta}$ yields $s \geqslant 2(\lambda - \delta)$. From $U \circ p = |h|$ and that h is harmonic, it follows that $\{z : U(z) = 0\}$ contains s distinct simple piecewise analytic curves $\gamma_n (n = 1, 2, \cdots, s)$ starting at the origin. Every pair of distinct γ_n and γ_m cannot intersect each other, otherwise in terms of the maximum principle of subharmonic functions, U vanishes identically in the domain surrounded by γ_n and γ_m and this contradicts the representation (7.3.7) of $U(z)$. And we can assume that γ_n end at a boundary point of D_0. Since $D_0 \cap \{z \in \mathbb{C} : U(z) > 0\}$ has at most q components, we get $q \geqslant s$. On the other hand, from the Denjoy-Carleman-Ahlfors Theorem for subharmonic functions (cf. Lemma 6.2.1) it follows that $q \leqslant 2(\lambda + \delta)$. This implies that $q = s = 2\lambda$, as δ is arbitrarily small.

Since h is harmonic on Σ, we can get a multi-valued analytic function T on Σ such that $h = \mathrm{Re} T$. The derivatives of the conjugate harmonic function of h with respect to the coordinates equal to the derivatives of h with respect to the suitable coordinates, that is to say that these derivatives are single-valued. For any point ξ on Σ a neighborhood Σ_0 of which is conformly mapped by p onto a disk B on $\mathbb{C}$, the derivatives of all branches of $T \circ p_0^{-1}(z)$ are equal, where p_0^{-1} is a branch of p^{-1}

from B onto Σ_0, and hence $\frac{\mathrm{d}T}{\mathrm{d}p} \circ p_0^{-1}(z)$ is single-valued in B and $\frac{\mathrm{d}T}{\mathrm{d}p}$ is single-valued at ξ. It is easily seen that $\psi = \frac{\mathrm{d}T}{\mathrm{d}p}$ is a single-valued meromorphic function on Σ, for p has only finitely many ramified points.

Next we want to prove in terms of ψ that p is ramified only over 0. Noting that each ramification point of p, that is, zero of $\mathrm{d}p$, is a zero of h of order at least 3 and $\mathrm{d}p$ has only simple zeros, we easily see that ψ vanishes at each ramification point of p. Therefore, ψ is analytic on Σ. We add the points of $p^{-1}(\infty)$ to Σ to get a compact Riemann surface $\widehat{\Sigma}$, and the points will be called the infinite points on $\widehat{\Sigma}$.

Take a large $R > 0$ such that p has no ramified points over $B = \{z : |z| > R\}$. Since p is a covering mapping, we have an analytic function $\tilde{H}(z)$ in a neighborhood of a point in $B = \{z : |z| > R\}$ such that $h \circ p^{-1}(z) = \mathrm{Re}\tilde{H}(z)$. We will get a multi-valued analytic function by continuing analytically $\tilde{H}(z)$ in $\{z : |z| > R\}$, which is denoted still by $\tilde{H}(z)$. Then we have $U(z) = |\mathrm{Re}\tilde{H}(z)|$ and so $0 \leqslant |\mathrm{Re}\tilde{H}(z)| \leqslant c|z|^{\lambda+1/4}$ in B. Now we expand $\tilde{H}(z)$ in B into the Puiseux series

$$\tilde{H}(z) = \sum_{k=-\infty}^{2\lambda} a_k z^{k/2}. \tag{7.3.8}$$

It follows that $a_{2\lambda} \neq 0$ from the assumption of that the set $\{z \in \mathbb{C} : U(z) > 0\}$ has precisely $q = 2\lambda$ tract and furthermore, we can write

$$U(z) = |\mathrm{Re}(a_{2\lambda} z^{\lambda}(1 + o(1)))|, \ z \to \infty.$$

Noting that $T = \tilde{H} \circ p$ yields that ψ has poles only in the infinite points of Σ whose total multiplicity is $q - 2$.

The same argument as the above implies that 0 is a zero of $U(z)$ with order at least q. Consider a neighborhood of 0 and as in the above discussion, we can get that the total multiplicity of zeros of ψ over 0 is at least $q - 2$. Since on $\widehat{\Sigma}$ the number of zeros of ψ equals the number of poles of ψ, therefore ψ has no other zeros than those over 0. This immediately implies that p is ramified only over 0, because ψ vanishes at each ramification point of p.

Thus we can continue $\tilde{H}(z)$ toward $\mathbb{C} \setminus \{0\}$ so that (7.3.8) holds in $\mathbb{C} \setminus \{0\}$. Employing the inequality $0 \leqslant U(z) \leqslant c|z|^{\lambda-\delta}, |z| < \eta$ to (7.3.8)yields

$$\tilde{H}(z) = a_{2\lambda} z^{\lambda}.$$

From this (7.3.6) follows. □

Now we are in position to prove Theorem 7.3.1.

Proof. From the Shea's result mentioned before Theorem 7.3.1 we can assume that $\mu(f) \geqslant 1/2$. First of all, we prove (7.3.3) and the results stated in the Nevanlinna's conjecture in the case when

$$m(r, f) = o(T(r, f)) \tag{7.3.9}$$

as $r \to \infty$, that is, ∞ is not a Valiron exceptional value of f. This together with (7.3.2) yields

$$\overline{N}(r,f) \sim N(r,f) \sim T(r,f). \tag{7.3.10}$$

In terms of Lemma 2.5.1 on the logarithmic derivative and (7.3.10), we have

$$T(r,f) = N(r,f) + o(T(r,f)) = \frac{N(r,f')}{2} + o(T(r,f)) \leqslant \frac{T(r,f')}{2} + o(T(r,f)),$$

and

$$T(r,f') \leqslant 2N(r,f) + m(r,f) + m(r,f'/f) \leqslant 2T(r,f) + o(T(2r,f)).$$

Consider a positive number β and a sequence of positive numbers $\{r_j\}$ tending to ∞ such that there exist two sequences $\{\varepsilon_j\}$ and $\{\tau_j\}$ with $\varepsilon_j \to 0$ and $\tau_j \to 0$ and for any small δ, we have

$$T(r,f) \leqslant (1+\tau_j)\left(\frac{r}{r_j}\right)^{\beta-\delta} T(r_j,f),\ \varepsilon_j r_j \leqslant r \leqslant \eta r_j,$$

$$T(r,f) \leqslant (1+\tau_j)\left(\frac{r}{r_j}\right)^{\beta+\delta} T(r_j,f),\ \tau r_j \leqslant r \leqslant \varepsilon_j^{-1} r_j,$$

where η and τ are two positive numbers only depending on δ.

For a fixed $\beta \in [\mu_*, \lambda^*]$ with $\beta > 0$, where λ^* and μ_* are respectively the Pólya order and Pólya lower order of $T(r,f)$ (for their definition, see the paragraph before Lemma 1.1.3), the Drasin and Shea Theorem [9] asserts the existence of a sequence $\{r_j\}$ of Pólya peak of order β, which obviously satisfies the above inequalities with $\eta < 1$ and $\tau > 1$.

Then

$$T(r_j,f') \sim 2T(r_j,f) \tag{7.3.11}$$

as $j \to \infty$.

Define a sequence of δ-subharmonic functions

$$U_j(z) = \frac{1}{T(r_j,f)} \log \frac{1}{|f'(r_j z)|}. \tag{7.3.12}$$

From (7.3.11) and Theorem 7.2.8, it follows that $\{r_j\}$ contains a subsequence, which we still denote by $\{r_j\}$, such that for a δ-subharmonic function U on $\mathbb{C}$, we have for each $r \in (0,\infty)$,

$$\lim_{j\to\infty} \int_0^{2\pi} \left| U_j(re^{i\theta}) - U(re^{i\theta}) \right| d\theta = 0$$

and for arbitrarily small $\delta > 0$, we have

$$0 \leqslant U(z) \leqslant c|z|^{\beta-\delta}, |z| < \eta \text{ and } 0 \leqslant U(z) \leqslant c|z|^{\beta+\delta}, \tau < |z|, \tag{7.3.13}$$

where η and τ depend on δ. The left side of above inequality comes from the following implication. Since $U_j^-(z) = \frac{1}{T(r_j,f)} \log^+ |f'(r_j z)|$, we have therefore

$$\begin{aligned} 0 \leqslant \int_0^{2\pi} U_j^-(re^{i\theta})d\theta &= \frac{1}{T(r_j,f)} m(r_j r, f') \\ &\leqslant \frac{1}{T(r_j,f)} (m(r_j r, f) + m(r_j r, f'/f)) \\ &= \frac{1}{T(r_j,f)} o(T(2r_j r, f)) \to 0, \end{aligned}$$

as $j \to \infty$. In terms of Theorem 7.2.4, U_j^- converges to 0 in the 1-Carleson measure. This together with the Maximal Principle for subharmonic functions implies that $U^- = 0$ and so $U \geqslant 0$. It is clear that

$$(\Delta U_j)^- = \frac{1}{T(r_j,f)} \sum_{f'(r_j z)=0} \delta_{r_j z}.$$

For arbitrary $r > 0$, we have

$$(\Delta U_j)^-(B(0,r)) = \frac{1}{T(r_j,f)} n_1(r r_j, f) \leqslant \frac{1}{T(r_j,f)} N_1(2 r r_j, f)(\log 2)^{-1} \to 0$$

as $j \to \infty$. This yields that the Riesz charge of U is a measure, denoted by 2μ, and therefore U is subharmonic.

In what follows, we want to characterize the behavior of U in terms of Lemma 7.2.3 and Lemma 7.3.1, through which we will complete the proof of Theorem 7.3.1. An application of the Denjoy-Carleman-Ahlfors Theorem to (7.3.13) yields that the set $\{z : U(z) > 0\}$ has a finite number of fine components, denoted by $E_1, E_2, \cdots, E_q$. By means of Lemma 7.2.4, U has the decomposition

$$U = \sum_{k=1}^{q} U_{E_k}, \tag{7.3.14}$$

where each U_{E_k} is subharmonic on $\mathbb{C}$.

Choose $q+1$ complex numbers $b_n (n = 1, 2, \cdots, q+1)$ such that b_n's are not Valiron exceptional values and all b_n-points of f are simple. Then

$$m(r, f - b_n) + m(r, (f - b_n)^{-1}) = o(T(r,f))$$

and it shows that

$$\frac{1}{T(r_j,f)} |\log |f(r_j z) - b_n|| \to 0$$

in $\mathscr{L}^1_{\text{loc}}$ with respect to the linear measure on $\{z : |z| = r\}$. For each n, the sequence of δ-subharmonic functions

$$\widehat{U}_{n,j}(z) = -\frac{1}{T(r_j,f)}\log\left|\left(\frac{1}{f-b_n}\right)'(r_jz)\right|$$
$$= -\frac{1}{T(r_j,f)}\log|f'(r_jz)| + \frac{2}{T(r_j,f)}\log|f(r_jz)-b_n|$$

has the same limit function $U(z)$ as $U_j(z)$ does.

Given an arbitrarily large $R>0$, by noting that U is unbounded in each E_k we can find circles $C_k \subset E_k$ such that

$$U(z) > M_U(R), z\in C_k, 1\leqslant k\leqslant q; \tag{7.3.15}$$

and $\{U_j\}$ and $\{\widehat{U}_{n,j}\}$ converges to U uniformly on C_k.

The next aim is to get a decomposition of U satisfying the assumptions of Lemma 7.2.3. To do this, we choose a point $z_k \in C_k$ and a subsequence of $\{r_j\}$, denoted still by $\{r_j\}$, such that

$$f(r_jz_k) = c_{k,j} \to c_k, 1\leqslant k\leqslant q, \tag{7.3.16}$$

as $j\to\infty$ and $c_k \in \widehat{\mathbb{C}}$. Take a complex number b from $\{b_1,\cdots,b_{q+1}\}$ such that $b\neq c_k (1\leqslant k\leqslant q)$. Set $a_k = (c_k-b)^{-1}$ ($a_k = 0$ if $c_k=\infty$), $a_{k,j} = (c_{k,j}-b)^{-1}$ and $F(z) = (f(z)-b)^{-1}$.

Define q sequences of δ-subharmonic functions

$$W_{k,j}(z) = \frac{1}{T(r_j,f)}\log\frac{1}{|F(r_jz)-a_{k,j}|}. \tag{7.3.17}$$

Since a_k's are finite and $a_{k,j}\to a_k$ as $j\to\infty$, the sequence $\{a_{k,j} : k=1,2\cdots,q; j=1,2,\cdots\}$ are bounded. This implies that

$$T(r,F-a_{k,j}) = T(r,F)+O(1) = T(r,f)+O(1)$$

uniformly in j. In terms of Theorem 7.2.8, there exists a subsequence, which we still denote by $\{r_j\}$, such that for a δ-subharmonic function W_k on $\mathbb{C}$, we have for each $r\in(0,\infty)$,

$$\lim_{j\to\infty}\int_0^{2\pi}\left|W_{k,j}(re^{i\theta})-W_k(re^{i\theta})\right|d\theta = 0$$

as $j\to\infty$. Noting the fact that $m(r,F-a_{k,j}) = m(r,F)+O(1) = o(T(r,f))$ and in terms of Theorem 7.2.9, we immediately attain

$$0\leqslant W_k\leqslant U,\ 1\leqslant k\leqslant q. \tag{7.3.18}$$

In terms of (7.3.15) and uniform convergence of $\{\widehat{U}_{n,j}\}$ to U on C_k, we have $|F'(z)| < \exp(-M_U(R)T(r_j,f))$, $z\in r_jC_k$ and hence

$$\begin{aligned}\log|F(z)-a_{k,j}| &= \log|F(z)-F(r_jz_k)|\\ &\leqslant \log\int_{\widehat{z,r_jz_k}}|F'(\zeta)||\mathrm{d}\zeta|\\ &\leqslant -M_U(R)T(r_j,f)+\log(2\pi R_kr_j),\ z\in r_jC_k,\end{aligned}$$

where R_k is the radius of C_k so that

$$W_{k,j}(z)\geqslant M_U(R)-\frac{1}{T(r_j,f)}\log(2\pi R_kr_j),\ z\in C_k.$$

This yields that

$$W_k(z)\geqslant M_U(R),\ z\in C_k. \tag{7.3.19}$$

Define q functions $u_k(z)=W_k(z)$ for $z\in E_k$ and $u_k(z)=0$ for $z\in\mathbb{C}\setminus E_k$. We want to prove that $u_k(1\leqslant k\leqslant q)$ satisfies the assumptions of Lemma 7.2.3. Since $\sum_{n\neq k}U_{E_n}=0$ for $z\in E_k$ and $W_k\leqslant\sum_{n\neq k}U_{E_n}$ for $z\in\mathbb{C}\setminus E_k$, we have

$$u_k=\left(W_k-\sum_{n\neq k}U_{E_n}\right)^+$$

and then u_k is δ-subharmonic. It is easy to see from (7.3.18) that

$$0\leqslant u_k\leqslant U,\ 1\leqslant k\leqslant q. \tag{7.3.20}$$

We want to prove that the support of u_k is E_k and connected. From Theorem 7.2.9 it follows that on each fine component $D_{k,m}$ of the set $\{z\in E_k: W_k(z)<U(z)\}$, $W_k\equiv t_{k,m}$, a real constant, and the fine component is also equal to a fine component of the set $\{z: U(z)>t_{k,m}\}$. In terms of (7.3.13) and Denjoy-Carleman-Ahlfors Theorem, for each t, the number of fine components of the set $\{z: U(z)>t\}$ is bounded from above and hence the set $\{z\in E_k: W_k(z)<U(z)\}$ has only a finite number of fine components. This implies that we have only finitely many $t_{k,m}$. Set $t_0=\min_{k,m}\{t_{k,m}\}$ and then $t_0\geqslant 0$.

Suppose that $t_0=0$ and then some $t_{k,m}=0$. In terms of the result in the final statement of Theorem 7.2.10, $D_{k,m}=E_k$, that is to say $W_k\equiv 0$ in E_k, but this contradicts (7.3.19), noting that $C_k\subset E_k$. We prove $t_0>0$.

It is clear that $u_k(z)=W_k(z)\geqslant t_0$ in the set $\{z\in E_k: W_k(z)<U(z)\}$, while in the set $\{z\in E_k: W_k(z)=U(z)\}$, $u_k(z)=W_k(z)=U(z)>0$. Thus $u_k(z)>0$ for $z\in E_k$, that is to say, the support of u_k is E_k.

In view of (7.3.13), $U(0)=0$ and hence the set $\{z: U(z)<t_0\}$ has a component $D(t_0)$ containing 0. Since the subharmonic function U is upper semi-continuous, $D(t_0)$ is an open set. The maximum principle for subharmonic functions yields that $D(t_0)$ must be simply connected. Set $D=D(t_0)$. If $z\in E_k\cap D$, then $U(z)<t_0$ and furthermore $z\notin\{z\in E_k: W_k(z)<U(z)\}$ and this deduces $u_k(z)=U(z)=U_{E_k}(z)$. An application of Lemma 7.2.4 to D and $U(z)$ yields that $u_k(z)$ is subharmonic in D and we have the following decomposition

$$U(z) = \sum_{k=1}^{q} u_k(z),\ z \in D. \tag{7.3.21}$$

Let μ_k be the Riesz charge of u_k and η_k be the Riesz charge of W_k. We want to prove that the restriction of μ_k and μ to D satisfies (7.2.2) in Lemma 7.2.3. From (7.3.21) it follows that $2\mu = \sum_{k=1}^{q} \mu_k$. Thus it is sufficient to prove that for each k, $\mu \geqslant \mu_k$. We know that $\mu[W_{k,j}]^+$ weakly converges to η_k^+ as $j \to \infty$ and

$$(\Delta W_{k,j})^+ = \frac{1}{T(r_j,f)} \sum_{f(r_j z)=b} \delta_{r_j z} = \frac{1}{2}(\Delta \widehat{U}_{n,j})^+$$

for some n with $b_n = b$, that is, $\widehat{U}_{n,j}(z)$ is the function in (7.3.12) with f' replaced by F', Recalling that each $\widehat{U}_{n,j}(z)$ has the limit function U, then we have $\mu = \eta_k^+$. From the definition of u_k, in view of Lemma 7.2.5 we have $\mu_k|_{E_k} = \eta_k|_{E_k}$ and $\mu_k|_{E_m} = 0$ for $m \neq k$. $E = \mathbb{C} \setminus \cup E_k = \{z : U(z) \leqslant 0\}$ is a Borel set and $u_k(z) = 0$ in E and $u_k(z) \geqslant 0$ on $\mathbb{C}$. Then using Lemma 7.2.2 yields $\mu_k|_E \leqslant 0$ and thus we get $\mu_k \leqslant \eta_k^+$ so that $\mu_k \leqslant \mu$. This immediately implies (7.2.2).

Now we can use Lemma 7.2.3 to get a Riemann surface Σ with a two-sheeted ramified covering $p : \Sigma \to D$ and a function h harmonic on Σ such that $U \circ p = |h|$. As in the proof of Lemma 7.3.1, we get $q = 2\beta$. Since β can be chosen to be arbitrary positive number between μ_* and λ^*, we immediately have $\mu_* = \lambda^* = q/2$. Furthermore, we have $\lambda = \lambda(f) = \mu(f) = q/2$. Hence we have proved result (1) in the Nevanlinna conjecture.

And it together with the Denjoy-Carleman-Ahlfors Theorem also produces that for each $t > 0$, $\{z : U(z) > t\}$ has at most q fine components, while $\{z : U(z) > 0\}$ has just q fine components. Then $\{z : U(z) > t\}$ has just q fine components. Since each fine component of the set $\{z \in E_k : W_k(z) < U(z)\}$ $(k = 1,2,\cdots,q)$ is a fine component of $\{z \in E_k : U(z) > t_k\}$ for some $t_k > 0$, therefore $D_k = \{z \in E_k : U(z) > t_k\}$ $(k = 1,2,\cdots,q)$ has only one fine component, that is, D_k is finely connected. It is easy to see that $u_k(z) = U(z) \leqslant t_k$ for $z \in E_k \setminus D_k$ and in terms of Theorem 7.2.9, $u_k(z) = W_k(z) = t_k$ for $z \in D_k$ so that $u_k(z) \leqslant t_k, z \in E_k$.

Employing (7.3.19) deduces that $u_k(z) = W_k(z) \geqslant M_U(R), z \in C_k$ so that $t_k \geqslant M_U(R)$. Now for $t = \min_k\{t_k\} > 0$ we define $D = D(t)$ which is the component of $\{z : U(z) < t\}$ containing 0. Then D contains the disk $B(0,R)$. The previous argument is used to this D, by noting that R can be chosen to be arbitrarily large, to show that the assumption of Lemma 7.3.1 is satisfied by U. Therefore, we have (7.3.6)

$$U(re^{i\theta}) = |a| r^{\lambda} |\cos \lambda(\theta - \theta_0)|$$

for some complex number a and some $\theta_0 \in [0, 2\pi)$.

Since in terms of (7.3.10)

$$\begin{aligned} m\left(r_j, \frac{1}{f'}\right) - m(r_j, f') &= N(r_j, f') - N\left(r_j, \frac{1}{f'}\right) + O(1) \\ &= 2T(r_j, f) + o(T(r_j, f)), \end{aligned}$$

we have

$$\frac{1}{2\pi}\int_0^{2\pi} U_j(e^{i\theta})d\theta = \frac{1}{T(r_j,f)}\left(m\left(r_j,\frac{1}{f'}\right) - m(r_j,f)\right) \to 2,$$

so that

$$\frac{1}{2\pi}\int_0^{2\pi} U(e^{i\theta})d\theta = 2,\ j\to\infty.$$

This immediately yields $|a| = \pi$, that is, we have

$$U(re^{i\theta}) = \pi r^{\lambda}|\cos\lambda(\theta-\theta_0)|. \tag{7.3.22}$$

Then for any fixed $t > 0$, we have

$$\frac{1}{2\pi}\int_0^{2\pi} U_j(te^{i\theta})d\theta \to \frac{1}{2\pi}\int_0^{2\pi} U(te^{i\theta})d\theta = 2t^{\lambda},\ j\to\infty.$$

On the other hand, we have

$$\begin{aligned}\frac{1}{2\pi}\int_0^{2\pi} U_j(te^{i\theta})d\theta &= \frac{1}{T(r_j,f)}\left(N(tr_j,f') - N\left(tr_j,\frac{1}{f'}\right) + O(1)\right)\\ &= \frac{2(1+o(1))T(tr_j,f)}{T(r_j,f)} + o(1).\end{aligned}$$

This implies that

$$T(tr_j,f)/T(r_j,f) \to t^{\lambda},\ \text{as } j\to\infty. \tag{7.3.23}$$

Since $\lambda^* = \mu_*$, every sequence of positive numbers tending to infinity satisfies the requirement of the sequence $\{r_j\}$ in the above discussion and so contains a subsequence, denoted by $\{r_j\}$, such that $U_j(z) \to U(z)$ in $\mathscr{L}^1_{\text{loc}}$ as $j\to\infty$ and $U(z)$ has the form (7.3.22). Furthermore we also have (7.3.23) and therefore every unbounded sequence of positive numbers contains a subsequence such that (7.3.23) holds. Thus it is not difficult to see that

$$\frac{T(tr,f)}{T(r,f)} \to t^{\lambda},\ r\to\infty.$$

Write $h(r) = T(r,f)r^{-\lambda}$, and the above limit yields that $h(tr)\sim h(r), r\to\infty$ uniformly with respect to $t\in[1,2]$.

Denote by X the set of all subharmonic functions of the form

$$u(re^{i\theta};\theta_0) = \pi r^{\lambda}|\cos\lambda(\theta-\theta_0)|, \theta_0\in[0,2\pi].$$

Set

$$U_t(z) = \frac{1}{t^{\lambda}h(t)}\log\frac{1}{|f'(tz)|},\ t>0.$$

The result we have previously obtained yields that for each sequence $\{t_j\}$ tending to infinity, $U_{t_j}(z)$ converges to an element in X in $\mathscr{L}^1_{\text{loc}}$. Letting $u_t(z) = u(z;\phi(t)) \in X$ be the closest element to U_t, then we have

$$d(U_t, u_t) = d(U_t, X) \to 0, t \to \infty$$

so that

$$U_t(z) = u(z;\phi(t)) + o(1),\ t \to \infty, \tag{7.3.24}$$

where $d(u,v)$ denotes the distance of u and v in $\mathscr{L}^1_{\text{loc}}$ and $o(1)$ stands for a function which tends to 0 in $\mathscr{L}^1_{\text{loc}}$ as $t \to \infty$ and thus

$$\log \frac{1}{|f'(tz)|} = \pi r^\lambda t^\lambda h(t) |\cos \lambda(\theta - \phi(t))| + o(1),\ z = re^{\mathrm{i}\theta}. \tag{7.3.25}$$

In terms of Theorem 7.2.4, (7.3.25) holds uniformly with respect to 1-Carleson measure, that is, uniformly with respect to θ outside a C_0^1 set E.

Now we prove that $\phi(t)$ satisfies the requirement of Theorem 7.3.1. It is sufficient to prove that

$$d(u_t, u_{ct}) \to 0,\ t \to \infty. \tag{7.3.26}$$

In fact, if (7.3.26) holds, then we find a $r > 0$ and a θ_0 such that $u_t(re^{\mathrm{i}\theta_0}) - u_{ct}(re^{\mathrm{i}\theta_0}) \to 0$ as $t \to \infty$. This immediately implies that

$$\lambda(\phi(ct) - \phi(t)) \to 0 \bmod \pi. \tag{7.3.27}$$

Now suppose that (7.3.26) fails. There exist a sequence $c_n \in [1,2]$ and a sequence $t_n \to \infty$ such that for some fixed $\varepsilon > 0$

$$d(u_{t_n}, u_{c_n t_n}) \geqslant \varepsilon > 0.$$

Since $c^{-\lambda}u(cz) = u(z)$ for all $u \in X$ and $c > 0$, we have

$$\begin{aligned} u_{c_n t_n}(z) &= U_{c_n t_n}(z) + o(1) \\ &= c_n^{-\lambda} U_{t_n}(c_n z) + o(1) \\ &= c_n^{-\lambda} u_{t_n}(c_n z) + o(1) \\ &= u_{t_n}(z) + o(1), \end{aligned}$$

which contradicts our assumption for c_n and t_n and hence (7.3.27) holds.

Therefore we have (7.3.3) in terms of (7.3.25) and $h(ct) \sim h(t)$ and (7.3.27).

Next we want to prove the asymptotic formula (7.3.4) and results (2) and (3) in the Nevanlinna conjecture. Consider the domain

$$V_k = \left\{ z = re^{\mathrm{i}\theta} : \frac{\pi}{2\lambda}(2k-1) \leqslant \theta - \phi(r) \leqslant \frac{\pi}{2\lambda}(2k+1) \right\}$$

$k = 1,2,\cdots,2\lambda$ and curve $\gamma(\vartheta) = \{re^{i\theta} : \theta - \phi(r) = \vartheta\}$ for $\vartheta \in [0,2\pi)$. Let Γ_ϑ be the curve which is constructed from $\gamma(\vartheta)$ by replacing the part of $\gamma(\vartheta)$ inside exceptional disks in E with parts of their circles so that Γ_ϑ is outside all the exceptional disks in E. It is clear that $\Gamma_\vartheta = \{re^{i\theta} : \theta - \phi(r) = \vartheta + o(1)\}$. Take a complex number z_k and a curve Γ from z_k to ∞ in V_k outside all exceptional disks in E and define

$$a_k = \int_\Gamma f'(\zeta)\mathrm{d}\zeta + f(z_k).$$

In terms of the asymptotic representation (7.3.3) of $\log|f'(z)|$, a_k is in fact independent of choice of Γ and z_k. For any point $z = re^{i\theta} \in V_k$ outside all the exceptional disks in E, it is easy to see that

$$\frac{1}{T(t,f)}\log^+\left|\frac{f(z_t)-a_k}{f'(z_t)}\right| \leqslant \frac{1}{T(t,f)}\log^+\int_{\Gamma_\vartheta(z_t)}\left|\frac{f'(\zeta)}{f'(z_t)}\right||\mathrm{d}\zeta| \to 0,\ t\to\infty,$$

where $\vartheta = \theta - \phi(r) \in [\frac{\pi}{2\lambda}(2k-1), \frac{\pi}{2\lambda}(2k+1)]$ and $\Gamma_\vartheta(z_t)$ is the part of Γ_ϑ from z_t to ∞ and $z_t \in \Gamma_\vartheta$, $|z_t| = tr$, $\arg z_t = \vartheta + \phi(tr) + o(1)$. Therefore,

$$\frac{1}{T(t,f)}\log\frac{1}{|f(z_t)-a_k|} - \frac{1}{T(t,f)}\log\frac{1}{|f'(z_t)|}$$

$$= \frac{1}{T(t,f)}\log^+\left|\frac{f'(z_t)}{f(z_t)-a_k}\right| - \frac{1}{T(t,f)}\log^+\left|\frac{f(z_t)-a_k}{f'(z_t)}\right| \to 0,\ t\to\infty$$

in $\mathscr{L}^1_{\mathrm{loc}}$ and furthermore, we have

$$\frac{1}{T(t,f)}\log\frac{1}{|f(tz)-a_k|} = U_t(z) + o(1),\ tz \in V_k,\ t\to\infty$$

in $\mathscr{L}^1_{\mathrm{loc}}$. This implies that

$$\log\frac{1}{|f(re^{i\theta})-a_k|} = \pi r^\lambda h(r)|\cos\lambda(\theta-\phi(r))| + o(r^\lambda h(r)), \tag{7.3.28}$$

uniformly for $z = re^{i\theta} \in V_k$ outside E as $r\to\infty$. From this it follows that for $a \in \widehat{\mathbb{C}}$,

$$\begin{aligned} m\left(r,\frac{1}{f-a}\right) &= \frac{1}{2}p(a)r^\lambda h(r)\int_{-\pi/2\lambda}^{\pi/2\lambda}\cos(\lambda\theta)\mathrm{d}\theta + o(r^\lambda h(r)) \\ &= \frac{p(a)}{\lambda}r^\lambda h(r) + o(r^\lambda h(r)) \\ &= \frac{p(a)}{\lambda}T(r,f) + o(T(r,f)),\ r\to\infty, \end{aligned}$$

where $p(a)$ is the number of a_k's equal to a and hence $\Delta(a,f) = \delta(a,f) = p(a)/\lambda$; the total sum of deficiencies is 2 and all deficient values are asymptotic values of $f(z)$.

In general case, we choose a complex number b such that $F=(f-b)^{-1}$ satisfies the above requirement for $f(z)$, that is, all b-points of $f(z)$ are simple and $\Delta(b,f)=\Delta(\infty,F)=0$. Then employing the above argument to F we get the results in the Nevanlinna's conjecture for F and further for f and $\Delta(\infty,f)=\delta(\infty,f)$.

Assume that $\delta(\infty,f)=0$ and so $\Delta(\infty,f)=0$. Therefore using the above argument yields (7.3.3) for $f(z)$.

We have completed the proof of Theorem 7.3.1. □

The following consequence is immediate from the final part of the above proof.

Corollary 7.3.1. *If $f(z)$ is a meromorphic function with the finite lower order. Then (7.3.1) is equivalent to (7.3.2).*

Below let us make a simple survey on the development of the Nevanlinna's conjecture. Dealing with the derivatives, Yang and Zhang [28] established the following

Theorem 7.3.2. *Let $f(z)$ be an entire function with finite lower order μ. If*

$$\sum_{j=-\infty}^{\infty}\sum_{a\neq 0,\infty}\delta(a,f^{(j)})=1,$$

then

$$\sum_{j=-\infty}^{\infty}p_j\leqslant\mu,$$

where p_j is the number of finite and non-zero deficient values of $f^{(j)}$. Furthermore, every deficient value of $f^{(j)}$ $(j=0,\pm1,\pm2,\cdots)$ is its a asymptotic value and the deficiency is the multiple of $\frac{1}{\mu}$.

It is natural to ask if the results stated in the Nevanlinna's conjecture is true when the derivatives are dealt with. Another approach develops the Nevanlinna's conjecture by considering a in (7.3.1) as small function with respect to f. For an entire function with finite lower order μ, Li and Ye [22] extended some of the results of Pfluger [23], that is to say, they proved that if $\lambda=\lambda(f)$ is not an integer, $\sum\delta(a,f)\leqslant 2-k(\lambda)$ where the sum is taken over all deficient small function of f and consequently if (7.3.1) holds for all small functions with respect to f, then $\lambda(f)=\mu(f)$ is a positive integer. Under (7.3.1) for all small functions with respect to f, Jin and Dai [19] [20] proved that the number of deficient small functions of entire function f does not exceed $\mu(f)$ and every deficiency is the multiple of $\frac{1}{\mu}$.

Eremenko and Sodin [12] solved the problem on the Nevanlinna conjecture dealing with the small functions. They proved that if $f(z)$ is a meromorphic function of the finite lower order and satisfies (7.3.1) for an at most countable number of small functions a of f, then the results in Theorem 7.3.1 holds without (7.3.3), where $a(z)$ is an asymptotic small function of f means that 0 is an asymptotic value of $f(z)-a(z)$.

Finally, let us take the singular directions of the function into account under the condition of the Nevanlinna's conjecture. According to the method which was used by F. Nevanlinna to study his conjecture, we consider the case when $N_1(r,f)\equiv 0$, that is, $f(z)$ has no multiple points. Since the Schwarzian derivative of f has poles only at multiple points of f,

$$S_f = \frac{f'''}{f'} - \frac{3}{2}\left(\frac{f''}{f'}\right)^2$$

is an entire function and from the lemma on logarithmic derivative, $T(r, S_f) = m(r, S_f) = O(\log r)$. This yields that S_f is a polynomial and from Theorem 3.7.5 we have that the Julia, Borel and T-directions coincide and the number of these singular directions equals to 2λ. However, the author do not know if this result is true for the general case and therefore this leads us ask the following conjecture.

Conjecture 7.3.1. *Under the condition of the Nevanlinna's conjecture, does the number of the singular directions of f equal to 2λ.*

References

1. Anderson, J. and Baernstein II, A., The size of the set on which a meromorphic function is large, Proc. London Math. Soc., **36**(1978), 518-539.
2. Arsove, M., Normal families of subharmonic functions, Proc. Amer. Math. Soc., (1)**7**(1956), 115-126.
3. Azarin, V. S., On the asymptotic behavior of subharmonic functions of finite order, Mat. Sbornik, **108**(1979), 147-167 (in Russian); English translation: Math. UssR Sbornik **36**(1980), 135-153.
4. Baernstein II, A., Proof of Edrei's spread conjecture, Proc. London Math. Soc., **26**(1973), 418-434.
5. Brelot, M., On topologies and boundaries in potential theory, Lect. Notes Math., **175**, Spronger-Verlag, Berlin, 1971.
6. Conway, J. B., Functions of one complex variable, Springer-Verlag,1978.
7. Doob, J., Potential Theory and Its Probabilistic Counterpart, Springer-Verlag, Berlin, 1984.
8. Drasin, D., Proof of a conjecture of F. Nevalinna concerning functions which have deficiency sum two, Acta Math. **158**(1987), 1-94.
9. Drasin, D. and Shea, D. F., Pólya peaks and the oscillation of positive functions, Proc. Amer. Math. Soc., (2)**34**(1972), 403-411.
10. Eremenko, A. E., A new proof of Drasin's theorem on meromorphic functions of finite order with maximal deficiency sum, I and II, Teorija Funktsii, Funkts. Anal. i Prilozh. Soviet Math. (6)**52**(1990), 3522-3529; 3397-3403.
11. Eremenko, A. E., Meromorphic functions with small ramification, Indiana University Math. J., (4)**42**(1993), 1193-1218.
12. Eremenko A. E. and Sodin, M. L., Meromorphic functions of finite order with maximal deficiency sum, Teoriya Funktsii, Funkts. Anal. i Prilozh., **55**(1991), 84-95; J. Soviet Math., **59**(1992), 643-651.
13. Eremenko A. E. and Sodin, M. L., The value distribution of meromorphic functions and meromorphic curves from the point of view of potential theory, St. Petersburg Math. J., (1)**3**(1992), 109-136.
14. Eremenko A. E. and Sodin, M. L., Iterations of rational functions and the distribution of the values of the Poincaré functions, J. Soviet Math., **58**(1992), 504-509.
15. Fugleds, B., Asymptotic paths for subharmonic functions and polygonal connectedness of fine domains, Université Paris 6. Séminaire de Théorie du Potential, **5**(1980), 1-20.

16. Grishin, A. Ph., On the sets of regular growth of entire functions, Teorija Funktsii, Funkts. Anal. i Prilozh. Soviet Math., **40**(1983), 36-47 (in Russian).
17. Hayman, W. K. and Kennedy, P. B., Subharmonic functions I, Academic Press, London, 1976.
18. Hewitt, E. and Stromberg, K., Real and Absract Analysis, Graduate Texts in Mathematics 25, Springer-Verlage, 1975.
19. Jin, L. and Dai, C. J., On F. Nevanlinna's conjecture of deficient functions I, A department report, East China Normal University, 1982 (in Chinese).
20. Jin, L. and Dai, C. J., On F. Nevanlinna's conjecture of deficient functions II, A department report, East China Normal University, 1982 (in Chinese).
21. Landkof, N. S., Foundations of modern potential theory, "Nauka", Moscow, 1966; English transl., Springer-Verlag, Berlin and New York, 1972.
22. Li, Q. Z. and Ye, Y. S., On the deficiency sum of deficient functrions and F. Nevanlinna's conjecture, Adv. in Mah., (2)**14**(1985), 168-174.
23. Pfluger, A., Zur Defektrelation ganzer Funktionen endlicher Ordnung, Coment. Math. Heiv., **19**(1946), 91-104.
24. Ransford, T., Potential theory in the complex plane, Cambridge Univ. Press, 1995.
25. Shea, D., On the frequency of multiple values of a meromorphic function of small order, Michigan Math. J., **32**(1985), 109-116.
26. Tsuji, M., Potential theory in modern function theory, Maruzen Co. LTD. Tokyo, 1959.
27. Weitsman, A., Meromorphic functions with maximal deficeincy sum and a conjecture of F. Nevanlinna, Acta Math., **123**(1969), 115-139.
28. Yang, L. and Zhang, G. H., Deficient values and asymptotic values of entire functions, Science in China, **1**(1979), 1-11 (in Chinese).

Index